Molecular Biology

Second Edition

The INSTANT NOTES series

Series editor
B.D. Hames
School of Biochemistry and Molecular Biology, University of Leeds, Leeds, UK

Animal Biology
Genetics
Microbiology
Chemistry for Biologists
Immunology
Biochemistry 2nd edition
Molecular Biology 2nd edition
Neuroscience
Psychology
Developmental Biology
Plant Biology
Ecology 2nd edition

Forthcoming titles
Bioinformatics

The INSTANT NOTES Chemistry series
Consulting editor: Howard Stanbury

Organic Chemistry
Inorganic Chemistry
Physical Chemistry
Medicinal Chemistry

Forthcoming titles
Analytical Chemistry

Molecular Biology

Second Edition

P.C. Turner, A.G. McLennan,
A.D. Bates & M.R.H. White

School of Biological Sciences,
University of Liverpool, Liverpool, UK

First published 1997
Second edition published 2000
Reprinted 2001, 2003, 2004

A CIP catalogue record for this book is available from the British Library.

ISBN 1 85996 152 5

Garland Science/BIOS Scientific Publishers
4 Park Square, Milton Park, Abingdon, Oxon OX14 4RN, UK
World Wide Web home page: http/www.bios.co.uk/

Published in the United States of America, its dependent territories and Canada by Springer-Verlag, New York Inc., 175 Fifth Avenue, New York, NY 10010-7858, in association with BIOS Scientific Publishers

Published in Hong Kong, Taiwan, Cambodia, Korea, The Philippines, Brunei, Laos and Macau only by Springer-Verlag Hong Kong Ltd, Unit 1702, Tower 1, Enterprise Square, 9 Sheung Yuet Road, Kowloon Bay, Kowloon, Hong Kong, in association with BIOS Scientific Publishers

Production Editor: Andrea Bosher
Typeset by Phoenix Photosetting, Chatham, Kent
Printed and bound in Great Britain by Biddles Ltd, King's Lynn, Norfolk

Contents

[a] Contributed by Dr M. Bennett, Department of Veterinary Pathology, University of Liverpool, Leahurst, Neston, South Wirral L64 7TE, UK.
[b] Contributed by Dr C. Green, School of Biological Sciences, University of Liverpool, PO Box 147, Liverpool L69 7ZB, UK.

ABBREVIATIONS

ADP	adenosine 5′-diphosphate
AIDS	acquired immune deficiency syndrome
AMP	adenosine 5′-monophosphate
ARS	autonomously replicating sequence
ATP	adenosine 5′-triphosphate
BAC	bacterial artificial chromosome
BER	base excision repair
bp	base pairs
BRF	TFIIB-related factor
BUdR	bromodeoxyuridine
bZIP	basic leucine zipper
CDK	cyclin-dependent kinase
cDNA	complementary DNA
CHEF	contour clamped homogeneous electric field
CJD	Creutzfeld–Jakob disease
CRP	cAMP receptor protein
CSF-1	colony-stimulating factor-1
CTD	carboxyl-terminal domain
Da	Dalton
dNTP	deoxynucleoside triphosphate
ddNTP	dideoxynucleoside triphosphate
DMS	dimethyl sulfate
DNA	deoxyribonucleic acid
DNase	deoxyribonuclease
DOP-PCR	degenerate oligonucleotide primer PCR
dsDNA	double-stranded DNA
EDTA	ethylenediamine tetraacetic acid
EF	elongation factor
ENU	ethylnitrosourea
ER	endoplasmic reticulum
ESI	electrospray ionization
ETS	external transcribed spacer
FADH	reduced flavin adenine dinucleotide
FIGE	field inversion gel electrophoresis
β-gal	β-galactosidase
GMO	genetically modified organism
GTP	guanosine 5′-triphosphate
HIV	human immunodeficiency virus
HLH	helix–loop–helix
hnRNA	heterogeneous nuclear RNA
hnRNP	heterogeneous nuclear ribonucleoprotein
HSP	heat-shock protein
HSV-1	herpes simplex virus-1
ICE	interleukin-1 β converting enzyme
IF	initiation factor
IHF	integration host factor
IPTG	isopropyl-β-D-thiogalactopyranoside
IS	insertion sequence
ITS	internal transcribed spacer
JAK	Janus activated kinase
kb	kilobase pairs in duplex nucleic acid, kilobases in single-stranded nucleic acid
kDa	kiloDalton
LAT	latency-associated transcript
LINES	long interspersed elements
LTR	long terminal repeat
MALDI	matrix-assisted laser desorption/ionization
MCS	multiple cloning site
MMS	methylmethane sulfonate
MMTV	mouse mammary tumor virus
mRNA	messenger RNA
NAD^+	nicotinamide adenine dinucleotide
NER	nucleotide excision repair
NLS	nuclear localization signal
NMN	nicotinamide mononucleotide
NMR	nuclear magnetic resonance
nt	nucleotide
NTP	nucleoside triphosphate
ORC	origin recognition complex
ORF	open reading frame
PAGE	polyacrylamide gel electrophoresis
PAP	poly(A) polymerase
PCNA	proliferating cell nuclear antigen
PCR	polymerase chain reaction
PDGF	platelet-derived growth factor
PFGE	pulsed field gel electrophoresis
PTH	phenylthiohydantoin
RACE	rapid amplification of cDNA ends
RBS	ribosome-binding site
RER	rough endoplasmic reticulum
RF	replicative form

RFLP	restriction fragment length polymorphism
RNA	ribonucleic acid
RNA Pol I	RNA polymerase I
RNA Pol II	RNA polymerase II
RNA Pol III	RNA polymerase III
RNase A	ribonuclease A
RNase H	ribonuclease H
RNP	ribonucleoprotein
ROS	reactive oxygen species
RP-A	replication protein A
rRNA	ribosomal RNA
RT	reverse transcriptase
RT–PCR	reverse transcriptase–polymerase chain reaction
SAM	*S*-adenosylmethionine
SDS	sodium dodecyl sulfate
SINES	short interspersed elements
SL1	selectivity factor 1
snoRNP	small nucleolar RNP
SNP	single nucleotide polymorphism
snRNA	small nuclear RNA
snRNP	small nuclear ribonucleoprotein
SRP	signal recognition particle
Ssb	single-stranded binding protein
SSCP	single stranded conformational polymorphism
ssDNA	single-stranded DNA
STR	simple tandem repeat
SV40	simian virus 40
TAF	TBP-associated factor
TBP	TATA-binding protein
α-TIF	α-*trans*-inducing factor
Tris	tris(hydroxymethyl)amino-methane
tRNA	transfer RNA
UBF	upstream binding factor
UCE	upstream control element
URE	upstream regulatory element
UV	ultraviolet
VNTR	variable number tandem repeat
X-gal	5-bromo-4-chloro-3-indolyl-β-D-galactopyranoside
XP	xeroderma pigmentosum
YAC	yeast artificial chromosome
YEp	yeast episomal plasmid

Preface to the Second Edition

To assess how to improve *Instant Notes in Molecular Biology* for the second edition, we studied the first edition reader's comments carefully and were pleasantly surprised to discover how little was deemed to have been omitted and how few errors had been brought to our attention. Thus, the problem facing us was how to add a number of fairly disparate items and topics without substantially affecting the existing structure of the book. We therefore chose to fit new material into existing topics as far as possible, only creating new topics where absolutely necessary. A superficial comparison might therefore suggest that little has changed in the second edition, but we have included, updated or extended the following areas: proteomics, LINES/SINES, signal transduction, BACs, Z-DNA, gene gun, genomics, DNA fingerprinting, DNA chips, microarrays, RFLPs, genetic polymorphism, genome sequencing projects, SSCP, automated DNA sequencing, positional cloning, chromosome jumping, PFGE, multiplex DNA amplification, RT-PCR, quantitative PCR, PCR screening, PCR mutagenesis, degenerate PCR and transgenic animals. In addition, three completely new topics have been added. Arguably, no molecular biology text should omit a discussion of Crick's central dogma and it now forms the basis of Topic D5 – The flow of genetic information. Two other rapidly expanding and essential subjects are The cell cycle and Apoptosis, each of which, we felt, deserved its own topic. These have been added to Section E on DNA replication and Section S on Tumor viruses and oncogenes, respectively. Finally, in keeping with the ethos of the first edition that *Instant Notes in Molecular Biology* should be used as a study guide and revision aid, we have added approximately 100 multiple choice questions grouped in section order. This single improvement will, we feel, greatly enhance the educational utility of the book.

Acknowledgments

We thank all those readers of the first edition who took the trouble to return their comments and suggestions, without which the second edition would have been less improved, Will Sansom, Andrea Bosher and Jonathan Ray at BIOS who kept encouraging us and finally our families, who once again had to suffer during the periods of (re)writing.

Preface to the First Edition

The last 20 years have witnessed a revolution in our understanding of the processes responsible for the maintenance, transmission and expression of genetic information at the molecular level – the very basis of life itself. Of the many technical advances on which this explosion of knowledge has been based, the ability to remove a specific fragment of DNA from an organism, manipulate it in the test tube, and return it to the same or a different organism must take pride of place. It is around this essence of recombinant DNA technology, or genetic engineering to give it its more popular title, that the subject of molecular biology has grown. Molecular biology seeks to explain the relationships between the structure and function of biological molecules and how these relationships contribute to the operation and control of biochemical processes. Of principal interest are the macromolecules and macromolecular complexes of DNA, RNA and protein and the processes of replication, transcription and translation. The new experimental technologies involved in manipulating these molecules are central to modern molecular biology. Not only does it yield fundamental information about the molecules, but it has tremendous practical applications in the development of new and safe products such as therapeutics, vaccines and foodstuffs, and in the diagnosis of genetic disease and in gene therapy.

An inevitable consequence of the proliferation of this knowledge is the concomitant proliferation of comprehensive, glossy textbooks, which, while beautifully produced, can prove somewhat overwhelming in both breadth and depth to first and second year undergraduate students. With this in mind, *Instant Notes in Molecular Biology* aims to deliver the core of the subject in a concise, easily assimilated form designed to aid revision. The book is divided into 19 sections containing 70 topics. Each topic consists of a 'Key Notes' panel, with extremely concise statements of the key points covered. These are then amplified in the main part of the topic, which includes simple and clear black and white figures, which may be easily understood and reproduced. To get the best from this book, material should first be learnt from the main part of the topic; the Key Notes can then be used as a rapid revision aid. Whilst there is a reasonably logical order to the topics, the book is designed to be 'dipped into' at any point. For this reason, numerous cross-references are provided to guide the reader to related topics.

The contents of the book have been chosen to reflect both the major techniques used and the conclusions reached through their application to the molecular analysis of biological processes. They are based largely on the molecular biology courses taught by the authors to first and second year undergraduates on a range of biological science degree courses at the University of Liverpool. Section A introduces the classification of cells and macromolecules and outlines some of the methods used to analyze them. Section B considers the basic elements of protein structure and the relationship of structure to function. The structure and physico-chemical properties of DNA and RNA molecules are discussed in Section C, including the complex concepts involved in the supercoiling of DNA. The organization of DNA into the intricate genomes of both prokaryotes and eukaryotes is covered in Section D. The related subjects of mutagenesis, DNA replication, DNA recombination and the repair of DNA damage are considered in Sections E and F.

Section G introduces the technology available for the manipulation of DNA sequences. As described above, this underpins much of our detailed understanding of the molecular mechanisms of cellular processes. A simple DNA cloning scheme is used to introduce the basic methods. Section H describes a number of the more sophisticated cloning vectors which are used for a variety of purposes. Section I considers the use of DNA libraries in the isolation of new gene sequences, while Section J covers more complex and detailed methods, including DNA sequencing and the analysis of cloned sequences. This section concludes with a discussion of some of the rapidly expanding applications of gene cloning techniques.

The basic principles of gene transcription in prokaryotes are described in Section K, while Section L gives examples of some of the sophisticated mechanisms employed by bacteria to control specific gene expression. Sections M and N provide the equivalent, but necessarily more complex, story of transcription in eukaryotic cells. The processing of newly transcribed RNA into mature molecules is detailed in Section O, and the roles of these various RNA molecules in the translation of the genetic code into protein sequences are described in Sections P and Q. The contributions that prokaryotic and eukaryotic viruses have made to our understanding of molecular information processing are detailed in Section R. Finally, Section S shows how the study of viruses, combined with the knowledge accumulated from many other areas of molecular biology is now leading us to a detailed understanding of the processes involved in the development of a major human affliction – cancer.

This book is not intended to be a replacement for the comprehensive mainstream textbooks; rather, it should serve as a direct complement to your lecture notes to provide a sound grounding in the subject. The major texts, some of which are listed in the Further Reading section at the end of the book, can then be consulted for more detail on topics specific to the particular course being studied. For those of you whose fascination and enthusiasm for the subject has been sufficiently stimulated, the reading list also directs you to some more detailed and advanced articles to take you beyond the scope of this book. Inevitably, there have had to be omissions from *Instant Notes in Molecular Biology* and we are sure each reader will spot a different one. However, many of these will be covered in other titles in the Instant Notes series, such as the companion volume, *Instant Notes in Biochemistry.*

Phil Turner, Sandy McLennan, Andy Bates and Mike White

Acknowledgments

We would like to acknowledge the support and understanding of our families for those many lost evenings and weekends when we could all have been in the pub instead of drafting and redrafting manuscripts. We are also indebted to our colleagues Malcolm Bennett and Chris Green for their contributions to the chapters on bacteriophages, viruses and oncogenes. Our thanks, too, go to the series editor, David Hames, and to Jonathan Ray, Rachel Robinson and Lisa Mansell of BIOS Scientific Publishers for providing prompt and helpful advice when required and for keeping the pressure on us to finish the book on time.

A1 CELLULAR CLASSIFICATION

Key Notes

Eubacteria

Structurally defined as prokaryotes, these cells have a plasma membrane, usually enclosed in a rigid cell wall, but no intracellular compartments. They have a single, major circular chromosome. They may be unicellular or multicellular. *Escherichia coli* is the best studied eubacterium.

Archaea

The Archaea are structurally defined as prokaryotes but probably branched off from the eukaryotes after their common ancestor diverged from the eubacteria. They tend to inhabit extreme environments. They are biochemically closer to eubacteria in some ways but to eukaryotes in others. They also have some biochemical peculiarities.

Eukaryotes

Cells of plants, animals, fungi and protists possess well-defined subcellular compartments bounded by lipid membranes (e.g. nuclei, mitochondria, endoplasmic reticulum). These organelles are the sites of distinct biochemical processes and define the eukaryotes.

Differentiation

In most multicellular eukaryotes, groups of cells differentiate during development of the organism to provide specialized functions (e.g. as in liver, brain, kidney). In most cases, they contain the same DNA but transcribe different genes. Like all other cellular processes, differentiation is controlled by genes. Co-ordination of the activities of different cell types requires communication between them.

Related topics

Subcellular organelles (A2)
Prokaryotic and eukaryotic chromosome structure (Section D)
Bacteriophages and eukaryotic viruses (Section R)

Eubacteria

The **Eubacteria** are one of two subdivisions of the **prokaryotes**. Prokaryotes are the simplest living cells, typically 1–10 μm in diameter, and are found in all environmental niches from the guts of animals to acidic hot springs. Classically, they are defined by their structural organization (*Fig. 1*). They are bounded by a **cell (plasma) membrane** comprising a lipid bilayer in which are embedded proteins that allow the exit and entry of small molecules. Most prokaryotes also have a rigid cell wall outside the plasma membrane which prevents the cell from swelling or shrinking in environments where the osmolarity differs significantly from that inside the cell. The cell interior (**cytoplasm** or **cytosol**) usually contains a single, circular chromosome compacted into a **nucleoid** and attached to the membrane (see Topic D1), and often plasmids [small deoxyribonucleic acid (DNA) molecules with limited genetic information, see Topic G2], ribonucleic acid (RNA), ribosomes (the sites of protein synthesis, see Section Q) and most of the proteins which perform the metabolic reactions of the cell. Some of these proteins are attached to the plasma membrane, but there are no distinct **subcel-**

lular organelles as in **eukaryotes** to compartmentalize different parts of the metabolism. The surface of a prokaryote may carry **pili**, which allow it to attach to other cells and surfaces, and **flagella**, whose rotating motion allows the cell to swim. Most prokaryotes are unicellular; some, however, have multicellular forms in which certain cells carry out specialized functions. The **Eubacteria** differ from the **Archaea** mainly in their biochemistry. The eubacterium *Escherichia coli* has a **genome size** (DNA content) of 4600 kilobase pairs (kb) which is sufficient genetic information for about 3000 proteins. Its molecular biology has been studied extensively. The genome of the simplest bacterium, *Mycoplasma genitalium*, has only 580 kb of DNA and encodes just 470 proteins. It has a very limited metabolic capacity.

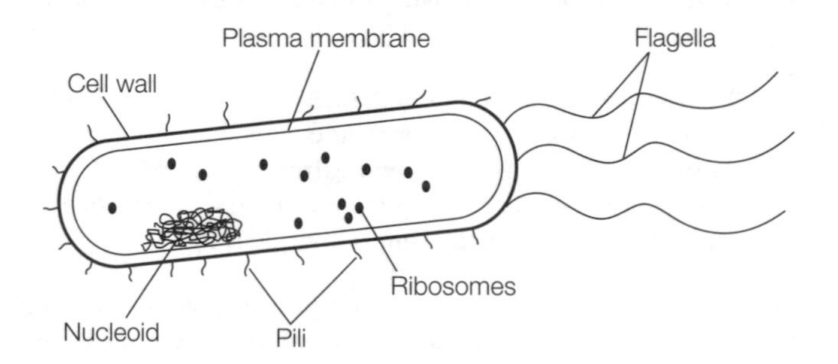

Fig. 1. Schematic diagram of a typical prokaryotic cell.

Archaea

The **Archaea**, or **archaebacteria**, form the second subdivision of the prokaryotes and tend to inhabit extreme environments. Structurally, they are similar to eubacteria. However, on the basis of the evolution of their ribosomal RNA (rRNA) molecules (see Topic O1), they appear as different from the eubacteria as both groups of prokaryotes are from the eukaryotes and display some unusual biochemical features, for example ether in place of ester linkages in membrane lipids (see Topic A3). The 1740 kb genome of the archaeon *Methanococcus jannaschii* encodes a maximum of 1738 proteins. Comparisons reveal that those involved in energy production and metabolism are most like those of eubacteria while those involved in replication, transcription and translation are more similar to those of eukaryotes. It appears that the Archaea and the eukaryotes share a common evolutionary ancestor which diverged from the ancestor of the Eubacteria.

Eukaryotes

Eukaryotes are classified taxonomically into four kingdoms comprising **animals**, **plants**, **fungi** and **protists** (algae and protozoa). Structurally, eukaryotes are defined by their possession of membrane-enclosed organelles (*Fig. 2*) with specialized metabolic functions (see Topic A2). Eukaryotic cells tend to be larger than prokaryotes: 10–100 μm in diameter. They are surrounded by a plasma membrane, which can have a highly convoluted shape to increase its surface area. Plants and many fungi and protists also have a rigid cell wall. The cytoplasm is a highly organized gel that contains, in addition to the organelles and ribosomes, an array of protein fibers called the **cytoskeleton** which controls the shape and movement of the cell and which organizes many of its metabolic functions. These fibers include **microtubules**, made of **tubulin**, and **microfilaments**, made of **actin** (see Topic A4). Many eukaryotes are multicellular, with groups of cells undergoing **differentiation** during development to form the specialized tissues of the whole organism.

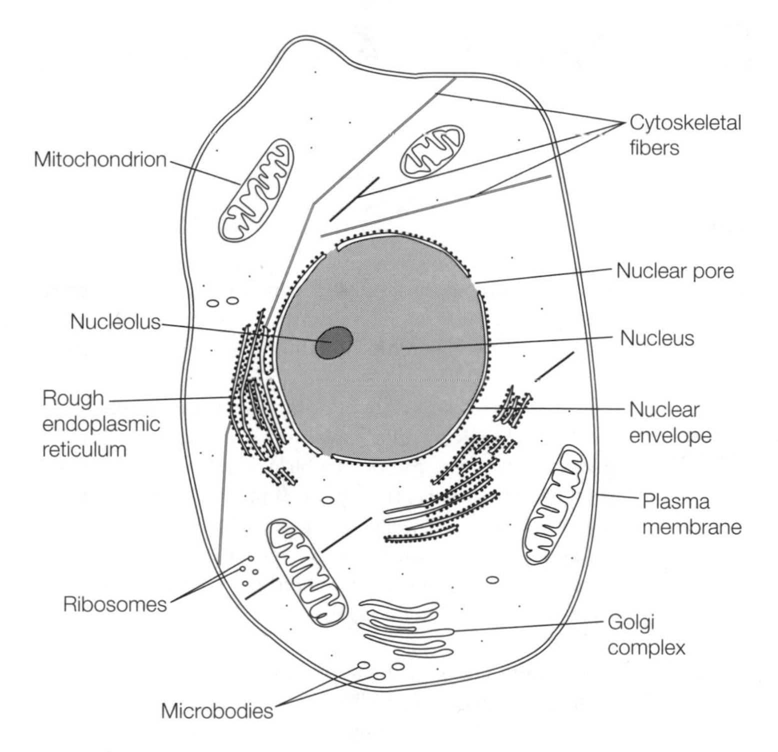

Fig. 2. Schematic diagram of a typical eukaryotic cell.

Differentiation

When a cell divides, the daughter cells may be identical in every way, or they may change their patterns of gene expression to become functionally different from the parent cell.

Among prokaryotes and lower eukaryotes, the formation of spores is an example of such **cellular differentiation** (see Topic L3). Among complex multicellular eukaryotes, embryonic cells differentiate into highly specialized cells, for example muscle, nerve, liver and kidney. In all but a few exceptional cases, the DNA content remains the same, but the genes which are transcribed have changed. Differentiation is regulated by developmental control genes (see Topic N2). Mutations in these genes result in abnormal body plans, such as legs in the place of antennae in the fruit fly *Drosophila*. Studying such gene mutations allows the process of embryonic development to be understood. In multicellular organisms, co-ordination of the activities of the various tissues and organs is controlled by communication between them. This involves signaling molecules such as neurotransmitters, hormones and growth factors which are secreted by one tissue and act upon another through specific cell-surface receptors.

A2 SUBCELLULAR ORGANELLES

Key Notes

Nuclei

The membrane-bound nucleus contains the bulk of the cellular DNA in multiple chromosomes. Transcription of this DNA and processing of the RNA occurs here. Nucleoli are contained within the nucleus.

Mitochondria and chloroplasts

Mitochondria are the site of cellular respiration where nutrients are oxidized to CO_2 and water, and adenosine 5′-triphosphate (ATP) is generated. They are derived from prokaryotic symbionts and retain some DNA, RNA and protein synthetic machinery, though most of their proteins are encoded in the nucleus. Photosynthesis takes place in the chloroplasts of plants and eukaryotic algae. Chloroplasts have a basically similar structure to mitochondria but with a thylakoid membrane system containing the light-harvesting pigment chlorophyll.

Endoplasmic reticulum

The smooth endoplasmic reticulum is a cytoplasmic membrane system where many of the reactions of lipid biosynthesis and xenobiotic metabolism are carried out. The rough endoplasmic reticulum has attached ribosomes engaged in the synthesis of membrane-targeted and secreted proteins. These proteins are carried in vesicles to the Golgi complex for further processing and sorting.

Microbodies

The lysosomes contain degradative, hydrolytic enzymes; the peroxisomes contain enzymes which destroy certain potentially dangerous free radicals and hydrogen peroxide; the glyoxysomes of plants carry out the reactions of the glyoxylate cycle.

Organelle isolation

After disruption of the plasma membrane, the subcellular organelles can be separated from each other and purified by a combination of differential centrifugation and density gradient centrifugation (both rate zonal and isopycnic). Purity can be assayed by measuring organelle-specific enzymes.

Related topics

Cellular classification (A1)
rRNA processing and ribosomes (O1)
Translational control and post-translational events (Q4)

Nuclei

The eukaryotic nucleus carries the genetic information of the cell in multiple chromosomes, each containing a single DNA molecule (see Topics D2 and D3). The nucleus is bounded by a lipid double membrane, the nuclear envelope, containing pores which allow passage of moderately large molecules (see Topic A1, *Fig. 2*). Transcription of RNA takes place in the nucleus (see Section M) and the processed RNA molecules (see Section O) pass into the cytoplasm where translation takes place (see Section Q). **Nucleoli** are bodies within the nucleus where rRNA is synthesized and ribosomes are partially assembled (see Topics M2 and O1).

Mitochondria and chloroplasts

Cellular respiration, that is the oxidation of nutrients to generate energy in the form of adenosine 5′-triphosphate (ATP), takes place in the **mitochondria**. These organelles are roughly 1–2 μm in diameter and there may be 1000–2000 per cell. They have a smooth outer membrane and a convoluted inner membrane that forms protrusions called **cristae** (see Topic A1, *Fig. 2*). They contain a small circular DNA molecule, mitochondrial-specific RNA and ribosomes on which some mitochondrial proteins are synthesized. However, the majority of mitochondrial (and chloroplast) proteins are encoded by nuclear DNA and synthesized in the cytoplasm. These latter proteins have specific **signal sequences** that target them to the mitochondria (see Topic Q4). The **chloroplasts** of plants are the site of photosynthesis, the light-dependent assimilation of CO_2 and water to form carbohydrates and oxygen. Though larger than mitochondria, they have a similar structure except that, in place of cristae, they have a third membrane system (the **thylakoids**) in the inner membrane space. These contain chlorophyll, which traps the light energy for photosynthesis. Chloroplasts are also partly genetically independent of the nucleus. Both mitochondria and chloroplasts are believed to have evolved from prokaryotes which had formed a symbiotic relationship with a primitive nucleated eukaryote.

Endoplasmic reticulum

The **endoplasmic reticulum** is an extensive membrane system within the cytoplasm and is continuous with the nuclear envelope (see Topic A1, *Fig. 2*). Two forms are visible in most cells. The **smooth** endoplasmic reticulum carries many membrane-bound enzymes, including those involved in the biosynthesis of certain lipids and the oxidation and detoxification of foreign compounds (**xenobiotics**) such as drugs. The **rough** endoplasmic reticulum (**RER**) is so-called because of the presence of many ribosomes. These ribosomes specifically synthesize proteins intended for secretion by the cell, such as plasma or milk proteins, or those destined for the plasma membrane or certain organelles. Apart from the plasma membrane proteins, which are initially incorporated into the RER membrane, these proteins are translocated into the interior space (**lumen**) of the RER where they are modified, often by **glycosylation** (see Topic Q4). The lipids and proteins synthesized on the RER are transported in specialized **transport vesicles** to the **Golgi complex**, a stack of flattened membrane vesicles which further modifies, sorts and directs them to their final destinations (see Topic A1, *Fig. 2*).

Microbodies

Lysosomes are small membrane-bound organelles which bud off from the Golgi complex and which contain a variety of digestive enzymes capable of degrading proteins, nucleic acids, lipids and carbohydrates. They act as recycling centers for macromolecules brought in from outside the cell or from damaged organelles. Some metabolic reactions which generate highly reactive free radicals and hydrogen peroxide are confined within organelles called **peroxisomes** to prevent these species from damaging cellular components. Peroxisomes contain the enzyme catalase, which destroys hydrogen peroxide:

$$2H_2O_2 \rightarrow 2H_2O + O_2$$

Glyoxysomes are specialized plant peroxisomes which carry out the reactions of the glyoxylate cycle. Lysosomes, peroxisomes and glyoxysomes are collectively known as **microbodies**.

Organelle isolation

The plasma membrane of eukaryotes can be disrupted by various means including osmotic shock, controlled mechanical shear or by certain nonionic detergents. Organelles displaying large size and density differences, for example nuclei and mitochondria, can be separated from each other and from other organelles by **differential centrifugation** according to the value of their **sedimentation coefficients** (see Topic A4). The **cell lysate** is centrifuged at a speed which is high enough to sediment only the heaviest organelles, usually the nuclei. The supernatant containing all the other organelles is removed then centrifuged at a higher speed to sediment the mitochondria, and so on (*Fig. 1a*). This technique is also used to fractionate suspensions containing cell types of different sizes, for example red cells, white cells and platelets in blood. These crude preparations of cells, nuclei and mitochondria usually require further purification by **density gradient centrifugation**. This is also used to separate organelles of similar densities. In **rate zonal centrifugation**, the mixture is layered on top of a pre-formed concentration (and, therefore, density) gradient of a suitable medium in a centrifuge tube. Upon centrifugation, bands or zones of the different components sediment at different rates depending on their sedimentation coefficients, and separate (*Fig. 1b*). The purpose of the density gradient of the supporting medium is to prevent convective mixing of the components after separation (i.e. to provide stability) and to ensure linear sedimentation rates of the components (it compensates for the acceleration of the components as they move further down the tube). In **equilibrium (isopycnic) centrifugation**, the density gradient extends to a density higher than that of one or more components of the mixture so that these components come to equilibrium at a point equal to their own density and stop moving. In this case, the density gradient can either be pre-formed, and the sample layered on top, or self-forming, in which case the sample may be mixed with the gradient material (*Fig. 1c*). Density gradients are made from substances such as sucrose, Ficoll (a synthetic polysaccharide), metrizamide (a synthetic iodinated heavy compound) or cesium chloride (CsCl), for separation of nucleic acids (see Topics C2 and G2). Purity of the subcellular fraction can be determined using an electron microscope or by assaying enzyme activities known to be associated specifically with particular organelles, for example succinate dehydrogenase in mitochondria.

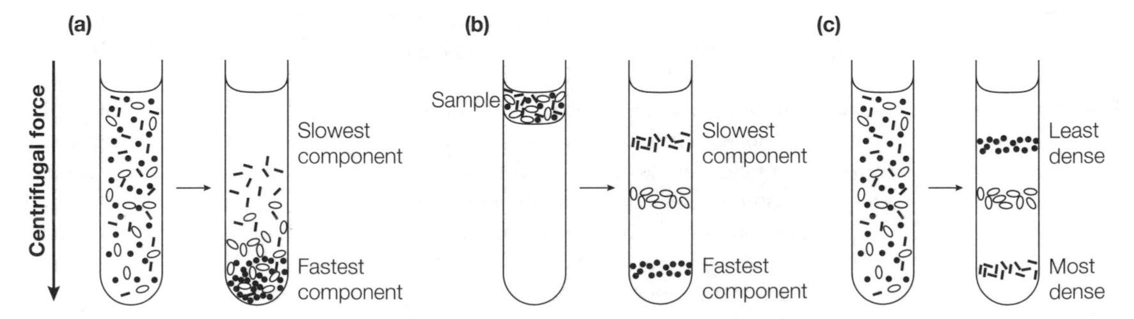

Fig. 1. Centrifugation techniques. (a) Differential, (b) rate zonal and (c) isopycnic (equilibrium).

A3 MACROMOLECULES

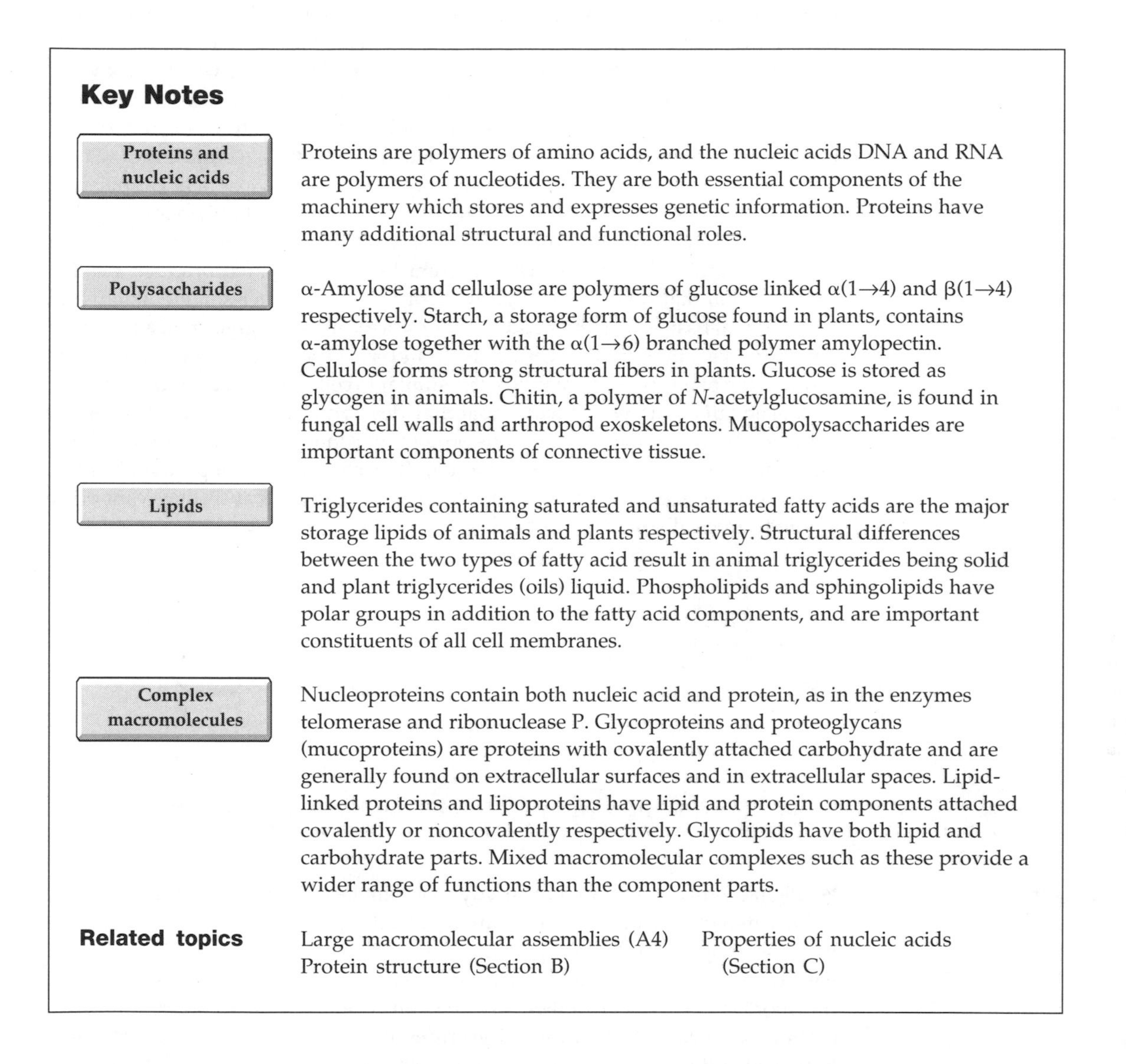

Key Notes

Proteins and nucleic acids

Proteins are polymers of amino acids, and the nucleic acids DNA and RNA are polymers of nucleotides. They are both essential components of the machinery which stores and expresses genetic information. Proteins have many additional structural and functional roles.

Polysaccharides

α-Amylose and cellulose are polymers of glucose linked α(1→4) and β(1→4) respectively. Starch, a storage form of glucose found in plants, contains α-amylose together with the α(1→6) branched polymer amylopectin. Cellulose forms strong structural fibers in plants. Glucose is stored as glycogen in animals. Chitin, a polymer of *N*-acetylglucosamine, is found in fungal cell walls and arthropod exoskeletons. Mucopolysaccharides are important components of connective tissue.

Lipids

Triglycerides containing saturated and unsaturated fatty acids are the major storage lipids of animals and plants respectively. Structural differences between the two types of fatty acid result in animal triglycerides being solid and plant triglycerides (oils) liquid. Phospholipids and sphingolipids have polar groups in addition to the fatty acid components, and are important constituents of all cell membranes.

Complex macromolecules

Nucleoproteins contain both nucleic acid and protein, as in the enzymes telomerase and ribonuclease P. Glycoproteins and proteoglycans (mucoproteins) are proteins with covalently attached carbohydrate and are generally found on extracellular surfaces and in extracellular spaces. Lipid-linked proteins and lipoproteins have lipid and protein components attached covalently or noncovalently respectively. Glycolipids have both lipid and carbohydrate parts. Mixed macromolecular complexes such as these provide a wider range of functions than the component parts.

Related topics

Large macromolecular assemblies (A4)
Protein structure (Section B)
Properties of nucleic acids (Section C)

Proteins and nucleic acids

Proteins are polymers of amino acids linked together by peptide bonds. The structures of the amino acids and of proteins are dealt with in detail in Section B. Proteins have both structural and functional roles. The nucleic acids DNA and RNA are polymers of nucleotides, which themselves consist of a nitrogenous base, a pentose sugar and phosphoric acid. Their structures are detailed in Section C. There are three main types of cellular RNA: messenger RNA (mRNA), ribosomal RNA (rRNA) and transfer RNA (tRNA). Nucleic acids are involved in the storage and processing of genetic information (see Topic D5), but the expression of this information requires proteins.

Polysaccharides

Polysaccharides are polymers of simple sugars covalently linked by glycosidic bonds. They function mainly as nutritional sugar stores and as structural materials. **Cellulose** and **starch** are abundant components of plants. Both are glucose polymers, but differ in the way the glucose monomers are linked. Cellulose is a linear polymer with β(1→4) linkages (*Fig. 1a*) and is a major structural component of the plant cell wall. About 40 parallel chains form horizontal sheets which stack vertically above one another. The chains and sheets are held together by hydrogen bonds (see Topic A4) to produce tough, insoluble fibers. Starch is a sugar store and is found in large intracellular granules which can be hydrolyzed quickly to release glucose for metabolism. It contains two components: α-amylose, a linear polymer with α(1→4) linkages (*Fig. 1b*), and amylopectin, a branched polymer with additional α(1→6) linkages. With up to 10^6 glucose residues, amylopectins are among the largest molecules known. The different linkages in starch produce a coiled conformation that cannot pack tightly, hence starch is water-soluble. Fungi and some animal tissues (e.g. liver and muscle) store glucose as **glycogen**, a branched polymer like amylopectin. **Chitin** is found in fungal cell walls and in the exoskeleton of insects and crustacea. It is similar to cellulose, but the monomer unit is *N*-acetylglucosamine. **Mucopolysaccharides** (glycosaminoglycans) form the gel-like solutions in which the fibrous proteins of connective tissue are embedded. Determination of the structures of large polysaccharides is complicated because they are heterogeneous in size and composition and because they cannot be studied genetically like nucleic acids and proteins.

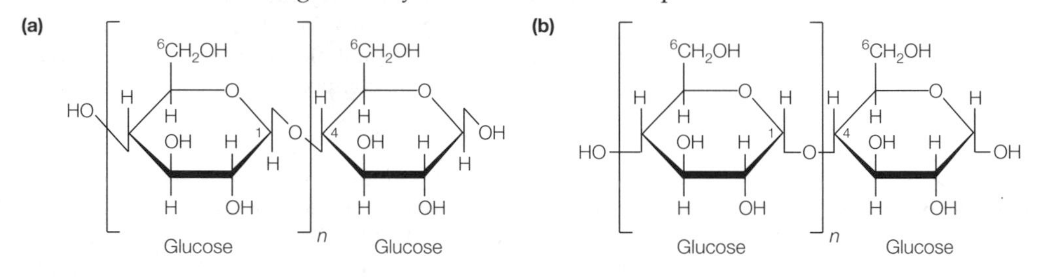

Fig. 1. The structures of (a) cellulose, with β(1→4) linkages and (b) starch α-amylose, with α(1→4) linkages. Carbon atoms 1, 4 and 6 are labeled. Additional α(1→6) linkages produce branches in amylopectin and glycogen.

Lipids

While individual lipids are not strictly macromolecules, many are built up from smaller monomeric units and they are involved in many macromolecular assemblies (see Topic A4). Large lipid molecules are predominantly hydrocarbon in nature and are poorly soluble in water. Some are involved in the storage and transport of energy while others are key components of membranes, protective coats and other cell structures. **Glycerides** have one, two or three long-chain fatty acids esterified to a molecule of glycerol. In animal triglycerides, the fatty acids have no double bonds (saturated) so the chains are linear, the molecules can pack tightly and the resulting fats are solid. Plant oils contain unsaturated fatty acids with one or more double bonds. The angled structures of these chains prevent close packing so they tend to be liquids at room temperature. Membranes contain **phospholipids**, which consist of glycerol esterified to two fatty acids and phosphoric acid. The phosphate is also usually esterified to a small molecule such as serine, ethanolamine, inositol or choline (*Fig. 2*). Membranes also contain **sphingolipids** such as **ceramide**, in which the long-chain amino alcohol sphingosine has a fatty acid linked by an amide bond. Attachment of phosphocholine to a ceramide produces **sphingomyelin**.

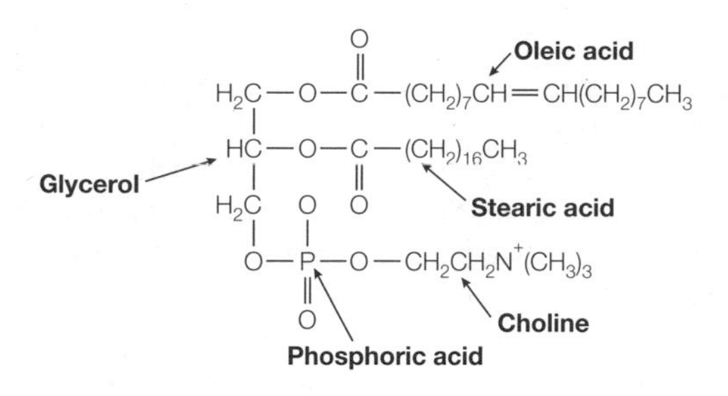

Fig. 2. A typical phospholipid: phosphatidylcholine containing esterified stearic and oleic acids.

Complex macromolecules

Many macromolecules contain covalent or noncovalent associations of more than one of the major classes of large biomolecules. This can greatly increase the functionality or structural capabilities of the resulting complex. For example, nearly all enzymes are proteins, but some have a noncovalently attached RNA component which is essential for catalytic activity. Associations of nucleic acid and protein are known as **nucleoproteins**. Examples are **telomerase**, which is responsible for replicating the ends of eukaryotic chromosomes (See Topics D3 and E3) and **ribonuclease P**, an enzyme which matures transfer RNA (tRNA). In telomerase, the RNA acts as a template for telomere DNA synthesis, while in ribonuclease P, the RNA contains the catalytic site of the enzyme. Ribonuclease P is an example of a **ribozyme** (see Topic O2).

Glycoproteins contain both protein and carbohydrate (between <1% and >90% of the weight) components; glycosylation is the commonest form of post-translational modification of proteins (see Topic Q4). The carbohydrate is always covalently attached to the surface of the protein, never the interior, and is often variable in composition, causing microheterogeneity (*Fig. 3*). This has made glycoproteins difficult to study. Glycoproteins have functions that span the entire range of protein activities, and are usually found extracellularly. They are important components of cell membranes and mediate cell–cell recognition.

Proteoglycans (mucoproteins) are large complexes (>10^7 Da) of protein and mucopolysaccharide found in bacterial cell walls and in the extracellular space in connective tissue. Their sugar units often have sulfate groups, which makes

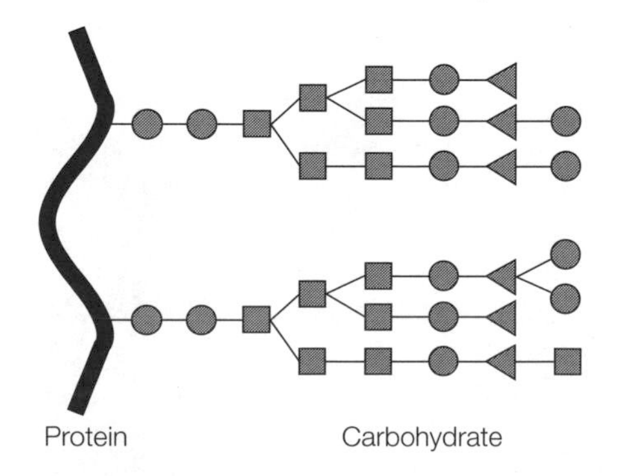

Fig. 3. Glycoprotein structure. The different symbols represent different monosaccharide units (e.g. galactose, N*-acetylglucosamine).*

them highly hydrated. This, coupled with their lengths (> 1000 units), produces solutions of high viscosity. Proteoglycans act as lubricants and shock absorbers in extracellular spaces.

Lipid-linked proteins have a covalently attached lipid component. This is usually a fatty acyl (e.g. myristoyl or palmitoyl) or isoprenoid (e.g. farnesyl or geranylgeranyl) group. These groups serve to anchor the proteins in membranes through hydrophobic interactions with the membrane lipids and also promote protein–protein associations (see Topic A4).

In **lipoproteins**, the lipids and proteins are linked noncovalently. Because lipids are poorly soluble in water, they are transported in the blood as lipoproteins. These are basically particles of triglycerides and cholesterol esters coated with a layer of phospholipids, cholesterol and protein (the **apolipoproteins**). The structures of the apolipoproteins are such that their hydrophobic amino acids face towards the lipid interior of the particles while the charged and polar amino acids (see Topic B1) face outwards into the aqueous environment. This renders the particles soluble.

Glycolipids, which include cerebrosides and gangliosides, have covalently linked lipid and carbohydrate components, and are especially abundant in the membranes of brain and nerve cells.

A4 Large macromolecular assemblies

Key Notes

Protein complexes

The eukaryotic cytoskeleton consists of various protein complexes: microtubules (made of tubulin), microfilaments (made of actin and myosin) and intermediate filaments (containing various proteins). These organize the shape and movement of cells and subcellular organelles. Cilia and flagella are also composed of microtubules complexed with dynein and nexin.

Nucleoprotein

Bacterial 70S ribosomes comprise a large 50S subunit, with 23S and 5S RNA molecules and 31 proteins, and a small 30S subunit, with a 16S RNA molecule and 21 proteins. Eukaryotic 80S ribosomes have 60S (28S, 5.8S and 5S RNAs) and 40S (18S RNA) subunits. Chromatin contains DNA and the basic histone proteins. Viruses are also nucleoprotein complexes.

Membranes

Membrane phospholipids and sphingolipids form bilayers with the polar groups on the exterior surfaces and the hydrocarbon chains in the interior. Membrane proteins may be peripheral or integral and act as receptors, enzymes, transporters or mediators of cellular interactions.

Noncovalent interactions

A large number of weak interactions hold macromolecular assemblies together. Charge–charge, charge–dipole and dipole–dipole interactions involve attractions between fully or partially charged atoms. Hydrogen bonds and hydrophobic interactions which exclude water are also important.

Related topics

Macromolecules (A3)
Chromatin structure (D2)
rRNA processing and ribosomes (O1)
Bacteriophages and eukaryotic viruses (Section R)

Protein complexes

Cellular architecture contains many large complexes of the different classes of macromolecules with themselves or with each other. Many of the major structural and locomotory elements of the cell consist of protein complexes. The **cytoskeleton** is an array of protein filaments which organizes the shape and motion of cells and the intracellular distribution of subcellular organelles. **Microtubules** are long polymers of tubulin, a 110 kiloDalton (kDa) globular protein (*Fig. 1*, see Topic B2). These are a major component of the cytoskeleton and of eukaryotic **cilia** and **flagella**, the hair-like structures on the surface of many cells which whip to move the cell or to move fluid across the cell surface. Cilia also contain the proteins nexin and dynein.

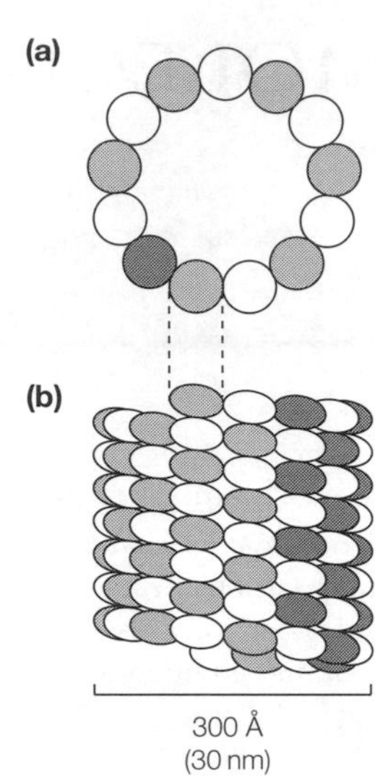

Fig. 1. Schematic diagram showing the (a) cross-sectional and (b) surface pattern of tubulin α and β subunits in a microtubule (see Topic B2). From: BIOCHEMISTRY, *4/E by Stryer © 1995 by Lubert Stryer. Used with permission of W.H. Freeman and Company. [After J.A. Snyder and J.R. McIntosh (1976)* Ann. Rev. Biochem. ***45****, 706. With permission from the* Annual Review of Biochemistry, *Volume 45, © 1976, by Annual Reviews Inc.]*

Microfilaments consisting of the protein **actin** form contractile assemblies with the protein **myosin** to cause cytoplasmic motion. Actin and myosin are also major components of muscle fibers. The **intermediate filaments** of the cytoskeleton contain a variety of proteins including **keratin** and have various functions, including strengthening cell structures. In all cases, the energy for motion is provided by the coupled hydrolysis of ATP or guanosine 5′-triphosphate (GTP).

Nucleoprotein

Nucleoproteins comprise both nucleic acid and protein. **Ribosomes** are large cytoplasmic **ribonucleoprotein** complexes which are the sites of protein synthesis (see Section Q). Bacterial 70S ribosomes have large (50S) and small (30S) subunits with a total mass of 2.5×10^6 Da. (The **S value**, e.g. 50S, is the numerical value of the **sedimentation coefficient**, *s*, and describes the rate at which a macromolecule or particle sediments in a centrifugal field. It is determined by both the mass and shape of the molecule or particle; hence S values are not additive.) The 50S subunit contains 23S and 5S RNA molecules and 31 different proteins while the 30S subunit contains a 16S RNA molecule and 21 proteins. Eukaryotic 80S ribosomes have 60S (with 28S, 5.8S and 5S RNAs) and 40S (with 18S RNA) subunits. Under the correct conditions, mixtures of the rRNAs and proteins will self-assemble in a precise order into functional ribosomes *in vitro*. Thus, all the information for ribosome structure is inherent in the structures of the components. The RNAs are not simply frameworks for the assembly of the ribosomal proteins, but participate in both the binding of the messenger RNA and in the catalysis of peptide bond synthesis (see Topic Q2).

Chromatin is the material from which eukaryotic chromosomes are made. It is a **deoxyribonucleoprotein** complex made up of roughly equal amounts of DNA and small, basic proteins called **histones** (see Topic D2). These form a repeating unit called a **nucleosome**. Correct assembly of nucleosomes and many other protein complexes requires assembly proteins, or **chaperones**. Histones neutralize the repulsion between the negative charges of the DNA sugar–phosphate backbone and allow the DNA to be tightly packaged within the chromosomes. **Viruses** are another example of nucleoprotein complexes. They are discussed in Section R.

Membranes

When placed in an aqueous environment, phospholipids and sphingolipids naturally form a **lipid bilayer** with the polar groups on the outside and the nonpolar hydrocarbon chains on the inside. This is the structural basis of all biological membranes. Such membranes form cellular and organellar boundaries and are selectively permeable to uncharged molecules. The precise lipid composition varies from cell to cell and from organelle to organelle. Proteins are also a major component of cell membranes (*Fig. 2*). **Peripheral** membrane proteins are loosely bound to the outer surface or are anchored via a lipid or **glycosyl phosphatidylinositol** anchor and are relatively easy to remove. **Integral** membrane proteins are embedded in the membrane and cannot be removed without destroying the membrane. Some protrude from the outer or inner surface of the membrane while **transmembrane** proteins span the bilayer completely and have both extracellular and intracellular **domains** (see Topic B2). The transmembrane regions of these proteins contain predominantly hydrophobic amino acids. Membrane proteins have a variety of functions, for example:

- receptors for signaling molecules such as hormones and neurotransmitters;
- enzymes for degrading extracellular molecules before uptake of the products;
- pores or channels for the selective transport of small, polar ions and molecules;
- mediators of cell–cell interactions (mainly glycoproteins).

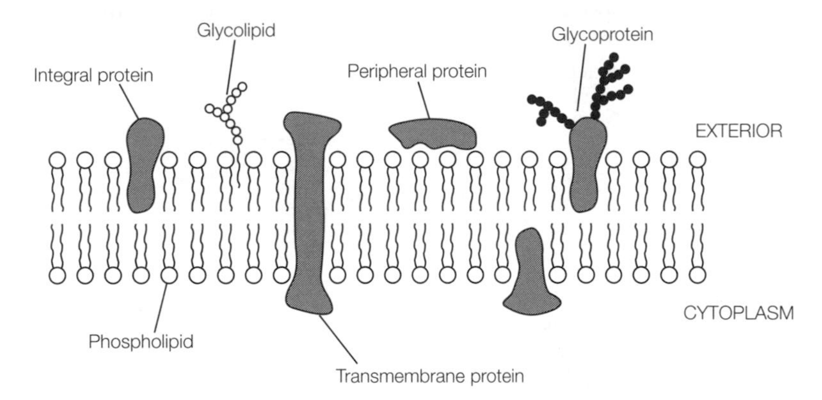

Fig. 2. Schematic diagram of a plasma membrane showing the major macromolecular components.

Noncovalent interactions

Most macromolecular assemblies are held together by a large number of different noncovalent interactions. **Charge–charge** interactions (**salt bridges**) operate between ionizable groups of opposite charge at physiological pH, for example between the negative phosphates of DNA and the positive lysine and arginine side chains of DNA-binding proteins such as histones (see Topic D2). **Charge–dipole** and **dipole–dipole** interactions are weaker and form when either or both of the participants is a dipole due to the asymmetric distribution of charge in the molecule (*Fig. 3a*). Even uncharged groups like methyl groups

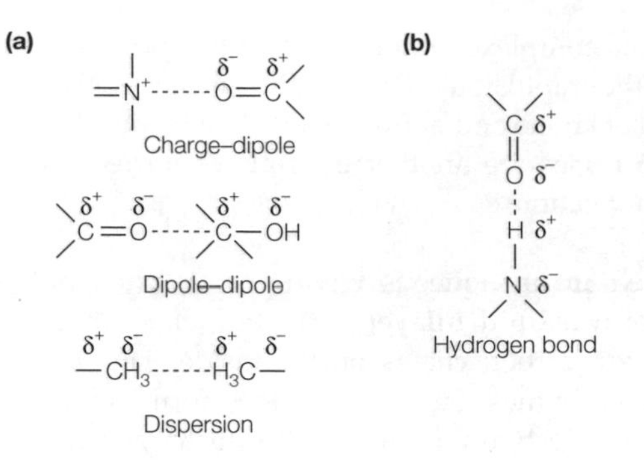

Fig. 3. Examples of (a) van der Waals forces and (b) a hydrogen bond.

can attract each other weakly through transient dipoles arising from the motion of their electrons (**dispersion forces**).

Noncovalent associations between electrically neutral molecules are known collectively as **van der Waals forces**. **Hydrogen bonds** are of great importance. They form between a covalently bonded hydrogen atom on a donor group (e.g. –O-H or –N-H) and a pair of nonbonding electrons on an acceptor group (e.g. :O=C– or :N–) (*Fig. 3b*). Hydrogen bonds and other interactions involving dipoles are directional in character and so help define macromolecular shapes and the specificity of molecular interactions. The presence of uncharged and nonpolar substances, for example lipids, in an aqueous environment tends to force a highly ordered structure on the surrounding water molecules. This is energetically unfavorable as it reduces the entropy of the system. Hence, nonpolar molecules tend to clump together, reducing the overall surface area exposed to water. This attraction is termed a **hydrophobic** (water-hating) interaction and is a major stabilizing force in protein–protein and protein–lipid interactions and in nucleic acids.

B1 AMINO ACIDS

Key Notes

Structure	The 20 common amino acids found in proteins have a chiral α-carbon atom linked to a proton, amino and carboxyl groups, and a specific side chain which confers different physical and chemical properties. They behave as zwitterions in solution. Nonstandard amino acids in proteins are formed by post-translational modification.
Charged side chains	Glutamic acid and aspartic acid have additional carboxyl groups and usually impart a negative charge to proteins. Lysine has an ε-amino group, arginine a guanidino group and histidine an imidazole group. These three basic amino acids generally impart a positive charge to proteins.
Polar uncharged side chains	Serine and threonine have hydroxyl groups, asparagine and glutamine have amide groups and cysteine has a thiol group.
Nonpolar aliphatic side chains	Glycine is the simplest amino acid with no side chain. Proline is a secondary amino acid (imino acid). Alanine, valine, leucine and isoleucine have hydrophobic alkyl groups. Methionine has a thioether sulfur atom.
Aromatic side chains	Phenylalanine, tyrosine and tryptophan have bulky aromatic side chains which absorb ultraviolet light.
Related topics	Protein structure and function (B2) Mechanism of protein synthesis (Q2)

Structure

Proteins are polymers of L-amino acids. Apart from proline, all of the 20 amino acids found in proteins have a common structure in which a carbon atom (the α-carbon) is linked to a carboxyl group, a primary amino group, a proton and a **side chain** (R) which is different in each amino acid (*Fig. 1*). Except in glycine, the α-carbon atom is asymmetric – it has four chemically different groups attached. Thus, amino acids can exist as pairs of optically active stereoisomers (D- and L-). However, only the L-isomers are found in proteins. Amino acids are dipolar ions (**zwitterions**) in aqueous solution and behave as both acids and bases (they are **amphoteric**). The side chains differ in size, shape, charge and chemical reactivity, and are responsible for the differences in the properties of different proteins (*Fig. 2*). A few proteins contain nonstandard amino acids, such as 4-hydroxyproline and 5-hydroxylysine in collagen. These are formed by **post-translational modification** of the parent amino acids proline and lysine (see Topic Q4).

Charged side chains

Taking pH 7 as a reference point, several amino acids have ionizable groups in their side chains which provide an extra positive or negative charge at this

Fig. 1. General structure of an L-amino acid. The R group is the side chain.

pH. The 'acidic' amino acids, **aspartic acid** and **glutamic acid**, have additional carboxyl groups which are usually ionized (negatively charged). The 'basic' amino acids have positively charged groups – **lysine** has a second amino group attached to the ε-carbon atom while **arginine** has a guanidino group. The imidazole group of **histidine** has a pK_a near neutrality. Reversible protonation of this group under physiological conditions contributes to the catalytic mechanism of many enzymes. Together, acidic and basic amino acids can form important salt bridges in proteins (see Topic A4).

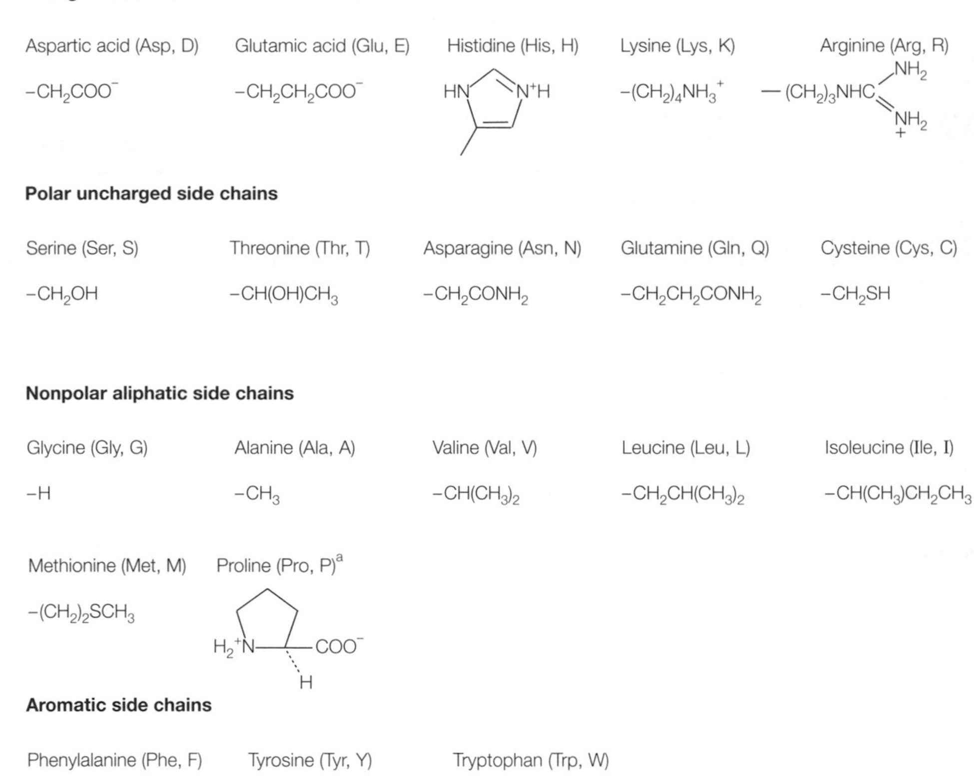

Aromatic side chains

Phenylalanine (Phe, F)

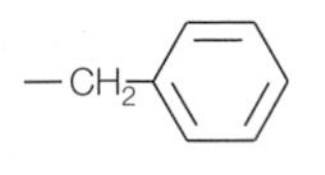

Tyrosine (Tyr, Y)

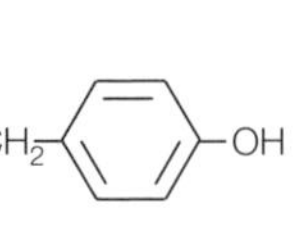

Tryptophan (Trp, W)

$—CH_2$

NH

Fig. 2. Side chains (R) of the 20 common amino acids. The standard three-letter abbreviations and one-letter code are shown in brackets. [a]The full structure of proline is shown as it is a secondary amino acid.

Polar uncharged side chains

These contain groups that form hydrogen bonds with water. Together with the charged amino acids, they are often described as **hydrophilic** ('water-loving'). **Serine** and **threonine** have hydroxyl groups, while **asparagine** and **glutamine** are the amide derivatives of aspartic and glutamic acids. **Cysteine** has a thiol (sulfhydryl) group which often oxidizes to **cystine**, in which two cysteines form a structurally important disulfide bond (see Topic B2).

Nonpolar aliphatic side chains

Glycine has a hydrogen atom in place of a side chain and is optically inactive. **Proline** is unusual in being a secondary amino (or imino) acid. **Alanine**, **valine**, **leucine** and **isoleucine** have **hydrophobic** ('water-hating') alkyl groups for side chains and participate in hydrophobic interactions in protein structure (see Topic A4). **Methionine** has a sulfur atom in a thioether link within its alkyl side chain.

Aromatic side chains

Phenylalanine, **tyrosine** and **tryptophan** have bulky hydrophobic side chains. Their aromatic structure accounts for most of the ultraviolet (UV) absorbance of proteins, which absorb maximally at 280 nm. The phenolic hydroxyl group of tyrosine can also form hydrogen bonds.

B2 Protein structure and function

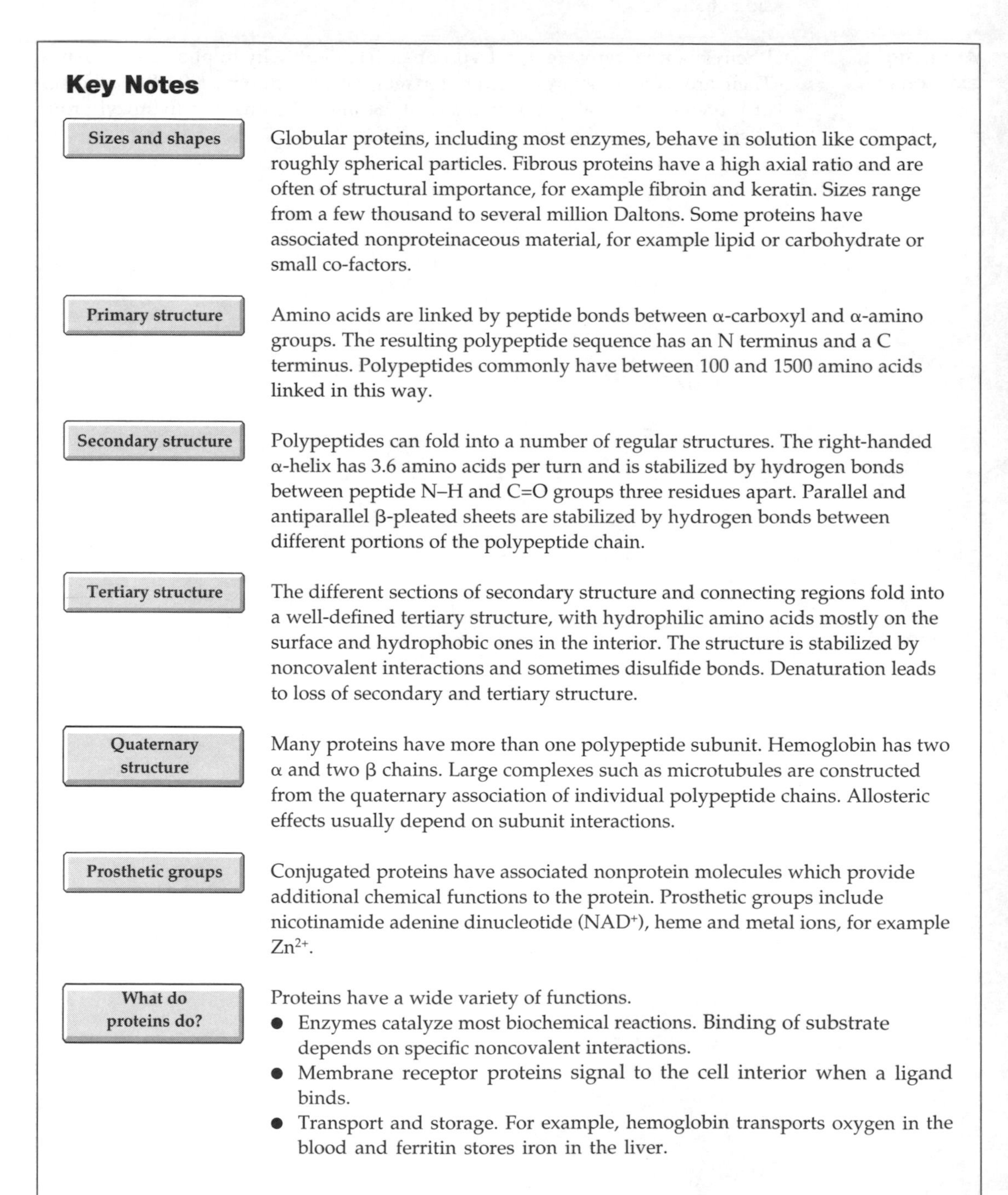

Key Notes

Sizes and shapes

Globular proteins, including most enzymes, behave in solution like compact, roughly spherical particles. Fibrous proteins have a high axial ratio and are often of structural importance, for example fibroin and keratin. Sizes range from a few thousand to several million Daltons. Some proteins have associated nonproteinaceous material, for example lipid or carbohydrate or small co-factors.

Primary structure

Amino acids are linked by peptide bonds between α-carboxyl and α-amino groups. The resulting polypeptide sequence has an N terminus and a C terminus. Polypeptides commonly have between 100 and 1500 amino acids linked in this way.

Secondary structure

Polypeptides can fold into a number of regular structures. The right-handed α-helix has 3.6 amino acids per turn and is stabilized by hydrogen bonds between peptide N–H and C=O groups three residues apart. Parallel and antiparallel β-pleated sheets are stabilized by hydrogen bonds between different portions of the polypeptide chain.

Tertiary structure

The different sections of secondary structure and connecting regions fold into a well-defined tertiary structure, with hydrophilic amino acids mostly on the surface and hydrophobic ones in the interior. The structure is stabilized by noncovalent interactions and sometimes disulfide bonds. Denaturation leads to loss of secondary and tertiary structure.

Quaternary structure

Many proteins have more than one polypeptide subunit. Hemoglobin has two α and two β chains. Large complexes such as microtubules are constructed from the quaternary association of individual polypeptide chains. Allosteric effects usually depend on subunit interactions.

Prosthetic groups

Conjugated proteins have associated nonprotein molecules which provide additional chemical functions to the protein. Prosthetic groups include nicotinamide adenine dinucleotide (NAD^+), heme and metal ions, for example Zn^{2+}.

What do proteins do?

Proteins have a wide variety of functions.

- Enzymes catalyze most biochemical reactions. Binding of substrate depends on specific noncovalent interactions.
- Membrane receptor proteins signal to the cell interior when a ligand binds.
- Transport and storage. For example, hemoglobin transports oxygen in the blood and ferritin stores iron in the liver.

- Collagen and keratin are important structural proteins. Actin and myosin form contractile muscle fibers.
- Casein and ovalbumin are nutritional proteins providing amino acids for growth.
- The immune system depends on antibody proteins to combat infection.
- Regulatory proteins such as transcription factors bind to and modulate the functions of other molecules, for example DNA.

Domains, motifs, families and evolution

Domains form semi-independent structural and functional units within a single polypeptide chain. Domains are often encoded by individual exons within a gene. New proteins may have evolved through new combinations of exons and, hence, protein domains. Motifs are groupings of secondary structural elements or amino acid sequences often found in related members of protein families. Similar structural motifs are also found in proteins which have no sequence similarity. Protein families arise through gene duplication and subsequent divergent evolution of the new genes.

Related topics

Macromolecules (A3)
Large macromolecular assemblies (A4)
Amino acids (B1)
Protein synthesis (Section Q)

Sizes and shapes

Two broad classes of protein may be distinguished. **Globular proteins** are folded compactly and behave in solution more or less as spherical particles; most enzymes are globular in nature. **Fibrous proteins** have very high axial ratios (length/width) and are often important structural proteins, for example silk fibroin and keratin in hair and wool. Molecular masses can range from a few thousand Daltons (Da) (e.g. the hormone insulin with 51 amino acids and a molecular mass of 5734 Da) to at least 5 million Daltons in the case of the enzyme complex pyruvate dehydrogenase. Some proteins contain bound nonprotein material, either in the form of small **prosthetic groups**, which may act as co-factors in enzyme reactions, or as large associations (e.g. the lipids in **lipoproteins** or the carbohydrate in **glycoproteins**, see Topic A3).

Primary structure

The α-carboxyl group of one amino acid is covalently linked to the α-amino group of the next amino acid by an amide bond, commonly known as a **peptide bond** when in proteins. When two amino acid **residues** are linked in this way the product is a **dipeptide**. Many amino acids linked by peptide bonds form a **polypeptide** (*Fig. 1*). The repeating sequence of α-carbon atoms and peptide bonds provides the **backbone** of the polypeptide while the different amino acid **side chains** confer functionality on the protein. The amino acid at one end of a polypeptide has an unattached α-amino group while the one at the other end has a free α-carboxyl group. Hence, polypeptides are directional, with an **N terminus** and a **C terminus**. Sometimes the N terminus is **blocked** with, for example, an acetyl group. The sequence of amino acids from the N to the C terminus is the **primary structure** of the polypeptide. Typical sizes for single polypeptide chains are within the range 100–1500 amino acids, though longer and shorter ones exist.

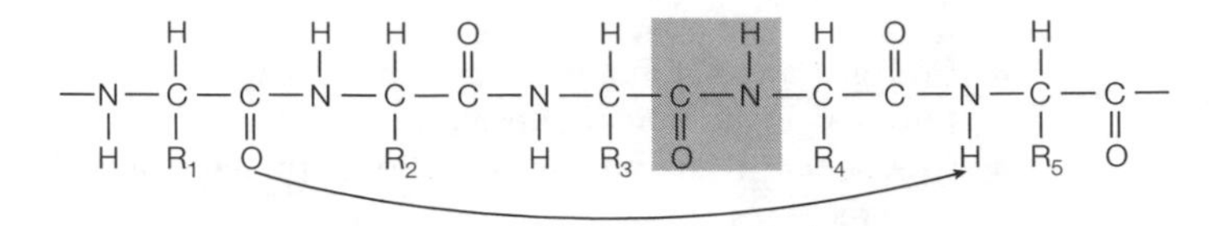

Fig. 1. Section of a polypeptide chain. The peptide bond is boxed. In the α-helix, the CO group of amino acid residue n is hydrogen-bonded to the NH group of residue n + 4 (arrowed).

Secondary structure

The highly polar nature of the C=O and N–H groups of the peptide bonds gives the C–N bond partial double bond character. This makes the peptide bond unit rigid and planar, though there is free rotation between adjacent peptide bonds. This polarity also favors hydrogen bond formation between appropriately spaced and oriented peptide bond units. Thus, polypeptide chains are able to fold into a number of regular structures which are held together by these hydrogen bonds. The best known **secondary structure** is the **α-helix** (*Fig. 2a*). The polypeptide backbone forms a right-handed helix with 3.6 amino acid residues per turn such that each peptide N–H group is hydrogen bonded to the C=O group of the peptide bond three residues away (*Fig. 1*). Sections of α-helical secondary structure are often found in globular proteins and in some fibrous proteins. The **β-pleated sheet** (β-sheet) is formed by hydrogen bonding of the peptide bond N–H and C=O groups to the complementary groups of another section of the polypeptide chain (*Fig. 2b*). Several sections of polypeptide chain may be involved side-by-side, giving a sheet structure with the side chains (R) projecting alternately above and below the sheet. If these sections run in the same direction (e.g. N terminus→C terminus), the sheet is **parallel**; if they alternate N→C and C→N, then the sheet is **antiparallel**. β-Sheets are strong and rigid and are important in structural proteins, for example silk fibroin. The connective tissue protein **collagen** has an unusual **triple helix** secondary structure in which three polypeptide chains are intertwined, making it very strong.

Tertiary structure

The way in which the different sections of α-helix, β-sheet, other minor secondary structures and connecting loops fold in three dimensions is the **tertiary structure** of the polypeptide (*Fig. 3*). The nature of the tertiary structure is inherent in the primary structure and, given the right conditions, most polypeptides will fold spontaneously into the correct tertiary structure as it is generally the lowest energy conformation for that sequence. However, *in vivo*, correct folding is often assisted by proteins called **chaperones** which help prevent misfolding of new polypeptides before their synthesis (and primary structure) is complete. Folding is such that amino acids with hydrophilic side chains locate mainly on the exterior of the protein where they can interact with water or solvent ions, while the hydrophobic amino acids become buried in the interior from which water is excluded. This gives overall stability to the structure. Various types of noncovalent interaction between side chains hold the tertiary structure together: van der Waals forces, hydrogen bonds, electrostatic salt bridges between oppositely charged groups (e.g. the ε-NH_3^+ group of lysine and the side chain COO^- groups of aspartate or glutamate) and hydrophobic interactions between the nonpolar side chains of the aliphatic and aromatic amino acids (see Topic A4). In addition, covalent disulfide bonds can form between two cysteine residues which may be far apart in the primary structure but close together in the folded tertiary structure. Disruption of secondary and

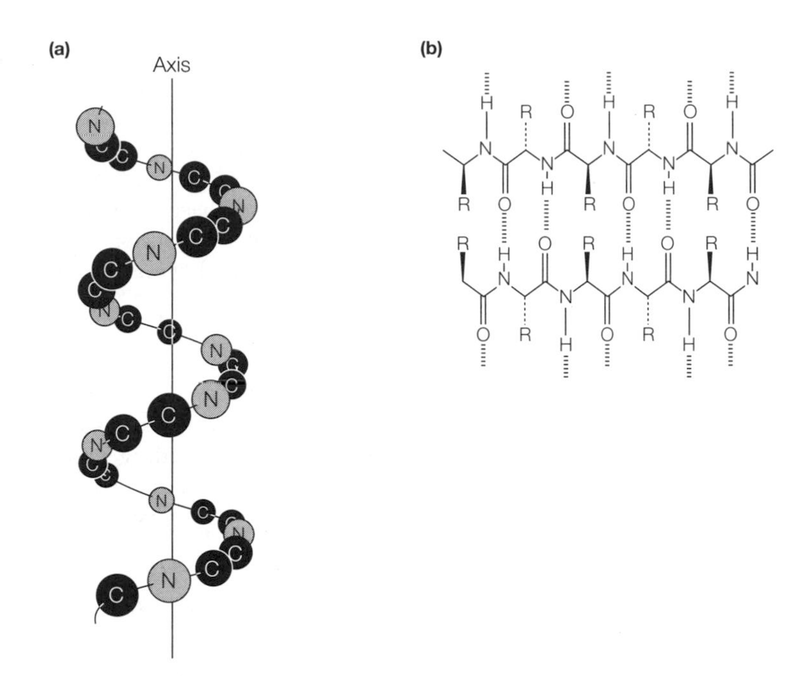

Fig. 2. (a) α-Helix secondary structure. Only the α-carbon and peptide bond carbon and nitrogen atoms of the polypeptide backbone are shown for clarity. (b) Section of a β-sheet secondary structure.

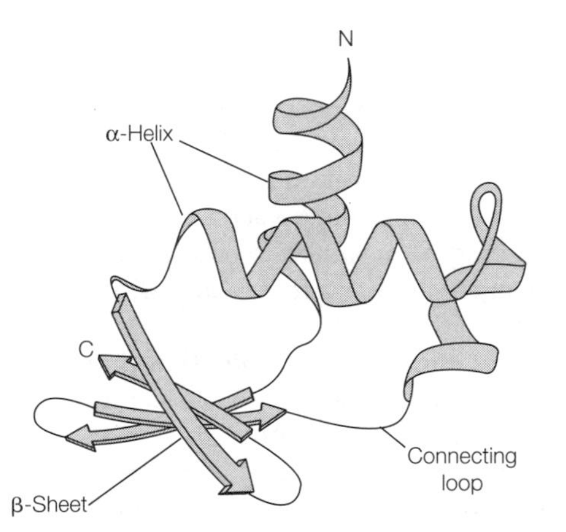

Fig. 3. Schematic diagram of a section of protein tertiary structure.

tertiary structure by heat or extremes of pH leads to **denaturation** of the protein and formation of a **random coil** conformation.

Quaternary structure

Many proteins are composed of two or more polypeptide chains (**subunits**). These may be identical or different. **Hemoglobin** has two α-globin and two β-globin chains ($\alpha_2\beta_2$). The same forces which stabilize tertiary structure hold these subunits together, including disulfide bonds between cysteines on separate polypeptides. This level of organization is known as the **quaternary structure** and has certain consequences. First, it allows very large protein molecules to be made. Tubulin is a dimeric protein made up of two small, nonidentical α and β subunits. Upon hydrolysis of tubulin-bound GTP, these dimers can polymerize into structures containing many hundreds of α and β subunits (see Topic A4, *Fig. 1*). These are the **microtubules** of the cytoskeleton. Secondly, it can provide greater functionality to a protein by combining different activities into a single entity, as in the fatty acid synthase complex. Often, the interactions between the subunits are modified by the binding of small molecules and this can lead to the **allosteric** effects seen in enzyme regulation.

Prosthetic groups

Many **conjugated** proteins contain covalently or noncovalently attached small molecules called **prosthetic groups** which give chemical functionality to the protein that the amino acid side chains cannot provide. Many of these are **co-factors** in enzyme-catalyzed reactions. Examples are nicotinamide adenine dinucleotide (NAD^+) in many dehydrogenases, pyridoxal phosphate in transaminases, heme in hemoglobin and cytochromes, and metal ions, for example Zn^{2+}. A protein without its prosthetic group is known as an **apoprotein**.

What do proteins do?

- **Enzymes.** Apart from a few catalytically active RNA molecules (see Topic O2), all enzymes are proteins. These can enhance the rate of biochemical reactions by several orders of magnitude. Binding of the **substrate** involves various noncovalent interactions with specific amino acid side chains, including van der Waals forces, hydrogen bonds, salt bridges and hydrophobic forces. Specificity of binding can be extremely high, with only a single substrate binding (e.g. glucose oxidase binds only glucose), or it can be group-specific (e.g. hexokinase binds a variety of hexose sugars). Side chains can also be directly involved in catalysis, for example by acting as nucleophiles, or proton donors or abstractors.
- **Signaling**. Receptor proteins in cell membranes can bind **ligands** (e.g. hormones) from the extracellular medium and, by virtue of the resulting conformational change, initiate reactions within the cell in response to that ligand. Ligand binding is similar to substrate binding but the ligand usually remains unchanged. Some hormones are themselves small proteins, such as insulin and growth hormone.
- **Transport and storage**. **Hemoglobin** transports oxygen in the red blood cells while **transferrin** transports iron to the liver. Once in the liver, iron is stored bound to the protein **ferritin**. Dietary fats are carried in the blood by **lipoproteins**. Many other molecules and ions are transported and stored in a protein-bound form. This can enhance solubility and reduce reactivity until they are required.
- **Structure and movement**. **Collagen** is the major protein in skin, bone and connective tissue, while hair is made mainly from **keratin**. There are also many structural proteins within the cell, for example in the **cytoskeleton**. The major muscle proteins **actin** and **myosin** form sliding filaments which are the basis of muscle contraction.
- **Nutrition**. **Casein** and **ovalbumin** are the major proteins of milk and eggs, respectively, and are used to provide the amino acids for growth of developing offspring. Seed proteins also provide nutrition for germinating plant embryos.
- **Immunity**. **Antibodies**, which recognize and bind to bacteria, viruses and other foreign material (the **antigen**) are proteins.

- **Regulation**. **Transcription factors** bind to and modulate the function of DNA. Many other proteins modify the functions of other molecules by binding to them.

Domains, motifs, families and evolution

Many proteins are composed of structurally independent units, or **domains**, that are connected by sections with limited higher order structure within the same polypeptide. The connections can act as hinges to permit the individual domains to move in relation to each other, and breakage of these connections by limited proteolysis can often separate the domains, which can then behave like independent globular proteins. The active site of an enzyme is sometimes formed in a groove between two domains, which wrap around the substrate. Domains can also have a specific function such as binding a commonly used molecule, for example ATP. When such a function is required in many different proteins, the same domain structure is often found. In eukaryotes, domains are often encoded by discrete parts of genes called **exons** (see Topic O3). Therefore, it has been suggested that, during evolution, new proteins were created by the duplication and rearrangement of domain-encoding exons in the genome to produce new combinations of binding sites, catalytic sites and structural elements in the resulting new polypeptides. In this way, the rate of evolution of new functional proteins may have been greatly increased.

Structural motifs (also known as **supersecondary structures**) are groupings of secondary structural elements that frequently occur in globular proteins. They often have functional significance and can represent the essential parts of binding or catalytic sites that have been conserved during the evolution of protein families from a common ancestor. Alternatively, they may represent the best solution to a structural–functional requirement that has been arrived at independently in unrelated proteins. A common example is the βαβ **motif** in which the connection between two consecutive parallel strands of a β sheet is an α-helix (*Fig. 4*). Two overlapping βαβ motifs (βαβαβ) form a dinucleotide (e.g. NAD^+) binding site in many otherwise unrelated proteins. Sequence motifs consist of only a few conserved, functionally important amino acids rather than supersecondary structures.

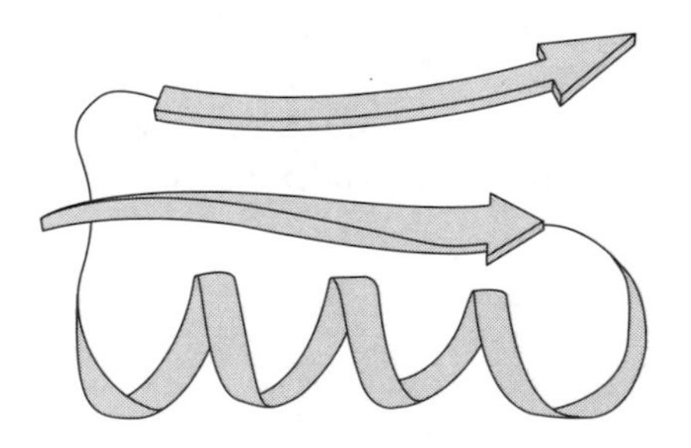

Fig. 4. Representation of a βαβ motif. The α-helix is shown as a coiled ribbon and the β-sheet segments as flat arrows.

Protein families arise through successive duplications and subsequent **divergent evolution** of an ancestral gene. Myoglobin, the oxygen-carrying protein in muscle, the α- and β-globin chains (and the minor δ chain) of adult hemoglobin and the γ-(gamma), ε-(epsilon) and ζ-(zeta) globins of embryonic and fetal hemoglobins are all related polypeptides within the **globin family** (*Fig. 5*). Their

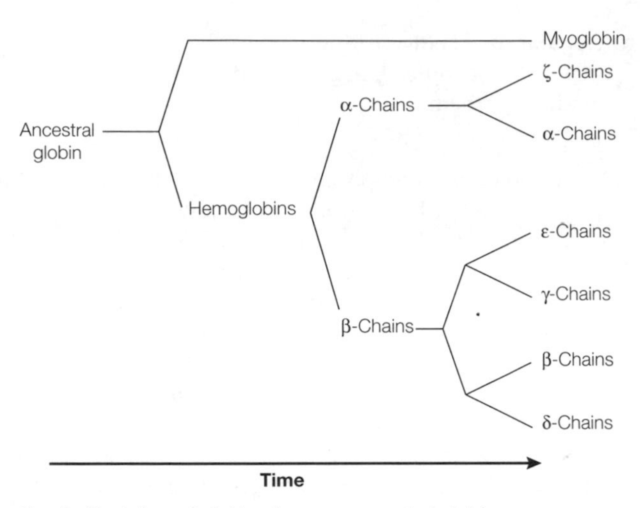

Fig. 5. Evolution of globins from an ancestral globin gene.

genes, and the proteins, are said to be **homologs**. Family members in different species that have retained the same function and carry out the same biochemical role (e.g. rat and mouse myoglobin) are **orthologs** while those that have evolved different, but often related functions (e.g. α-globin and β-globin) are **paralogs**. The degree of similarity between the amino acid sequences of orthologous members of a protein family in different organisms depends on how long ago the two organisms diverged from their common ancestor and on how important conservation of the sequence is for the function of the protein.

The function of a protein, whether structural or catalytic, is inherently related to its structure. As indicated above, similar structures and functions can also be achieved by **convergent evolution** whereby unrelated genes evolve to produce proteins with similar structures or catalytic activities. A good example is provided by the proteolytic enzymes subtilisin (bacterial) and chymotrypsin (animal). Even though their amino acid sequences are very different and they are composed of different structural motifs, they have evolved the same spatial orientation of the **catalytic triad** of active site amino acids — serine, histidine and aspartic acid — and use exactly the same catalytic mechanism to hydrolyse peptide bonds. Such proteins are termed **functional analogs.** Where similar structural motifs have evolved independently, the resulting proteins are **structural analogs**.

B3 PROTEIN ANALYSIS

Key Notes

Protein purification

Proteins are purified from crude cellular extracts by a combination of methods that separate according to different properties. Gel filtration chromatography separates by size. Ion-exchange chromatography, isoelectric focusing and electrophoresis take advantage of the different ionic charges on proteins. Hydrophobic interaction chromatography exploits differences in hydrophobicity. Affinity chromatography depends on the specific affinity between enzymes or receptors and ligands such as substrates or inhibitors. Overexpression of the protein greatly increases the yield, and inclusion of a purification tag in the recombinant can allow one-step purification.

Protein sequencing

After breaking a polypeptide down into smaller peptides using specific proteases or chemicals, the peptides are sequenced from the N terminus by sequential Edman degradation in an automated sequencer. The original sequence is recreated from the overlaps produced by cleavage with proteases of different specificities. Sequencing of the gene or complementary DNA (cDNA) for a protein is simpler.

Mass determination

Approximate molecular masses can be obtained by gel filtration chromatography and gel electrophoresis in the presence of sodium dodecyl sulfate. Mass spectrometry using electrospray ionization and matrix-assisted laser desorption/ionization techniques gives accurate masses for proteins of less than 100 kDa. Mass spectrometry also detects post-translational modifications.

X-ray crystallography and NMR

Many proteins can be crystallized and their three-dimensional structures determined by X-ray diffraction. The structures of small proteins in solution can also be determined by multi-dimensional nuclear magnetic resonance, particularly if the normal ^{12}C and ^{14}N are substituted by ^{13}C and ^{15}N.

Functional analysis

Functional analysis of a protein involves its isolation and study *in vitro* combined with a study of the behavior of a mutant organism in which the protein has been rendered nonfunctional by mutation or deletion of its gene. The function of a new protein can sometimes be predicted by comparing its sequence and structure to those of known proteins.

Proteomics

Proteomics is the identification and analysis of the total protein complement expressed by any given cell type under defined conditions.

Related topics

Subcellular organelles (A2) Protein structure and function (B2)

Protein purification

A typical eukaryotic cell may contain thousands of different proteins, some abundant and some present in only a few copies. In order to study any individual protein, it must be purified away from other proteins and nonprotein molecules.

The principal properties that can be exploited to separate proteins from each other are size, charge, hydrophobicity and affinity for other molecules. Usually, a combination of procedures is used to give complete purification.

Size

Gel filtration chromatography employs columns filled with an aqueous suspension of beads containing pores of a particular size (a molecular sieve). When applied to the top of the column, protein molecules larger than the pores elute at the bottom relatively quickly, as they are excluded from the beads and so have a relatively small volume of buffer available through which to travel. Molecules smaller than the pore size can enter the beads and so have a larger volume of buffer through which they can diffuse. Thus, smaller proteins elute from the column after larger ones. The sedimentation rate (**S value**) of a protein in the high centrifugal field produced in an ultracentrifuge varies with its size and shape (see Topic A4). Ultracentrifugation can be used to separate proteins but is also a powerful analytical tool for studying protein structure.

Charge

Because of the presence of ionizable side chains on their surface, proteins carry a net charge. As these side chains are all titratable, there will exist for each protein a pH at which its net surface charge is zero – the **isoelectric point (pI)**. Proteins are least soluble at their pI. At all other pH values, the net charge can be exploited for the separation and purification of proteins by **electrophoresis** and **ion-exchange chromatography**. In electrophoresis, the protein mixture is applied to a supporting medium, usually a gel, and an electric field applied across the support. Depending on their net charge, proteins will travel at different rates towards the anode or cathode and can then be recovered from the gel after separation. In ion-exchange chromatography, ions that are electrostatically bound to an insoluble support (the ion exchanger) packed in a column are reversibly replaced with charged proteins from solution. Different proteins have different affinities for the ion exchanger and require different ionic strengths to displace them again. Usually, a salt gradient of increasing ionic strength is passed through the column and the bound proteins elute separately. In **isoelectric focusing**, a pH gradient is generated from a mixture of buffers known as **polyampholytes** by an electric field. Proteins migrate to positions corresponding to their isoelectric points in this gradient and form tight, focused bands as their net charge becomes zero and they stop moving (*Fig. 1*).

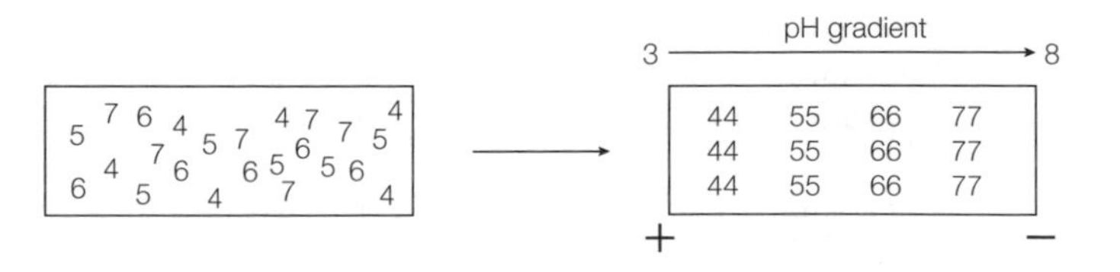

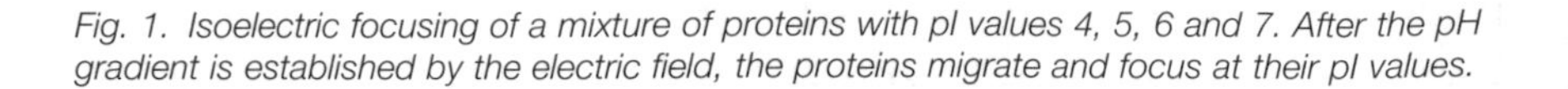
Fig. 1. Isoelectric focusing of a mixture of proteins with pI values 4, 5, 6 and 7. After the pH gradient is established by the electric field, the proteins migrate and focus at their pI values.

Hydrophobicity

Hydrophobic interactions between proteins and a column material containing aromatic or aliphatic alkyl groups are promoted in solutions of high ionic strength. By applying a gradient of decreasing ionic strength, proteins elute at different stages.

Affinity

The high specificities of enzyme–substrate, receptor–ligand and antibody–antigen interactions mean that these can be exploited to give very high degrees of purification. For example, a column made of a support to which the hormone insulin is covalently linked will specifically bind the insulin receptor and no other protein. The insulin receptor is a low abundance cell-surface protein responsible for transmitting the biological activity of insulin to the cell interior. Competitive inhibitors of enzymes are often used as affinity ligands as they are not degraded by the enzyme to be purified.

Recombinant techniques

Purification can be greatly simplified and the yields increased by overexpressing the recombinant protein in a suitable host cell (see Topics H1 and J6). For example, if the gene for the protein is inserted in a phage or plasmid **expression vector** under the control of a strong promoter (see Topic K3) and the vector is then transformed into *E. coli*, the protein can be synthesized to form up to 30% of the total protein of the cell. Purification can be aided further by adding a purification **tag**, such as six histidine codons, to the 5′- or 3′-end of the gene. In this case, the protein is synthesized with a hexahistidine tag at the N or C terminus, which allows one-step purification on an affinity column containing immobilized metal ions such as Ni^{2+}, which bind the histidine. The tag often has little effect on the function of the protein. Eukaryotic proteins may require expression in a eukaryotic host–vector system to achieve the correct post-translational modifications (see Topic H4).

Protein sequencing

An essential requirement for understanding how a protein works is a knowledge of its primary structure. The amino acid composition of a protein can be determined by hydrolyzing all the peptide bonds with acid (6 M HCl, 110°C, 24 h) and separating the resulting amino acids by chromatography. This indicates how many glycines and serines, etc., there are but does not give the actual sequence. Sequence determination involves splitting the protein into a number of smaller peptides using specific proteolytic enzymes or chemicals that break only certain peptide bonds (*Fig. 2*). For example, trypsin cleaves only after lysine (K) or arginine (R) and V8 protease only after glutamic acid (E). Cyanogen bromide cleaves polypeptides only after methionine residues.

Each peptide is then subjected to sequential **Edman degradation** in an automated protein sequencer. Phenylisothiocyanate reacts with the N-terminal amino acid which, after acid treatment, is released as the phenylthiohydantoin (PTH) derivative, leaving a new N terminus. The PTH-amino acid is identified by chromatography by comparison with standards and the cycle repeated to identify the next amino acid, and so on. The order of the peptides in the original protein can be deduced by sequencing peptides produced by proteases with different specificities and looking for the overlapping sequences (*Fig. 2*). This method is both laborious and expensive, so most proteins are now sequenced indirectly by sequencing the DNA of the gene or complementary DNA (cDNA) (see Topic J2) and deducing the protein sequence using the genetic code. This is simpler and faster but misses post-translational modifications (see Topic Q4). Direct sequencing is now usually confined to determination of the

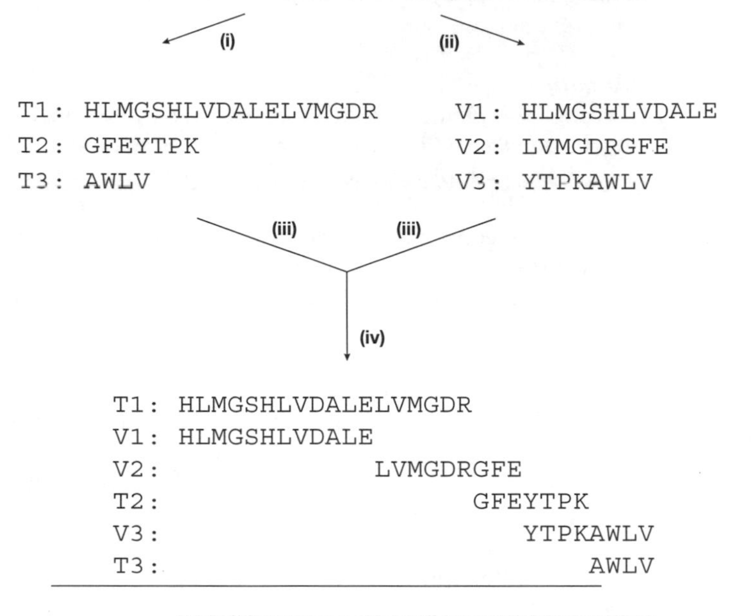

Fig. 2. Example of polypeptide sequence determination. (i) Cleave the polypeptide with trypsin; (ii) cleave with V8 protease; (iii) determine the sequences of tryptic (T1–T3) and V8 (V1–V3) peptides by Edman degradation; (iv) reassemble the original sequence from overlaps.

N-terminal or some limited internal sequence to provide information for the construction of an oligonucleotide or antibody **probe** which is then used to find the gene or cDNA (see Section I).

Mass determination

Gel filtration chromatography can be used to give the approximate molecular mass of a protein by comparing its elution time with that of known standards. Electrophoresis in polyacrylamide gels in the presence of the ionic detergent sodium dodecyl sulfate (SDS), which imparts a mass-dependent negative charge to all proteins from the ionized sulfate of the SDS, can also be used to determine the size of individual polypeptide chains (quaternary structure is lost) as their rate of movement in an electrical field now becomes dependent on mass rather than charge. These methods are cheap and easy though not particularly accurate (5–10% error). **Mass spectrometry** offers an extremely accurate method. Conventionally, molecules are vaporized and ionized by a beam of Xe or Ar atoms. The degree of deflection of the ions in an electromagnetic field is mass-dependent and can be measured. However, such methods have an upper mass limit of only a few kiloDaltons and are too destructive for protein analysis. Recent nondestructive ionization techniques have greatly extended this mass range. In **electrospray ionization (ESI)**, ions are formed by creating a fine spray of highly charged droplets of protein solution which are then vaporized to yield charged gaseous protein ions. In **matrix-assisted laser desorption/ionization (MALDI)** mass spectrometry, gas-phase ions are generated by the laser vaporization of a solid matrix containing the protein. Mass accuracies of 0.01% for proteins smaller than 100 kDa are now possible. Mass spectrometry is also helpful for detecting post-translational modifications and, most recently, for protein identification and sequencing.

X-ray crystallography and NMR

Because they have such well-defined three-dimensional (3-D) structures, many globular proteins have been crystallized. The 3-D structure can then be determined by **X-ray crystallography**. X-rays interact with the electrons in the matter through which they pass. By measuring the pattern of diffraction of a beam of X-rays as it passes through a crystal, the positions of the atoms in the crystal can be calculated. By crystallizing an enzyme in the presence of its substrate, the precise intermolecular interactions responsible for binding and catalysis can be seen. The structures of small globular proteins in solution can also be determined by two- or three-dimensional nuclear magnetic resonance (**NMR**) **spectroscopy**. In NMR, the relaxation of protons is measured after they have been excited by the radiofrequencies in a strong magnetic field. The properties of this relaxation depend on the relative positions of the protons in the molecule. The multi-dimensional approach is required for proteins to spread out and resolve the overlapping data produced by the large number of protons. Substituting ^{13}C and ^{15}N for the normal isotopes ^{12}C and ^{14}N in the protein also greatly improves data resolution by eliminating unwanted resonances. In this way, the structures of proteins up to about 30 kDa in size can be deduced. Where both X-ray and NMR methods have been used to determine the structure of a protein, the results usually agree well. This suggests that the measured structures are the true *in vivo* structures.

Functional analysis

The three dimensional structures of many proteins have now been determined. This structural information is of great value in the rational development of new drugs designed to bind specifically and with high affinity to target proteins. However, tertiary and quaternary structural determination is still a relatively costly and laborious procedure and lags well behind the availability of new protein primary structures predicted from genomic sequencing projects (see Topic J2). Thus, there is now great interest in computational methods that will allow the prediction of both structure and possible function from simple amino acid sequence information. These methods are based largely on the fact that there are only a limited number of supersecondary structures in nature and involve mapping of new protein sequences on to the known three dimensional structures of proteins with related amino acid sequences. However, caution must still be exerted in interpreting such results. For example, lysozyme and α-lactalbumin share high sequence (about 70%) and structural similarity and are clearly closely related, yet α-lactalbumin has lost the essential catalytic amino acid residues that allow lysozyme to hydrolyse carbohydrates and so it only binds sugars. It is not an enzyme, as might be predicted on the basis of sequence similarity. Thus, at least for the moment, understanding of the true function of a protein still requires its isolation and biochemical and structural characterization. Isolation is greatly aided by recombinant techniques. This has allowed the production of minor proteins that have never even been detected before by cloning and expression of new genes identified in genome sequencing projects (see Topic J2). Identification of all the other proteins with which a protein interacts in the cell is another important aspect of functional analysis.

Knowledge of the biochemical properties of an isolated protein does not necessarily tell you what that protein does in the cell. Additional genetic analysis is usually required. If the gene for the protein can be inactivated by mutagenesis or deleted by recombinant DNA techniques, then the **phenotype** of the resulting mutant can be studied. In conjunction with the biochemical information, the altered behaviour of the mutant cell can help to pinpoint the function of the protein *in vivo*. All 6000 or so protein-coding genes of the yeast *Saccharomyces cerevisiae* have now been individually deleted to produce a set of mutants that should help to define the role of all the proteins in this relatively simple eukaryote.

Proteomics

By analogy with the word genome, the term **transcriptome** has been coined to describe the complement of mRNAs transcribed from a cell's genome (see Topic K1) while the term **proteome** is used to describe the total set of proteins expressed from the transcriptome of a cell either during its lifetime or at any one time during its existence. The proteome includes all the variants of a single gene product that may result from the alternative splicing of transcribed RNA (see Topic O4) and from post-translational modifications of a single protein product (see Topic Q4). Thus, while the genome is relatively constant from cell to cell within an organism, the transcriptome and proteome vary widely between cell types (e.g. between brain and liver) and even within a single cell type during the cell cycle (see Topic E3) or when exposed to different external stimuli that produce changes in gene expression (e.g. hormones, growth factors, stress).

Proteomics is the study of the proteome using techniques of high resolution protein separation and identification. Currently, the best separation method is **two dimensional gel electrophoresis**. Proteins extracted from a cell or tissue are first separated according to charge in a narrow tube of polyacrylamide gel by **isoelectric focusing**. The gel is then rotated by 90° and the proteins electrophoresed into a slab of gel containing SDS, which further separates them by mass. A 2-D map of protein spots is thus created which can contain several thousand resolved protein species. Individual spots are then cut from the gel and treated with proteases such as trypsin to produce a set of peptides characteristic of that protein. The precise masses of each peptide in the sample are then determined by MALDI mass spectrometry to produce a **peptide mass fingerprint** of that protein. This is then compared to a database of predicted fingerprints constructed for all known proteins from the abundant DNA sequence information that is now available for many organisms. Many of these procedures can be automated with robots and so large numbers of proteins can be identified within the proteome of a given cell. Such identification is of crucial importance to our understanding of how cells function and of how function changes during disease and is of great interest to pharmaceutical companies in their quest for new drug targets.

C1 NUCLEIC ACID STRUCTURE

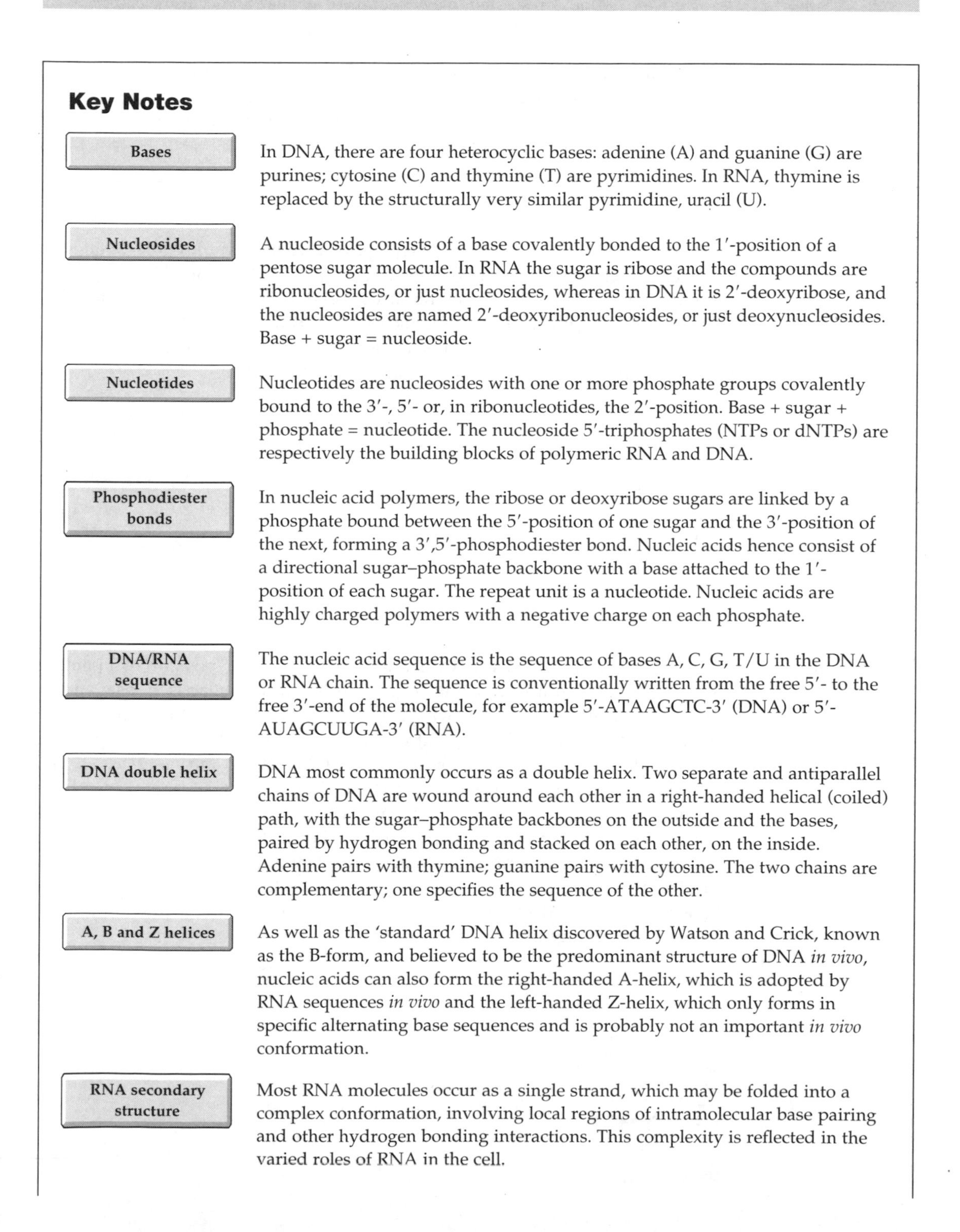

Key Notes

Bases

In DNA, there are four heterocyclic bases: adenine (A) and guanine (G) are purines; cytosine (C) and thymine (T) are pyrimidines. In RNA, thymine is replaced by the structurally very similar pyrimidine, uracil (U).

Nucleosides

A nucleoside consists of a base covalently bonded to the 1′-position of a pentose sugar molecule. In RNA the sugar is ribose and the compounds are ribonucleosides, or just nucleosides, whereas in DNA it is 2′-deoxyribose, and the nucleosides are named 2′-deoxyribonucleosides, or just deoxynucleosides. Base + sugar = nucleoside.

Nucleotides

Nucleotides are nucleosides with one or more phosphate groups covalently bound to the 3′-, 5′- or, in ribonucleotides, the 2′-position. Base + sugar + phosphate = nucleotide. The nucleoside 5′-triphosphates (NTPs or dNTPs) are respectively the building blocks of polymeric RNA and DNA.

Phosphodiester bonds

In nucleic acid polymers, the ribose or deoxyribose sugars are linked by a phosphate bound between the 5′-position of one sugar and the 3′-position of the next, forming a 3′,5′-phosphodiester bond. Nucleic acids hence consist of a directional sugar–phosphate backbone with a base attached to the 1′-position of each sugar. The repeat unit is a nucleotide. Nucleic acids are highly charged polymers with a negative charge on each phosphate.

DNA/RNA sequence

The nucleic acid sequence is the sequence of bases A, C, G, T/U in the DNA or RNA chain. The sequence is conventionally written from the free 5′- to the free 3′-end of the molecule, for example 5′-ATAAGCTC-3′ (DNA) or 5′-AUAGCUUGA-3′ (RNA).

DNA double helix

DNA most commonly occurs as a double helix. Two separate and antiparallel chains of DNA are wound around each other in a right-handed helical (coiled) path, with the sugar–phosphate backbones on the outside and the bases, paired by hydrogen bonding and stacked on each other, on the inside. Adenine pairs with thymine; guanine pairs with cytosine. The two chains are complementary; one specifies the sequence of the other.

A, B and Z helices

As well as the 'standard' DNA helix discovered by Watson and Crick, known as the B-form, and believed to be the predominant structure of DNA *in vivo*, nucleic acids can also form the right-handed A-helix, which is adopted by RNA sequences *in vivo* and the left-handed Z-helix, which only forms in specific alternating base sequences and is probably not an important *in vivo* conformation.

RNA secondary structure

Most RNA molecules occur as a single strand, which may be folded into a complex conformation, involving local regions of intramolecular base pairing and other hydrogen bonding interactions. This complexity is reflected in the varied roles of RNA in the cell.

Modified nucleic acids	Covalent modifications of nucleic acids have specific roles in the cell. In DNA, these are normally restricted to methylation of adenine and cytosine bases, but the range of modifications of RNA is much greater.

Related topics

Chemical and physical properties of nucleic acids (C2)
Spectroscopic and thermal properties of nucleic acids (C3)
DNA supercoiling (C4)
Prokaryotic and eukaryotic chromosome structure (Section D)

Bases

The **bases** of DNA and RNA are heterocyclic (carbon- and nitrogen-containing) aromatic rings, with a variety of substituents (*Fig. 1*). Adenine (A) and guanine (G) are **purines**, bicyclic structures (two fused rings), whereas cytosine (C), thymine (T) and uracil (U) are monocyclic **pyrimidines**. In RNA, the thymine base is replaced by uracil. Thymine differs from uracil only in having a methyl group at the 5-position, that is thymine is 5-methyluracil.

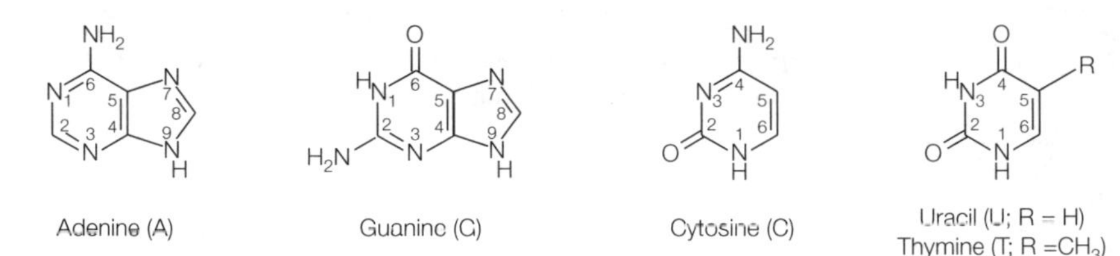

Fig. 1. Nucleic acid bases.

Nucleosides

In nucleic acids, the bases are covalently attached to the 1′-position of a pentose sugar ring, to form a **nucleoside** (*Fig. 2*). In RNA, the sugar is **ribose**, and in DNA, it is **2′-deoxyribose**, in which the hydroxyl group at the 2′-position is replaced by a hydrogen. The point of attachment to the base is the 1-position (*N*-1) of the pyrimidines and the 9-position (*N*-9) of the purines (*Fig. 1*). The numbers of the atoms in the ribose ring are designated 1′-, 2′-, etc., merely to distinguish them from the base atoms. The bond between the bases and the sugars is the **glycosylic (or glycosidic) bond**. If the sugar is ribose, the nucleosides (technically ribonucleosides) are adenosine, guanosine, cytidine and uridine. If the sugar is deoxyribose (as in DNA), the nucleosides (2′-deoxyribonucleosides) are deoxyadenosine, etc. Thymidine and deoxythymidine may be used interchangeably.

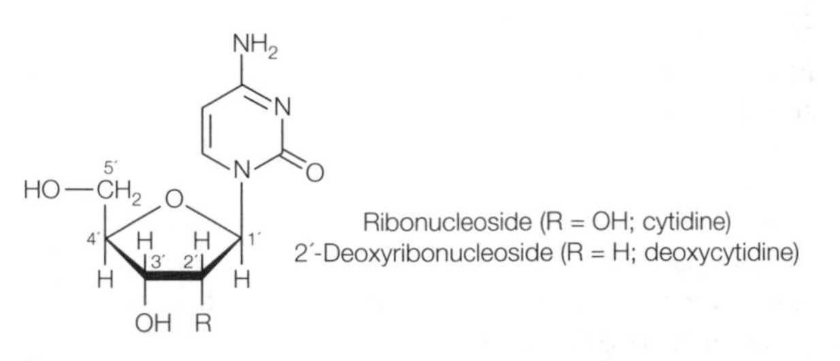

Fig. 2. Nucleosides.

Nucleotides

A **nucleotide** is a nucleoside with one or more phosphate groups bound covalently to the 3′-, 5′- or (in ribonucleotides only) the 2′-position. If the sugar is deoxyribose, then the compounds are termed **deoxynucleotides** (*Fig. 3*). Chemically, the compounds are phosphate esters. In the case of the 5′-position, up to three phosphates may be attached, to form, for example, adenosine 5′-triphosphate, or deoxyguanosine 5′-triphosphate, commonly abbreviated to ATP and dGTP respectively. In the same way, we have dCTP, UTP and dTTP (equivalent to TTP). 5′-Mono and -diphosphates are abbreviated as, for example, AMP and dGDP. Nucleoside 5′-triphosphates (NTPs), or deoxynucleoside 5′-triphosphates (dNTPs) are the building blocks of the polymeric nucleic acids. In the course of DNA or RNA synthesis, two phosphates are split off as pyrophosphate to leave one phosphate per nucleotide incorporated into the nucleic acid chain (see Topics E1 and K1). The repeat unit of a DNA or RNA chain is hence a nucleotide.

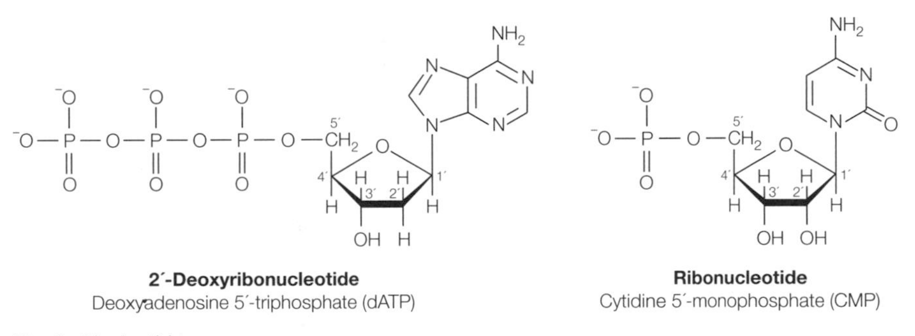

Fig. 3. Nucleotides.

Phosphodiester bonds

In a DNA or RNA molecule, deoxyribonucleotides or ribonucleotides respectively are joined into a polymer by the covalent linkage of a phosphate group between the 5′-hydroxyl of one ribose and the 3′-hydroxyl of the next (*Fig. 4*). This kind of bond or linkage is called a **phosphodiester bond**, since the phosphate is chemically in the form of a diester. A nucleic acid chain can hence be seen to have a direction. Any nucleic acid chain, of whatever length (unless it is circular; see Topic C4), has a free 5′-end, which may or may not have any attached phosphate groups, and a free 3′-end, which is most likely to be a free hydroxyl group. At neutral pH, each phosphate group has a single negative charge. This is why nucleic acids are termed acids; they are the anions of strong acids. Nucleic acids are thus **highly charged polymers**.

DNA/RNA sequence

Conventionally, the repeating monomers of DNA or RNA are represented by their single letters A, T, G, C or U. In addition, there is a convention to write the sequences with the 5′-end at the left. Hence a stretch of DNA sequence might be written 5′-ATAAGCTC-3′, or even just ATAAGCTC. An RNA sequence might be 5′-AUAGCUUGA-3′. Note that the directionality of the chain means that, for example, ATAAG is not the same as GAATA.

DNA double helix

DNA most commonly occurs in nature as the well-known **'double helix'**. The basic features of this structure were deduced by James Watson and Francis Crick in 1953. Two separate chains of DNA are wound around each other, each following a helical (coiling) path, resulting in a **right-handed** double helix (*Fig. 5a*). The negatively charged sugar–phosphate backbones of the molecules are on the outside, and the planar bases of each strand stack one above the

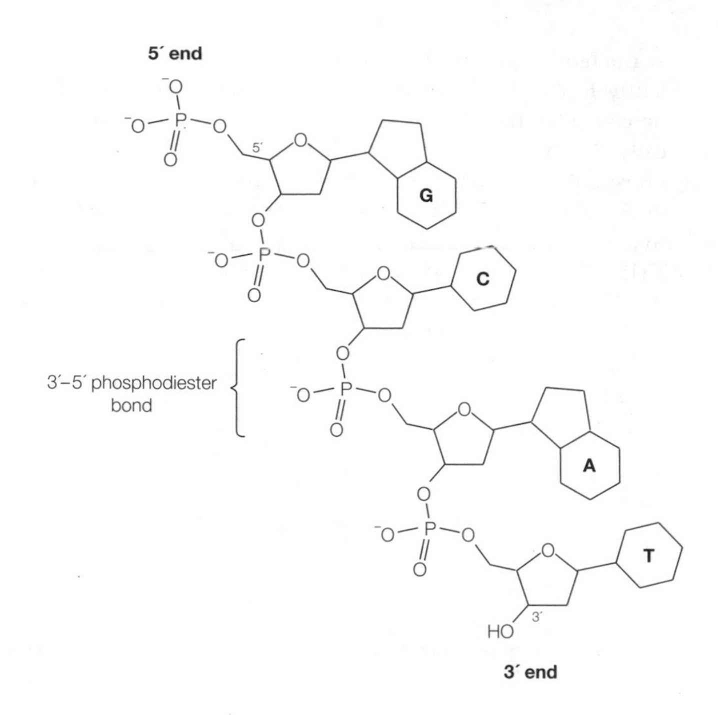

Fig. 4. Phosphodiester bonds and the covalent structure of a DNA strand.

other in the center of the helix (*Fig. 5b*). Between the backbone strands run the **major** and **minor grooves**, which also follow a helical path. The strands are joined noncovalently by hydrogen bonding between the bases on opposite strands, to form **base pairs**. There are around 10 base pairs per turn in the DNA double helix. The two strands are oriented in opposite directions (**antiparallel**) in terms of their 5′→3′ direction and, most crucially, the two strands are **complementary** in terms of sequence. This last feature arises because the structures of the bases and the constraints of the DNA backbone dictate that the bases hydrogen-bond to each other as purine–pyrimidine pairs which have very similar geometry and dimensions (*Fig. 6*). Guanine pairs with cytosine (three H-bonds) and adenine pairs with thymine (two H-bonds). Hence, any sequence can be accommodated within a regular double-stranded DNA structure. The sequence of one strand uniquely specifies the sequence of the other, with all that that implies for the mechanism of copying (replication) of DNA and the transcription of DNA sequence into RNA (see Topics E1 and K1).

A, B and Z helices

In fact, a number of different forms of nucleic acid double helix have been observed and studied, all having the basic pattern of two helically-wound antiparallel strands. The structure identified by Watson and Crick, and described above, is known as **B-DNA** (*Fig. 7a*), and is believed to be the idealized form of the structure adopted by virtually all DNA *in vivo*. It is characterized by a helical repeat of 10 bp/turn (although it is now known that 'real' B-DNA has a repeat closer to 10.5 bp/turn), by the presence of base pairs lying on the helix axis and almost perpendicular to it, and by having well-defined, deep major and minor grooves.

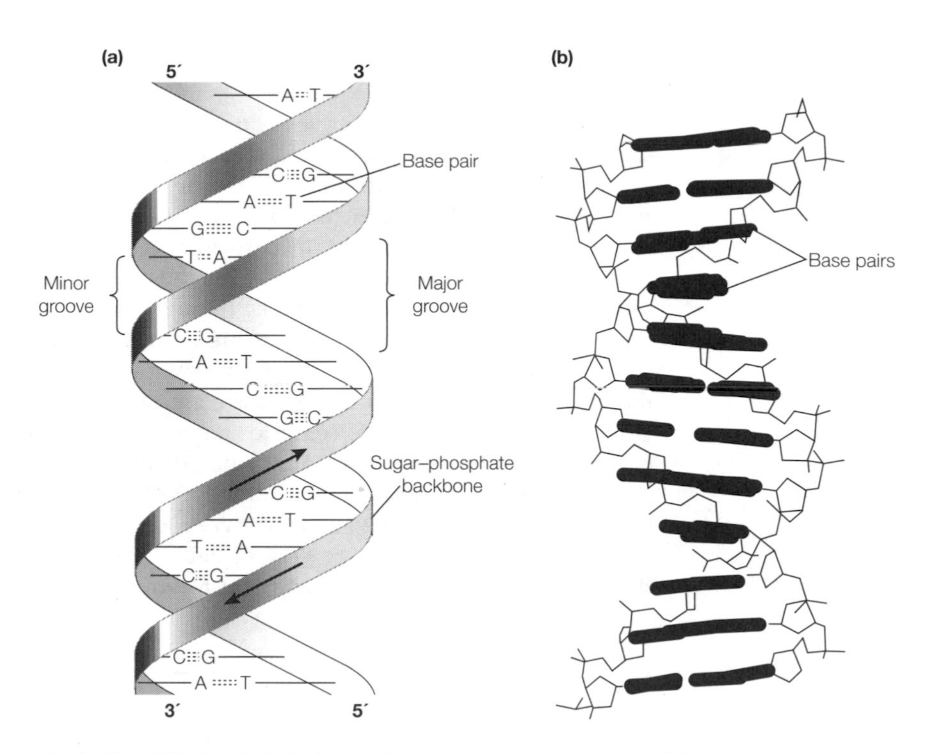

Fig. 5. The DNA double helix. (a) A schematic view of the structure; (b) a more detailed structure, highlighting the stacking of the base pairs (in bold).

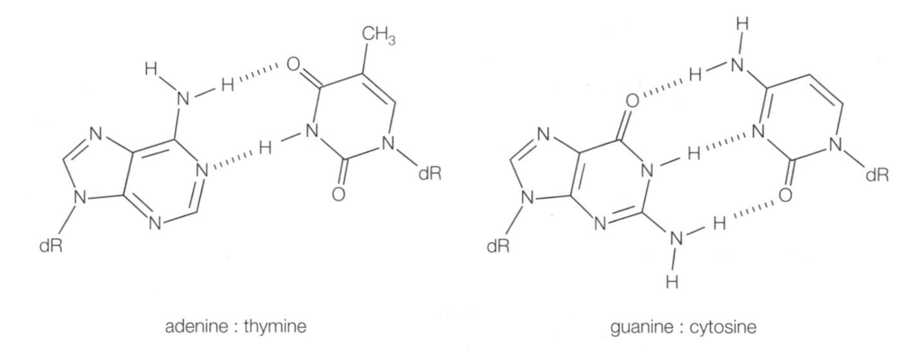

Fig. 6. The DNA base pairs. Hydrogen bonds are shown as dashed lines; dR = deoxyribose.

DNA can be induced to form an alternative helix, known as the **A-form** (*Fig. 7b*), under conditions of low humidity. The A-form is right-handed, like the B-form, but has a wider, more compressed structure in which the base pairs are tilted with respect to the helix axis, and actually lie off the axis (seen end-on, the A-helix has a hole down the middle). The helical repeat of the A-form is around 11 bp/turn. Although it may be that the A-form, or something close to it, is adopted by DNA *in vivo* under unusual circumstances, the major importance of the A-form is that it is the helix formed by RNA (see below), and by DNA-RNA hybrids; it turns out that it is impossible to fit the 2′-OH of RNA into the theoretically more stable B-form structure.

A further unusual helical structure can be formed by DNA. The left-handed **Z-DNA** (*Fig. 7c*) is stable in synthetic double stranded DNA consisting purely of alternating pyrimidine-purine sequence (such as 5´-CGCGCG-3´, with the same in the other strand, of course). This is because in this structure, the pyrimidine and the purine nucleotides adopt very different conformations, unlike in A- and B-form, where each nucleotide has essentially the same conformation and immediate environment. In particular, the purine nucleotides in the Z-form adopt the ***syn*** conformation, in which the purine base lies directly above the deoxyribose ring (imagine rotating through 180° around the glycosylic bond in *Fig. 3*; the nucleotides shown there are in the alternative ***anti*** conformation). The pyrimidine nucleotides, and all nucleotides in the A- and B- forms adopt the *anti* conformation. The Z-helix has a zig-zag appearance, with 12 bp/turn, although it probably makes sense to think of it as consisting of 6 'dimers of base pairs' per turn; the repeat unit along each strand is really a dinucleotide. Z-DNA does not easily form in normal DNA, even in regions of repeating CGCGCG, since the boundaries between the left-handed Z-form and the surrounding B-form would be very unstable. Although it has its enthusiasts, the Z-form is probably not a significant feature of DNA (or RNA) *in vivo*.

Table 1. Summary of the major features of A, B and Z nucleic acid helices

	A-form	B-form	Z-form
Helical sense	Right handed	Right handed	Left handed
Diameter	~2.6 nm	~2.0 nm	~1.8 nm
Base-pairs per helical turn (n)	11	10	12 (6 dimers)
Helical twist per bp (= $360/n$)	33°	36°	60° (per dimer)
Helix rise per bp (h)	0.26 nm	0.34 nm	0.37 nm
Helix pitch (= nh)	2.8 nm	3.4 nm	4.5 nm
Base tilt to helix axis	20°	6°	7°
Major groove	Narrow/deep	Wide/deep	Flat
Minor groove	Wide/shallow	Narrow/deep	Narrow/deep
Glycosylic bond	*anti*	*anti*	*anti* (pyr) *syn* (pur)

RNA secondary structure

RNA normally occurs as a single-stranded molecule, and hence it does not adopt a long regular helical structure like double-stranded DNA. RNA instead forms relatively globular conformations, in which local regions of helical structure are formed by **intramolecular** hydrogen bonding and base stacking within the single nucleic acid chain. These regions can form where one part of the RNA chain is complementary to another (see Topic P2, *Fig. 2*). This conformational variability is reflected in the more diverse roles of RNA in the cell, when compared with DNA. RNA structures range from short small nuclear RNAs, which help to mediate the splicing of pre-mRNAs in eukaryotic cells (see Topic O3), to large rRNA molecules, which form the structural backbone of the ribosomes and participate in the chemistry of protein synthesis (see Topic Q2).

Modified nucleic acids

The chemical modification of bases or nucleotides in nucleic acids is widespread, and has a number of specific roles. In cellular DNA, the modifications are restricted to the methylation of the *N*-6 position of adenine and the 4-amino group and the 5-position of cytosine (*Fig. 1*), although more complex modifications occur in some phage DNAs. These methylations have a role in restriction

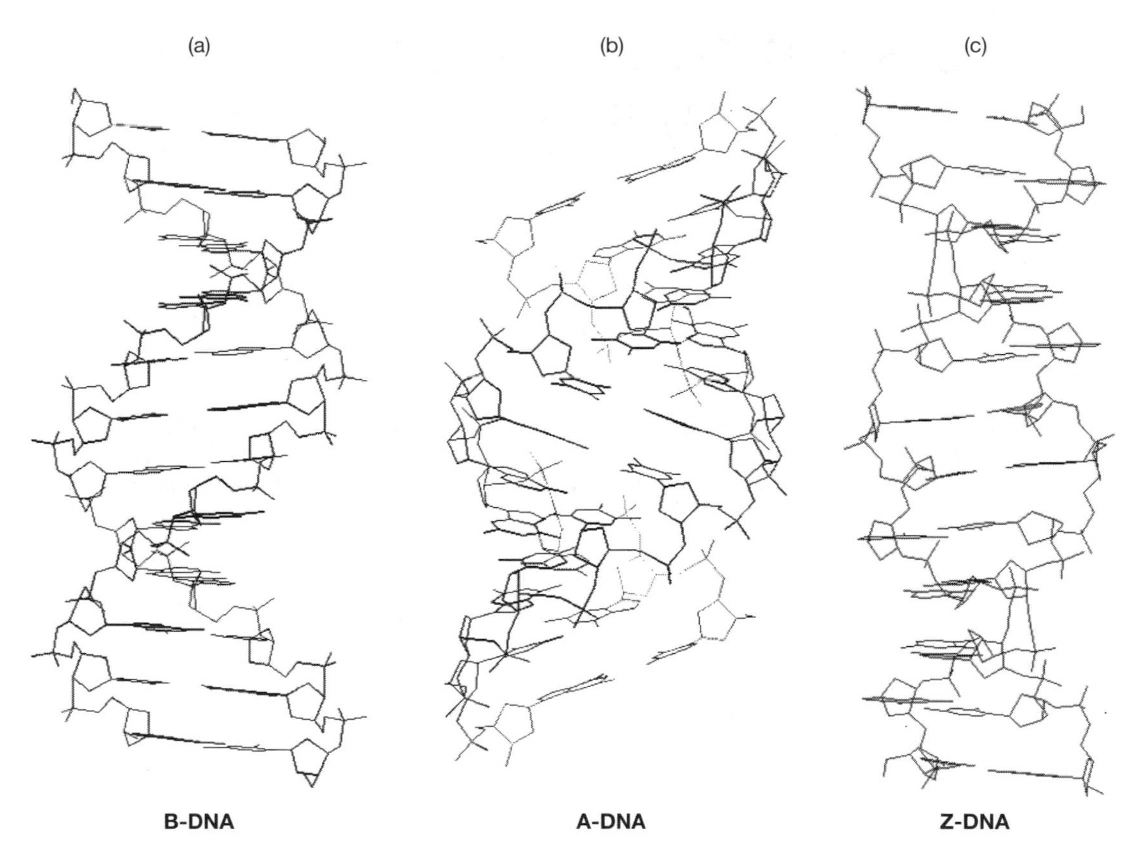

Fig. 7. The alternative helical forms of the DNA double helix.

modification (see Topic G3), base mismatch repair (see Topic F3) and eukaryotic genome structure (see Topic D3). A much more diverse range of modifications occurs in RNA after transcription, which again reflects the different roles of RNA in the cell. These are considered in more detail in Topics O3 and P2.

C2 CHEMICAL AND PHYSICAL PROPERTIES OF NUCLEIC ACIDS

Key Notes

Stability of nucleic acids

Although it might seem obvious that DNA double strands and RNA structures are stabilized by hydrogen bonding, this is not the case. H-bonds determine the specificity of the base pairing, but the stability of a nucleic acid helix is the result of hydrophobic and dipole–dipole interactions between the stacked base pairs.

Effect of acid

Highly acidic conditions may hydrolyze nucleic acids to their components: bases, sugar and phosphate. Moderate acid causes the hydrolysis of the purine base glycosylic bonds to yield apurinic acid. More complex chemistry has been developed to remove particular bases, and is the basis of chemical DNA sequencing.

Effect of alkali

High pH denatures DNA and RNA by altering the tautomeric state of the bases and disrupting specific hydrogen bonding. RNA is also susceptible to hydrolysis at high pH, by participation of the 2′-OH in intramolecular cleavage of the phosphodiester backbone.

Chemical denaturation

Some chemicals, such as urea and formamide, can denature DNA and RNA at neutral pH by disrupting the hydrophobic forces between the stacked bases.

Viscosity

DNA is very long and thin, and DNA solutions have a high viscosity. Long DNA molecules are susceptible to cleavage by shearing in solution – this process can be used to generate DNA of a specific average length.

Buoyant density

DNA has a density of around 1.7 g cm^{-3}, and can be analyzed and purified by its ability to equilibrate at its buoyant density in a cesium chloride density gradient formed in a centrifuge. The exact density of DNA is a function of its G+C content, and this technique may be used to analyze DNAs of different composition.

Related topics

Nucleic acid structure (C1)
Spectroscopic and thermal properties of nucleic acids (C3)

Stability of nucleic acids

At first sight, it might seem that the double helices of DNA and RNA secondary structure are stabilized by the hydrogen bonding between base pairs. In fact this is not the case. As in proteins (see Topic A4), the presence of H-bonds within a structure does not normally confer stability. This is because one must consider the *difference* in energy between, in the case of DNA, the single-stranded random coil state, and the double-stranded conformation. H-bonds between base pairs in double-stranded DNA merely replace what would be equally strong and energetically favorable H-bonds with water molecules in free solution, if the DNA were single-stranded. Hydrogen bonding contributes to the specificity required for base pairing in a double helix (double-stranded DNA will only form if the strands are complementary; see Topic C1), but it does not contribute to the overall stability of that helix. The root of this stability lies elsewhere, in the **stacking interactions** between the base pairs (see Topic C1, *Fig. 5b*). The flat surfaces of the aromatic bases cannot hydrogen-bond to water when they are in free solution, in other words they are **hydrophobic**. The hydrogen bonding network of bulk water becomes destabilized in the vicinity of a hydrophobic surface, since not all the water molecules can participate in full hydrogen bonding interactions, and they become more ordered. Hence it is energetically favorable to exclude water altogether from pairs of such surfaces by stacking them together; more water ends up in the bulk hydrogen-bonded network. This also maximizes the interaction between **charge dipoles** on the bases (see Topic A4). Even in single-stranded DNA, the bases have a tendency to stack on top of each other, but this stacking is maximized in double-stranded DNA, and the **hydrophobic effect** ensures that this is the most energetically favorable arrangement. In fact this is a simplified discussion of a rather complex phenomenon; a fuller explanation is beyond the scope of this book.

Effect of acid

In strong acid and at elevated temperatures, for example perchloric acid ($HClO_4$) at more than 100°C, nucleic acids are **hydrolyzed** completely to their constituents: bases, ribose or deoxyribose and phosphate. In more dilute mineral acid, for example at pH 3–4, the most easily hydrolyzed bonds are selectively broken. These are the glycosylic bonds attaching the purine bases to the ribose ring, and hence the nucleic acid becomes **apurinic** (see Topic F2). More complex chemistry has been developed which removes bases specifically, and cleaves the DNA or RNA backbone at particular bases. This forms the basis of the chemical DNA sequencing method developed by **Maxam and Gilbert**, which bears their name (see Topic J2).

Effect of alkali

DNA

Increasing pH above the physiological range (pH 7–8) has more subtle effects on DNA structure. The effect of alkali is to change the **tautomeric state** of the bases. This effect can be seen with reference to the model compound, cyclohexanone (*Fig. 1a*). The molecule is in equilibrium between the tautomeric keto and enol forms (1) and (2). At neutral pH, the compound is predominantly in the **keto** form (1). Increasing the pH causes a shift to the **enolate** form (3) when the molecule loses a proton, since the negative charge is most stably accommodated on the electronegative oxygen atom. In the same way, the structure of guanine (*Fig. 1b*) is also shifted to the enolate form at high pH, and analogous shifts take place in the structures of the other bases. This affects the specific hydrogen bonding between the base pairs, with the result that the double-stranded structure of the DNA breaks down; that is the DNA becomes **denatured** (*Fig. 1c*).

RNA

In RNA, the same denaturation of helical regions will take place at higher pH, but this effect is overshadowed by the susceptibility of RNA to **hydrolysis** in

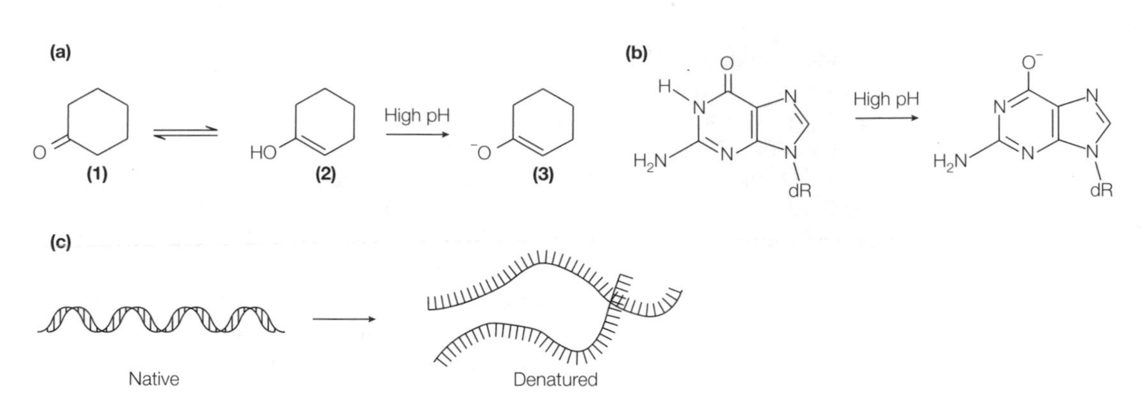

Fig. 1. The denaturation of DNA at high pH. (a) Alkali shifts the tautomeric ratio to the enolate form; (b) the tautomeric shift of deoxyguanosine; (c) the denaturation of double-helical DNA.

alkali. This comes about because of the presence of the 2′-OH group in RNA, which is perfectly positioned to participate in the cleavage of the RNA backbone by intramolecular attack on the phosphate of the phosphodiester bond (*Fig.* 2). This reaction is promoted by high pH, since ^{-}OH acts as a general base. The products are a free 5′-OH and a **2′,3′-cyclic phosphodiester**, which is subsequently hydrolyzed to either the 2′- or 3′-monophosphate. Even at neutral pH, RNA is much more susceptible to hydrolysis than DNA, which of course lacks the 2′-OH. This is a plausible reason why DNA may have evolved to incorporate 2′-deoxyribose, since its function requires extremely high stability.

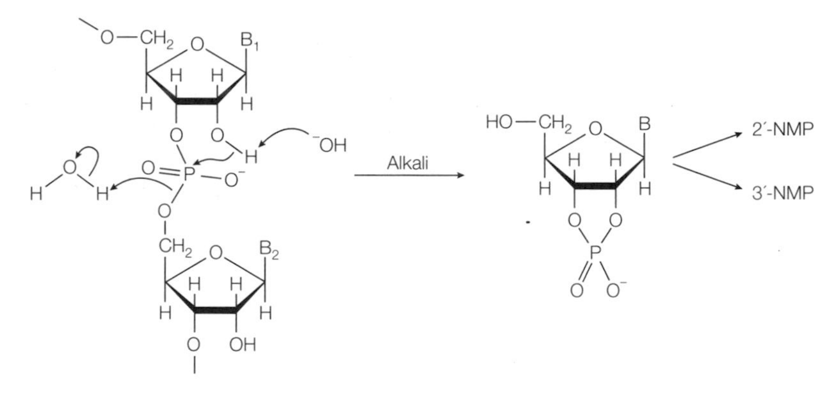

Fig. 2. Intramolecular cleavage of RNA phosphodiester bonds in alkali.

Chemical denaturation

A number of chemical agents can cause the denaturation of DNA or RNA at neutral pH, the best known examples being **urea** (H_2NCONH_2) and **formamide** ($HCONH_2$). A relatively high concentration of these agents (several molar) has the effect of disrupting the hydrogen bonding of the bulk water solution. This means that the energetic stabilization of the nucleic acid secondary structure, caused by the exclusion of water from between the stacked hydrophobic bases, is lessened and the strands become denatured.

Viscosity

Cellular DNA is very long and thin; technically, it has a high **axial ratio**. DNA is around 2 nm in diameter, and may have a length of micrometers, millimeters or even several centimeters in the case of eukaryotic chromosomes. To give a flavor of this, if DNA had the same diameter as spaghetti, then the *E. coli* chromosome (4.6 million base pairs) would have a length of around 1 km. In addition, DNA is a relatively stiff molecule; its stiffness would be similar to that of partly cooked spaghetti, using the same analogy. A consequence of this is that DNA solutions have a **high viscosity**. Furthermore, long DNA molecules can easily be damaged by **shearing** forces, or by **sonication** (high-intensity ultrasound), with a concomitant reduction in viscosity. Sensitivity to shearing is a problem if very large DNA molecules are to be isolated intact, although sonication may be used to produce DNA of a specified average length (see Topic D4). Note that neither shearing nor sonication denatures the DNA; they merely reduce the length of the double-stranded molecules in the solution.

Buoyant density

Analysis and purification of DNA can be carried out according to its density. In solutions containing high concentrations of a high molecular weight salt, for example 8 M **cesium chloride** (CsCl), DNA has a similar density to the bulk solution, around 1.7 g cm^{-3}. If the solution is centrifuged at very high speed, the dense cesium salt tends to migrate down the tube, setting up a **density gradient** (*Fig. 3*). Eventually the DNA sample will migrate to a sharp band at a position in the gradient corresponding to its own **buoyant density**. This technique is known as **equilibrium density gradient centrifugation** or **isopycnic centrifugation** (Greek for 'same density'; see Topic A2). Since, under these conditions, RNA pellets at the bottom of the tube and protein floats, this can be an effective way of purifying DNA away from these two contaminants (see Topic G2). However, the method is also analytically useful, since the precise buoyant density of the DNA (ρ) is a linear function of its **G+C content**:

$$\rho = 1.66 + 0.098 \times \text{Frac (G + C)}$$

Hence, the sedimentation of DNA may be used to determine its average G+C content or, in some cases, DNA fragments with different G+C contents from the bulk sequence can be separated from it (see Topic D4).

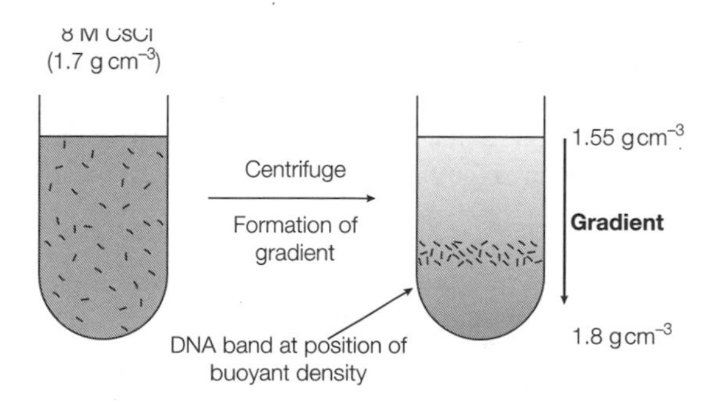

Fig. 3. Equilibrium density gradient centrifugation of DNA.

C3 SPECTROSCOPIC AND THERMAL PROPERTIES OF NUCLEIC ACIDS

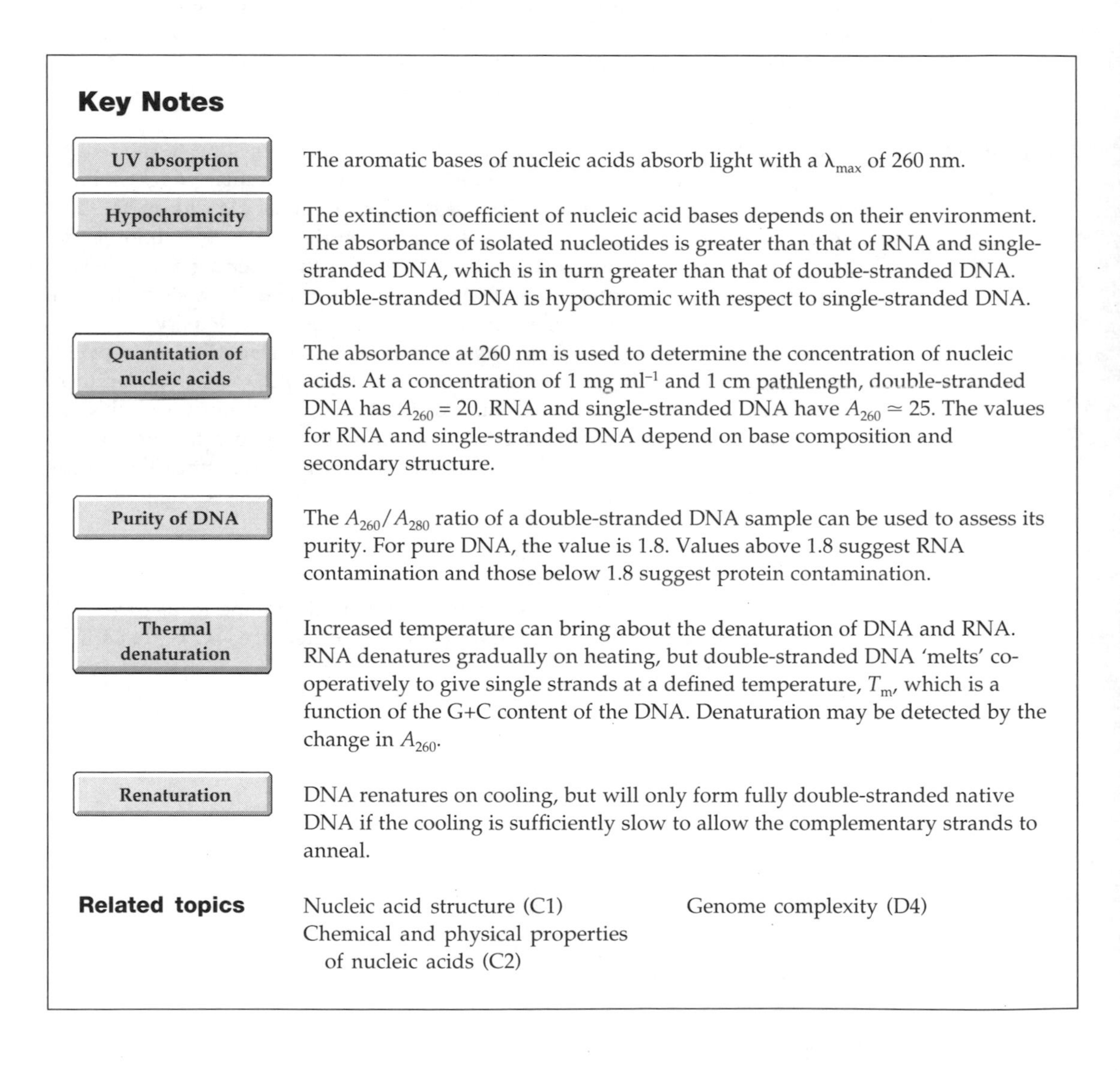

Key Notes

UV absorption — The aromatic bases of nucleic acids absorb light with a λ_{max} of 260 nm.

Hypochromicity — The extinction coefficient of nucleic acid bases depends on their environment. The absorbance of isolated nucleotides is greater than that of RNA and single-stranded DNA, which is in turn greater than that of double-stranded DNA. Double-stranded DNA is hypochromic with respect to single-stranded DNA.

Quantitation of nucleic acids — The absorbance at 260 nm is used to determine the concentration of nucleic acids. At a concentration of 1 mg ml^{-1} and 1 cm pathlength, double-stranded DNA has $A_{260} = 20$. RNA and single-stranded DNA have $A_{260} \simeq 25$. The values for RNA and single-stranded DNA depend on base composition and secondary structure.

Purity of DNA — The A_{260}/A_{280} ratio of a double-stranded DNA sample can be used to assess its purity. For pure DNA, the value is 1.8. Values above 1.8 suggest RNA contamination and those below 1.8 suggest protein contamination.

Thermal denaturation — Increased temperature can bring about the denaturation of DNA and RNA. RNA denatures gradually on heating, but double-stranded DNA 'melts' co-operatively to give single strands at a defined temperature, T_m, which is a function of the G+C content of the DNA. Denaturation may be detected by the change in A_{260}.

Renaturation — DNA renatures on cooling, but will only form fully double-stranded native DNA if the cooling is sufficiently slow to allow the complementary strands to anneal.

Related topics — Nucleic acid structure (C1); Chemical and physical properties of nucleic acids (C2); Genome complexity (D4)

UV absorption

Nucleic acids absorb UV light due to the conjugated aromatic nature of the bases; the sugar–phosphate backbone does not contribute appreciably to absorption. The wavelength of maximum absorption of light by both DNA and RNA is 260 nm ($\lambda_{max} = 260$ nm), which is conveniently distinct from the λ_{max} of protein (280 nm) (see Topic B1). The absorption properties of nucleic acids can be used for detection, quantitation and assessment of purity.

Hypochromicity

Although the λ_{max} for DNA or RNA bases is constant, the extinction coefficient depends on the environment of the bases. The absorbance at 260 nm (A_{260}) is greatest for isolated nucleotides, intermediate for single-stranded DNA (ssDNA) or RNA, and least for double-stranded DNA (dsDNA). This effect is caused by the fixing of the bases in a hydrophobic environment by stacking (see Topics C1 and C2). The classical term for this change in absorbance is **hypochromicity**, that is, dsDNA is **hypochromic** (from the Greek for 'less colored') relative to ssDNA. Alternatively, ssDNA may be said to be **hyperchromic** when compared with dsDNA.

Quantitation of nucleic acids

It is not convenient to speak of the molar extinction coefficient (ε) of a nucleic acid, since its value will depend on the length of the molecule in question; instead, extinction coefficients are usually quoted in terms of concentration in mg ml^{-1}: 1 mg ml^{-1} dsDNA has an A_{260} of 20. The corresponding value for RNA or ssDNA is approximately 25. The values for single-stranded DNA and RNA are approximate for two reasons; the values are the sum of absorbances contributed by the different bases (purines have a higher extinction coefficient than pyrimidines), and hence are sensitive to the base composition of the molecule. Double-stranded DNA has equal numbers of purines and pyrimidines, and so does not show this effect. The absorbance values also depend on the amount of secondary structure (double-stranded regions) in a given molecule, due to hypochromicity. In the case of a short oligonucleotide, where secondary structure will be minimal, it is usual to calculate the extinction coefficient based on the sum of values for individual nucleotides in the single strand.

Purity of DNA

The approximate purity of dsDNA preparations (see Topic G2) may be estimated by determination of the ratio of absorbance at 260 and 280 nm (A_{260}/A_{280}). The shape of the absorption spectrum (*Fig. 1*), as well as the extinction coefficient, varies with the environment of the bases such that pure dsDNA has an A_{260}/A_{280} of 1.8, and pure RNA one of around 2.0. Protein, of course, with λ_{max} = 280 nm has a 260/280 ratio of less than 1 (actually around 0.5). Hence, if a DNA sample has an A_{260}/A_{280} greater than 1.8, this suggests RNA contamination, whereas one less than 1.8 suggests protein in the sample.

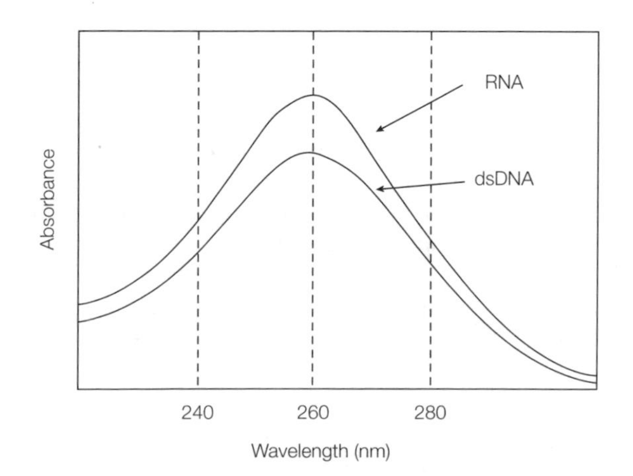

Fig. 1. The ultraviolet absorption spectra of RNA and double-stranded DNA solutions at equal concentration (mg ml^{-1}).

Thermal denaturation

A number of chemicals can bring about the denaturation of nucleic acids (see Topic C2). Heating also leads to the destruction of double-stranded hydrogen-bonded regions of DNA and RNA. The process of denaturation can be observed conveniently by the increase in absorbance as double-stranded nucleic acids are converted to single strands (*Fig. 2*).

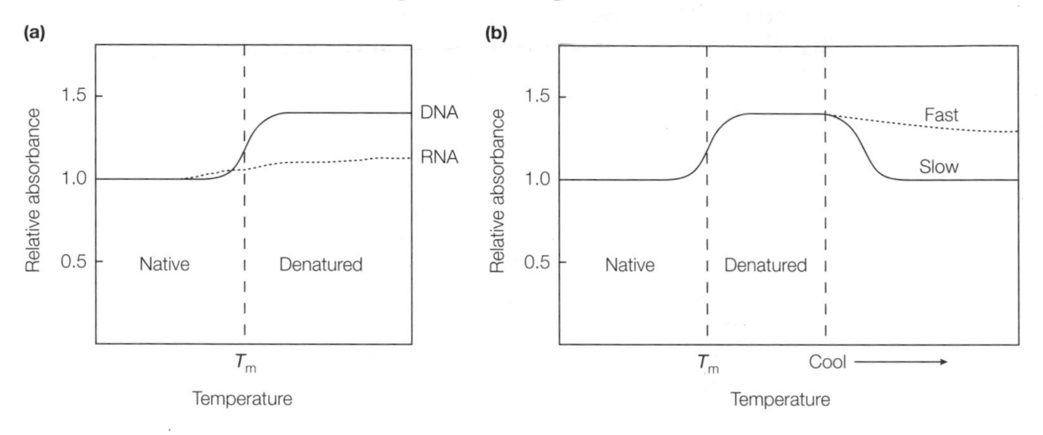

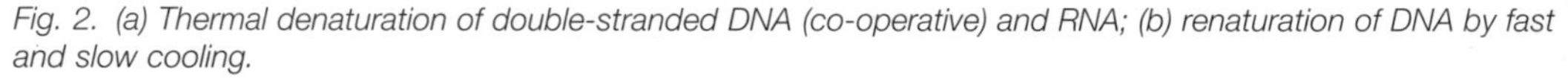
Fig. 2. (a) Thermal denaturation of double-stranded DNA (co-operative) and RNA; (b) renaturation of DNA by fast and slow cooling.

The thermal behaviors of dsDNA and RNA are very different. As the temperature is increased, the absorbance of an RNA sample gradually and erratically increases as the stacking of the bases in double-stranded regions is reduced. Shorter regions of base pairing will denature before longer regions, since they are more thermally mobile. In contrast, the **thermal denaturation**, or **melting**, of dsDNA is **co-operative**. The denaturation of the ends of the molecule, and of more mobile AT-rich internal regions, will destabilize adjacent regions of helix, leading to a progressive and concerted melting of the whole structure at a well-defined temperature corresponding to the mid-point of the smooth transition, and known as the **melting temperature** (**T_m**) (*Fig. 2a*). The melting is accompanied by a 40% increase in absorbance. T_m is a function of the G+C content of the DNA sample, and ranges from 80°C to 100°C for long DNA molecules.

Renaturation

The thermal denaturation of DNA may be reversed by cooling the solution. In this case, the rate of cooling has an influence on the outcome. Rapid cooling allows only the formation of local regions of dsDNA, formed by the base pairing or **annealing** of short regions of complementarity within or between DNA strands; the decrease in A_{260} is hence rather small (*Fig. 2b*). On the other hand, slow cooling allows time for the wholly complementary DNA strands to find each other, and the sample can become fully double-stranded, with the same absorbance as the original native sample. The renaturation of regions of complementarity between different nucleic acid strands is known as **hybridization**.

C4 DNA SUPERCOILING

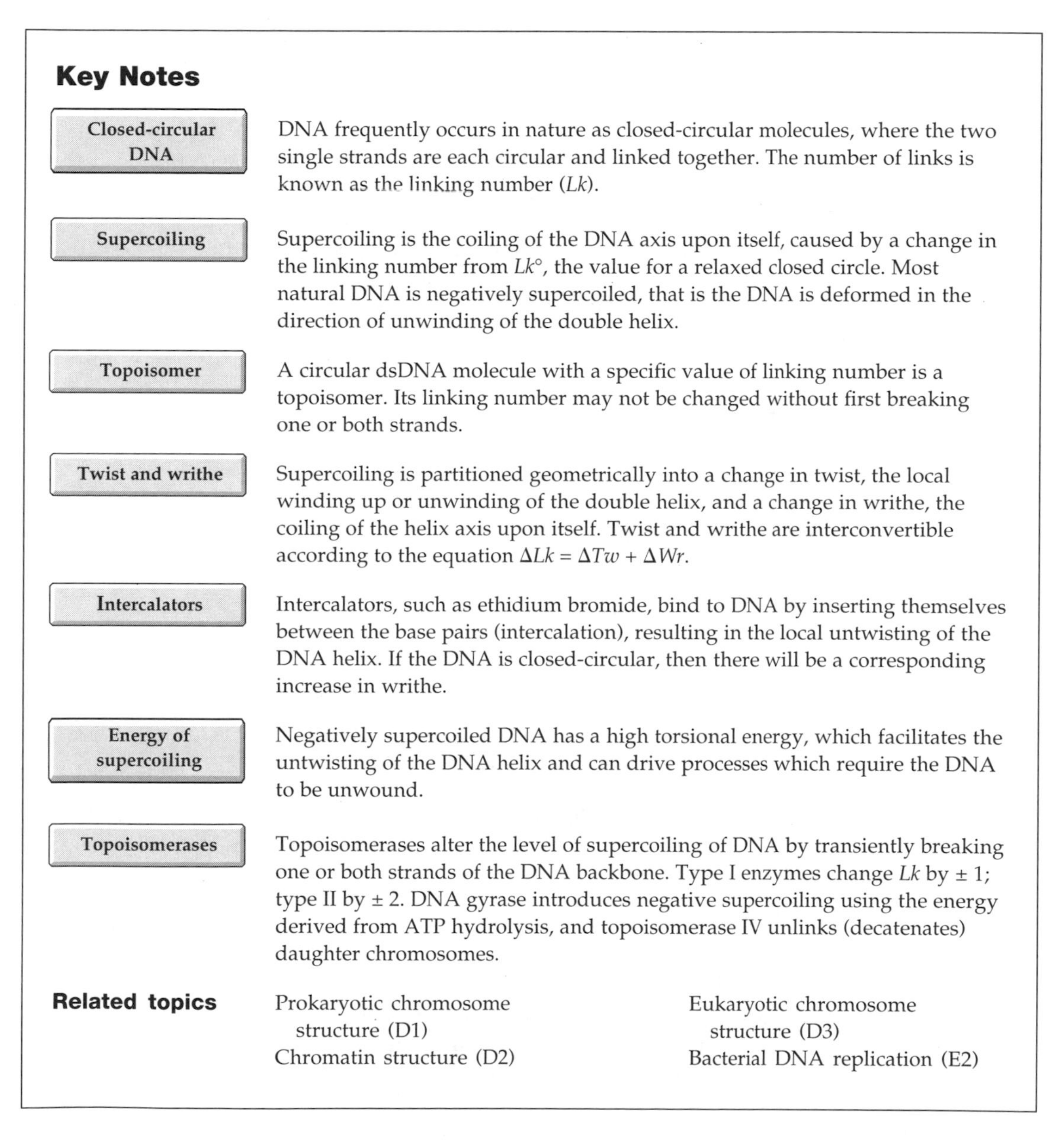

Key Notes

Closed-circular DNA — DNA frequently occurs in nature as closed-circular molecules, where the two single strands are each circular and linked together. The number of links is known as the linking number (*Lk*).

Supercoiling — Supercoiling is the coiling of the DNA axis upon itself, caused by a change in the linking number from $Lk°$, the value for a relaxed closed circle. Most natural DNA is negatively supercoiled, that is the DNA is deformed in the direction of unwinding of the double helix.

Topoisomer — A circular dsDNA molecule with a specific value of linking number is a topoisomer. Its linking number may not be changed without first breaking one or both strands.

Twist and writhe — Supercoiling is partitioned geometrically into a change in twist, the local winding up or unwinding of the double helix, and a change in writhe, the coiling of the helix axis upon itself. Twist and writhe are interconvertible according to the equation $\Delta Lk = \Delta Tw + \Delta Wr$.

Intercalators — Intercalators, such as ethidium bromide, bind to DNA by inserting themselves between the base pairs (intercalation), resulting in the local untwisting of the DNA helix. If the DNA is closed-circular, then there will be a corresponding increase in writhe.

Energy of supercoiling — Negatively supercoiled DNA has a high torsional energy, which facilitates the untwisting of the DNA helix and can drive processes which require the DNA to be unwound.

Topoisomerases — Topoisomerases alter the level of supercoiling of DNA by transiently breaking one or both strands of the DNA backbone. Type I enzymes change *Lk* by ± 1; type II by ± 2. DNA gyrase introduces negative supercoiling using the energy derived from ATP hydrolysis, and topoisomerase IV unlinks (decatenates) daughter chromosomes.

Related topics

Prokaryotic chromosome structure (D1)
Chromatin structure (D2)
Eukaryotic chromosome structure (D3)
Bacterial DNA replication (E2)

Closed-circular DNA

Many DNA molecules in cells consist of **closed-circular** double-stranded molecules, for example bacterial plasmids and chromosomes and many viral DNA molecules. This means that the two complementary single strands are each joined into circles, 5′ to 3′, and are twisted around one another by the helical path of the DNA. The molecule has no free ends, and the two single strands are linked together a number of times corresponding to the number of double-helical turns in the molecule. This number is known as the **linking number (*Lk*)**.

Supercoiling

A number of properties arise from this circular constraint of a DNA molecule. A good way to imagine these is to consider the DNA double helix as a piece of rubber tubing, with a line drawn along its length to enable us to follow its twisting. The tubing may be joined by a connector into a closed circle (*Fig. 1*). If we imagine a twisting of the DNA helix (tubing) followed by the joining of the ends, then the deformation so formed is locked into the system (*Fig. 1*). This deformation is known as **supercoiling**, since it manifests itself as a coiling of the DNA axis around itself in a higher-order coil, and corresponds to a change in linking number from the simple circular situation. If the twisting of the DNA is in the same direction as that of the double helix, that is the helix is twisted up before closure, then the supercoiling formed is **positive**; if the helix is untwisted, then the supercoiling is **negative**. Almost all DNA molecules in cells are on average negatively supercoiled. This is true even for linear DNAs such as eukaryotic chromosomes, which are constrained into large loops by interaction with a protein scaffold (see Topics D2 and E3).

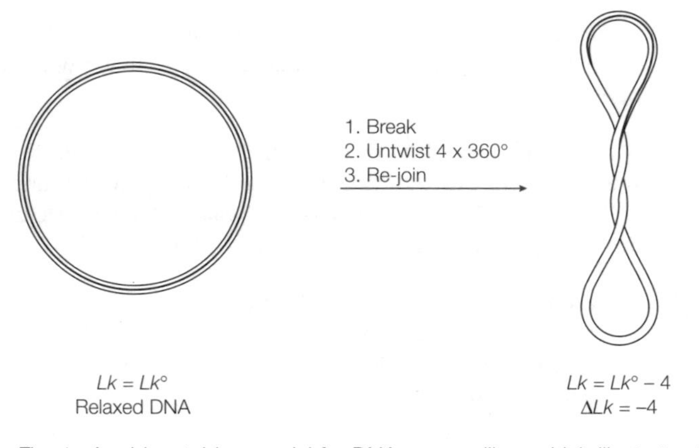

Fig. 1. A rubber tubing model for DNA supercoiling, which illustrates the change in DNA conformation. The changes in linking number are indicated (see text for details).

The level of supercoiling may be quantified in terms of the change in linking number (**ΔLk**) from that of the unconstrained (**relaxed**) closed-circular molecule (**$Lk°$**). This corresponds to the number of 360° twists introduced before ring closure. DNA when isolated from cells is commonly negatively supercoiled by around six turns per 100 turns of helix (1000 bp), that is $\Delta Lk/Lk° = -0.06$.

Topoisomer

The linking number of a closed-circular DNA is a **topological** property, that is one which cannot be changed without breaking one or both of the DNA backbones. A molecule of a given linking number is known as a **topoisomer**. Topoisomers differ from each other only in their linking number.

Twist and writhe

The conformation (geometry) of the DNA can be altered while the linking number remains constant. Two extreme conformations of a supercoiled DNA topoisomer may be envisaged (*Fig. 2*), corresponding to the partition of the supercoiling (ΔLk) completely into **writhe** (*Fig. 2a*) or completely into **twist** (*Fig. 2c*). The line on the rubber tubing model helps to keep track of local twisting of the DNA axis. The equilibrium situation lies between these two extremes (*Fig. 2b*), and corresponds to some change in both twist and writhe induced by supercoiling. This partition may be expressed by the equation:

$$\Delta Lk = \Delta Tw + \Delta Wr$$

ΔLk must be an integer, but ΔTw and ΔWr need not be. The topological change in supercoiling of a DNA molecule is partitioned into a conformational change of twist and/or a change of writhe.

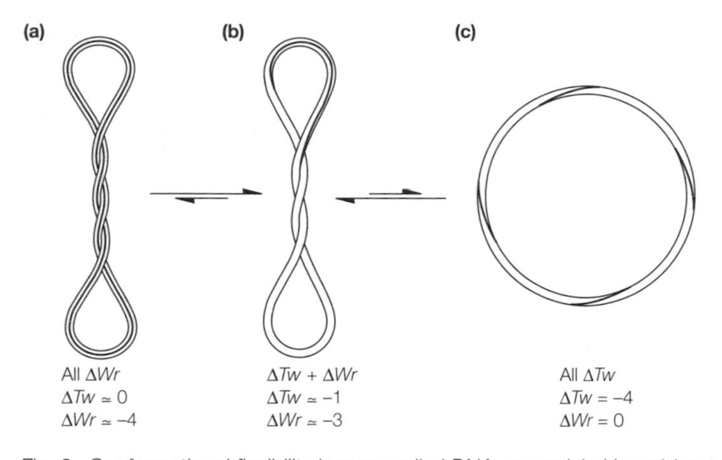

Fig. 2. Conformational flexibility in supercoiled DNA as modeled by rubber tubing. The changes in twist and writhe at constant linking number are shown (see text for details).

Intercalators

The geometry of a supercoiled molecule may be altered by any factor which affects the intrinsic twisting of the DNA helix. For example, an increase in temperature reduces the twist, and an increase in ionic strength may increase the twist. One important factor is the presence of an **intercalator**. The best-known example of an intercalator is **ethidium bromide** (*Fig. 3a*). This is a positively charged polycyclic aromatic compound, which binds to DNA by inserting itself between the base pairs (**intercalation**; *Fig. 3b*). In addition to a large increase in fluorescence of the ethidium bromide molecule on binding, which is the basis of its use as a stain of DNA in gels (see Topic G4), the binding causes a local unwinding of the helix by around 26°. This is a reduction in twist, and hence results in a corresponding increase in writhe in a closed-circular molecule. In a negatively supercoiled molecule, this corresponds to a decrease in the (negative) writhe of the molecule, and the shape of the molecule will be altered in the direction a→b→c in *Fig. 2*.

Energy of supercoiling

Supercoiling involves the introduction of **torsional stress** into DNA molecules. Supercoiled DNA hence has a higher energy than relaxed DNA. For negative supercoiling, this energy makes it easier for the DNA helix to be locally untwisted, or unwound. Negative supercoiling may thus facilitate processes which require the unwinding of the helix, such as **transcription initiation** or **replication** (see Topics E1 and K1).

Topoisomerases

Enzymes exist which regulate the level of supercoiling of DNA molecules; these are termed **topoisomerases**. To alter the linking number of DNA, the enzymes must transiently break one or both DNA strands, which they achieve by the attack of a tyrosine residue on a backbone phosphate, resulting in a temporary covalent attachment of the enzyme to one of the DNA ends via a **phosphotyrosine bond**. There are two classes of topoisomerase. **Type I** enzymes break one strand of the DNA, and change the linking number in steps of ± 1 by passing the other strand

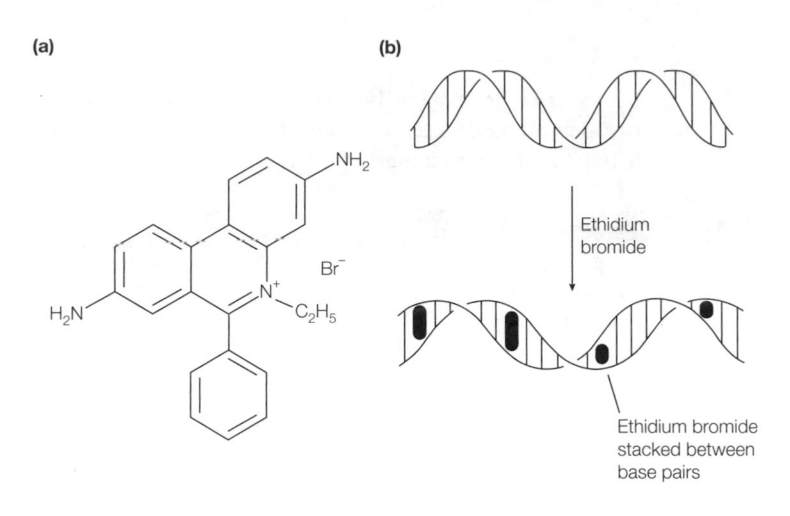

Fig. 3. (a) Ethidium bromide; (b) the process of intercalation, illustrating the lengthening and untwisting of the DNA helix.

through the break (*Fig. 4a*). **Type II** enzymes, which require the hydrolysis of ATP, break both strands of DNA and change the linking number in steps of ± 2, by the transfer of another double-stranded segment through the break (*Fig. 4b*). Most topoisomerases reduce the level of positive or negative supercoiling, that is they operate in the energetically favorable direction. However, **DNA gyrase**, a bacterial type II enzyme, uses the energy of ATP hydrolysis to introduce negative supercoiling into DNA hence removing positive supercoiling generated during replication (see Topic E2). Since their mechanism involves the passing of one double strand through another, type II topoisomerases are also able to unlink DNA molecules, such as daughter molecules produced in replication, which are linked (**catenated**) together (see Topic E2). In bacteria, this function is carried out by **topoisomerase IV**. Topoisomerases are essential enzymes in all organisms, being involved in replication, recombination and transcription (see Topics E1, F4 and K1). DNA gyrase and topoisomerase IV are the targets of **anti-bacterial drugs** in bacteria, and both type I and type II enzymes are the target of **anti-tumor agents** in humans.

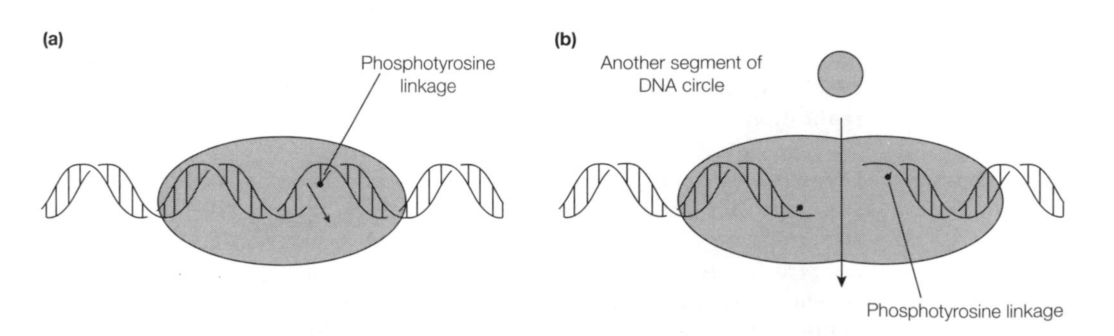

Fig. 4. The mechanisms of (a) type I and (b) type II topoisomerases (see text for details).

D1 PROKARYOTIC CHROMOSOME STRUCTURE

Key Notes

The *Escherichia coli* chromosome

The *E. coli* chromosome is a closed-circular DNA of length 4.6 million base pairs, which resides in a region of the cell called the nucleoid. In normal growth, the DNA is being replicated continuously.

DNA domains

The genome is organized into 50–100 large loops or domains of 50–100 kb in length, which are constrained by binding to a membrane–protein complex.

Supercoiling of the genome

The genome is negatively supercoiled. Individual domains may be topologically independent, that is they may be able to support different levels of supercoiling.

DNA-binding proteins

The DNA domains are compacted by wrapping around nonspecific DNA-binding proteins such as HU and H-NS (histone-like proteins). These proteins constrain about half of the supercoiling of the DNA. Other molecules such as integration host factor, RNA polymerase and mRNA may help to organize the nucleoid.

Related topics

DNA supercoiling (C4)
Chromatin structure (D2)
Genome complexity (D4)

The *Escherichia coli* chromosome

Prokaryotic genomes are exemplified by the *E. coli* chromosome. The bulk of the DNA in *E. coli* cells consists of a single closed-circular (see Topic C4) DNA molecule of length 4.6 million base pairs. The DNA is packaged into a region of the cell known as the **nucleoid**. This region has a very high DNA concentration, perhaps 30–50 mg ml^{-1}, as well as containing all the proteins associated with DNA, such as polymerases, repressors and others (see below). A fairly high DNA concentration in the test tube would be 1 mg ml^{-1}. In normal growth, the DNA is being replicated continuously and there may be on average around two copies of the genome per cell, when growth is at the maximal rate (see Topic E2).

DNA domains

Experiments in which DNA from *E. coli* is carefully isolated free of most of the attached proteins and observed under the electron microscope reveal one level of organization of the nucleoid. The DNA consists of 50–100 **domains** or **loops**, the ends of which are constrained by binding to a structure which probably consists of proteins attached to part of the cell membrane (*Fig. 1*). The loops are about 50–100 kb in size. It is not known whether the loops are static or dynamic, but one model suggests that the DNA may spool through sites of polymerase or other enzymic action at the base of the loops.

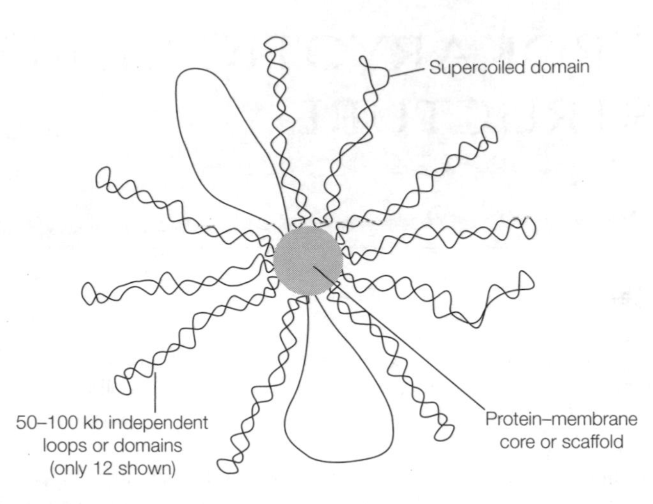

Fig. 1. A schematic view of the structure of the E. coli *chromosome (4600 kb) as visualized by electron microscopy. The thin line is the DNA double helix.*

Supercoiling of the genome

The *E. coli* chromosome as a whole is negatively supercoiled ($\Delta Lk/Lk° = -0.06$; see Topic C4), although there is some evidence that individual domains may be supercoiled independently. Electron micrographs indicate that some domains may not be supercoiled, perhaps because the DNA has become broken in one strand (see Topic C4), where other domains clearly do contain supercoils (*Fig. 1*). The attachment of the DNA to the protein–membrane scaffold may act as a barrier to rotation of the DNA, such that the domains may be **topologically independent**. There is, however, no real biochemical evidence for major differences in the level of supercoiling in different regions of the chromosome *in vivo*.

DNA-binding proteins

The looped DNA domains of the chromosome are constrained further by interaction with a number of DNA-binding proteins. The most abundant of these are **protein HU**, a small basic (positively charged) dimeric protein, which binds DNA nonspecifically by the wrapping of the DNA around the protein, and **H-NS** (formerly known as **protein H1**), a monomeric neutral protein, which also binds DNA nonspecifically in terms of sequence, but which seems to have a preference for regions of DNA which are intrinsically bent. These proteins are sometimes known as **histone-like** proteins (see Topic D2), and have the effect of compacting the DNA, which is essential for the packaging of the DNA into the nucleoid, and of stabilizing and constraining the supercoiling of the chromosome. This means that, although the chromosome when isolated has a $\Delta Lk/Lk° = -0.06$, that is approximately one supercoil for every 17 turns of DNA helix, approximately half of this is **constrained** as permanent wrapping of DNA around proteins such as HU (actually a form of writhing; see Topic C4). Only about half the supercoiling is **unconstrained** in the sense of being able to adopt the twisting and writhing conformations described in Topic C4. It has also been suggested that RNA polymerase and mRNA molecules, as well as site-specific DNA-binding proteins such as integration host factor (IHF), a homolog of HU, which binds to specific DNA sequences and bends DNA through 140°, may be important in the organization of the DNA domains. It may be that the organization of the nucleoid is fairly complex, although highly ordered DNA–protein complexes such as nucleosomes (see Topic D2) have not been detected.

D2 CHROMATIN STRUCTURE

Key Notes

Chromatin

Eukaryotic chromosomes each contain a long linear DNA molecule, which must be packaged into the nucleus. The name chromatin is given to the highly ordered DNA–protein complex which makes up the eukaryotic chromosomes. The chromatin structure serves to package and organize the chromosomal DNA, and is able to alter its level of packing at different stages of the cell cycle.

Histones

The major protein components of chromatin are the histones; small, basic (positively charged) proteins which bind tightly to DNA. There are four families of core histone, H2A, H2B, H3, H4, and a further family, H1, which has some different properties, and a distinct role. Individual species have a number of variants of the different histone proteins.

Nucleosomes

The nucleosome core is the basic unit of chromosome structure, consisting of a protein octamer containing two each of the core histones, with 146 bp of DNA wrapped 1.8 times in a left-handed fashion around it. The wrapping of DNA into nucleosomes accounts for virtually all of the negative supercoiling in eukaryotic DNA.

The role of H1

A single molecule of H1 stabilizes the DNA at the point at which it enters and leaves the nucleosome core, and organizes the DNA between nucleosomes. A nucleosome core plus H1 is known as a chromatosome. In some cases, H1 is replaced by a variant, H5, which binds more tightly, and is associated with DNA which is inactive in transcription.

Linker DNA

The linker DNA between the nucleosome cores varies between less than 10 and more than 100 bp, but is normally around 55 bp. The nucleosomal repeat unit is hence around 200 bp.

The 30 nm fiber

Chromatin is organized into a larger structure, known as the 30 nm fiber or solenoid, thought to consist of a left-handed helix of nucleosomes with approximately six nucleosomes per helical turn. Most chromatin exists in this form.

Higher order structure

On the largest scale, chromosomal DNA is organized into loops of up to 100 kb in the form of the 30 nm fiber, constrained by a protein scaffold, the nuclear matrix. The overall structure somewhat resembles that of the organizational domains of prokaryotic DNA.

Related topics

DNA supercoiling (C4)
Prokaryotic chromosome structure (D1)
Eukaryotic chromosome structure (D3)
Genome complexity (D4)

Chromatin

The total length of DNA in a eukaryotic cell depends on the species, but it can be thousands of times as much as in a prokaryotic genome, and is made up of a number of discrete bodies called **chromosomes** (46 in humans). The DNA in each chromosome is believed to be a single linear molecule, which can be up to several centimeters long (see Topic D3). All this DNA must be packaged into the **nucleus** (see Topic A2), a space of approximately the same volume as a bacterial cell; in fact, in their most highly condensed forms, the chromosomes have an enormously high DNA concentration of perhaps 200 mg ml^{-1} (see Topic D1). This feat of packing is accomplished by the formation of a highly organized complex of DNA and protein, known as **chromatin**, a nucleoprotein complex (see Topic A4). More than 50% of the mass of chromatin is protein. Chromosomes greatly alter their level of compactness as cells progress through the cell cycle (see Topics D3 and E3), varying between highly condensed chromosomes at **metaphase** (just before cell division), and very much more diffuse structures in **interphase**. This implies the existence of different levels of organization of chromatin.

Histones

Most of the protein in eukaryotic chromatin consists of **histones**, of which there are five families, or classes: H2A, H2B, H3 and H4, known as the **core histones**, and H1. The core histones are small proteins, with masses between 10 and 20 kDa, and H1 histones are a little larger at around 23 kDa. All histone proteins have a large positive charge; between 20 and 30% of their sequences consist of the basic amino acids, lysine and arginine (see Topic B1). This means that histones will bind very strongly to the negatively charged DNA in forming chromatin.

Members of the same histone class are very highly conserved between relatively unrelated species, for example between plants and animals, which testifies to their crucial role in chromatin. Within a given species, there are normally a number of closely similar variants of a particular class, which may be expressed in different tissues, and at different stages in development. There is not much similarity in sequence between the different histone classes, but structural studies have shown that the classes do share a similar tertiary structure (see Topic B2), suggesting that all histones are ultimately evolutionarily related (see Topic B3).

H1 histones are somewhat distinct from the other histone classes in a number of ways; in addition to their larger size, there is more variation in H1 sequences both between and within species than in the other classes. Histone H1 is more easily extracted from bulk chromatin, and seems to be present in roughly half the quantity of the other classes, of which there are very similar amounts. These facts suggest a specific and distinct role for histone H1 in chromatin structure.

Nucleosomes

A number of studies in the 1970s pointed to the existence of a basic unit of chromatin structure. Nucleases are enzymes which hydrolyze the phosphodiester bonds of nucleic acids. Exonucleases release single nucleotides from the ends of nucleic acid strands, whereas endonucleases cleave internal phosphodiester bonds. Treatment of chromatin with **micrococcal nuclease**, an endonuclease which cleaves double-stranded DNA, led to the isolation of DNA fragments with discrete sizes, in multiples of approximately 200 bp. It was discovered that each 200 bp fragment is associated with an **octamer** core of histone proteins, $(H2A)_2(H2B)_2(H3)_2(H4)_2$, which is why these are designated the **core histones**, and more loosely with one molecule of H1. The proteins protect the DNA from the action of micrococcal nuclease. More prolonged digestion with nuclease leads to the loss of H1 and yields a very resistant structure consisting of **146 bp** of DNA associated very tightly with the histone octamer. This structure is known as the **nucleosome core**, and is structurally very similar whatever the source of the chromatin.

The structure of the nucleosome core particle is now known in considerable detail, from structural studies culminating in X-ray crystallography. The histone octamer forms a wedge-shaped disk, around which the 146 bp of DNA is wrapped in 1.8 turns in a left-handed direction. *Figure 1* shows the basic features and the dimensions of the structure. The left-handed wrapping of the DNA around the nucleosome corresponds to negative supercoiling, that is the turns are **superhelical turns** (technically writhing; see Topic C4). Although eukaryotic DNA is negatively supercoiled to a similar level as that of prokaryotes, on average, virtually all the supercoiling is accounted for by wrapping in nucleosomes and there is no unconstrained supercoiling (see Topic D1).

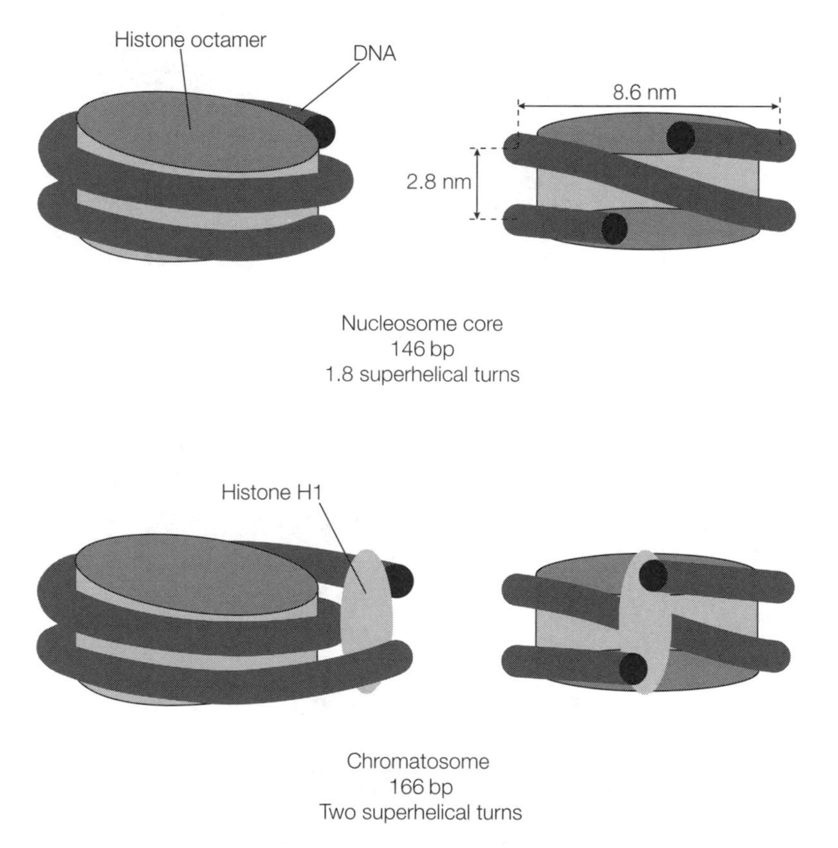

Fig. 1. A schematic view of the structure of a nucleosome core and a chromatosome.

The role of H1

One molecule of histone H1 binds to the nucleosome, and acts to stabilize the point at which the DNA enters and leaves the nucleosome core (*Fig. 1*). In the presence of H1, a further 20 bp of DNA is protected from nuclease digestion, making 166 bp in all, corresponding to two full turns around the histone octamer. A nucleosome core plus H1 is known as a **chromatosome**. The larger size of H1 compared with the core histones is due to the presence of an additional C-terminal tail, which serves to stabilize the DNA between the nucleosome cores. As stated above, H1 is more variable in sequence than the other histones and, in some cell types, it may be replaced by an extreme variant called histone H5, which binds chromatin particularly tightly, and is associated with DNA which is not undergoing transcription (see Topic D3).

Linker DNA

In electron micrographs of nucleosome core particles on DNA under certain conditions, an array, sometimes called the **'beads on a string'** structure is visible. This comprises globular particles (nucleosomes), connected by thin strands of DNA. This **linker DNA** is the additional DNA required to make up the 200 bp nucleosomal repeat apparent in the micrococcal nuclease experiments (see above). The average length of linker DNA between core particles is 55 bp, but the length varies between species and tissues from almost nothing to more than 100 bp (*Fig. 2*).

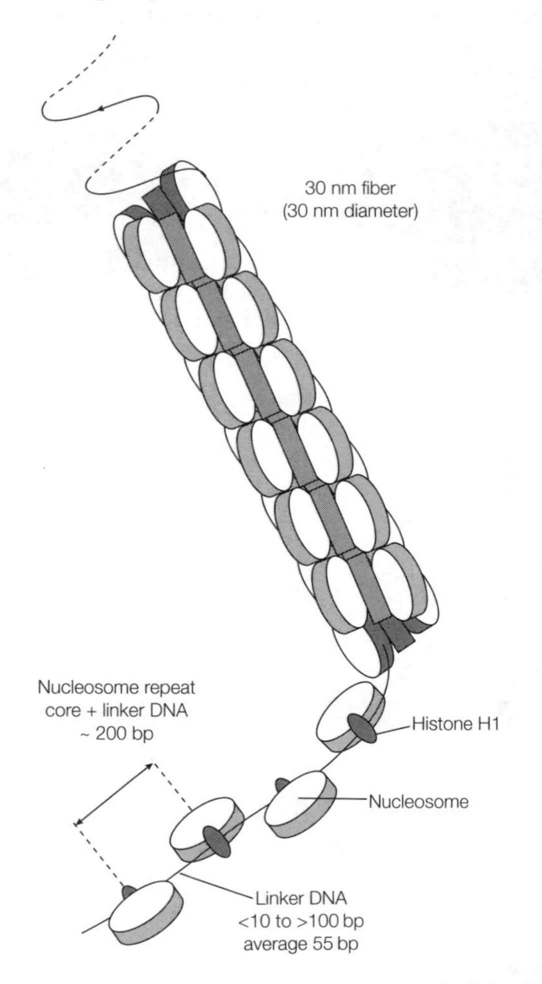

Fig. 2. An array of nucleosomes separated by linker DNA, and the 30 nm fiber.

The 30 nm fiber

The presence of histone H1 increases the organization of the 'beads on a string' to show a zig-zag structure in electron micrographs. With a change in the salt concentration, further organization of the nucleosomes into a **fiber** of 30 nm diameter takes place. Detailed studies of this process have suggested that the nucleosomes are wound into a higher order left-handed helix, dubbed a **solenoid**, with around six nucleosomes per turn (*Fig. 2*). However, there is still some conjecture about the precise organization of the fiber structure, including the path of the linker DNA and the way in which different linker lengths might be incorporated into what seems to be a very uniform structure. Most chromosomal DNA *in vivo* is packaged into the 30 nm fiber (see Topic D3).

Higher order structure

The organization of chromatin at the highest level seems rather similar to that of prokaryotic DNA (see Topic D1). Electron micrographs of chromosomes which have been stripped of their histone proteins show a looped domain structure, which is similar to that illustrated in Topic D1, *Fig. 1*. Even the size of the loops is approximately the same, up to around 100 kb of DNA, although there are many more loops in a eukaryotic chromosome. The loops are constrained by interaction with a protein complex known as the **nuclear matrix** (see Topic E4). The DNA in the loops is in the form of 30 nm fiber, and the loops form an array about 300 nm across (*Fig. 3*).

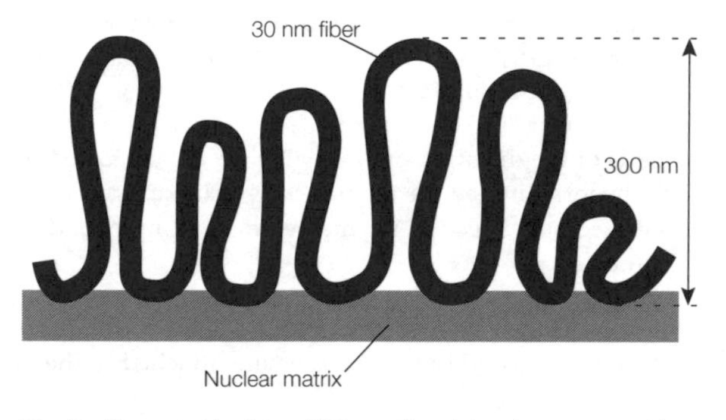

Fig. 3. The organization of 30 nm fiber into chromosomal loops.

D3 EUKARYOTIC CHROMOSOME STRUCTURE

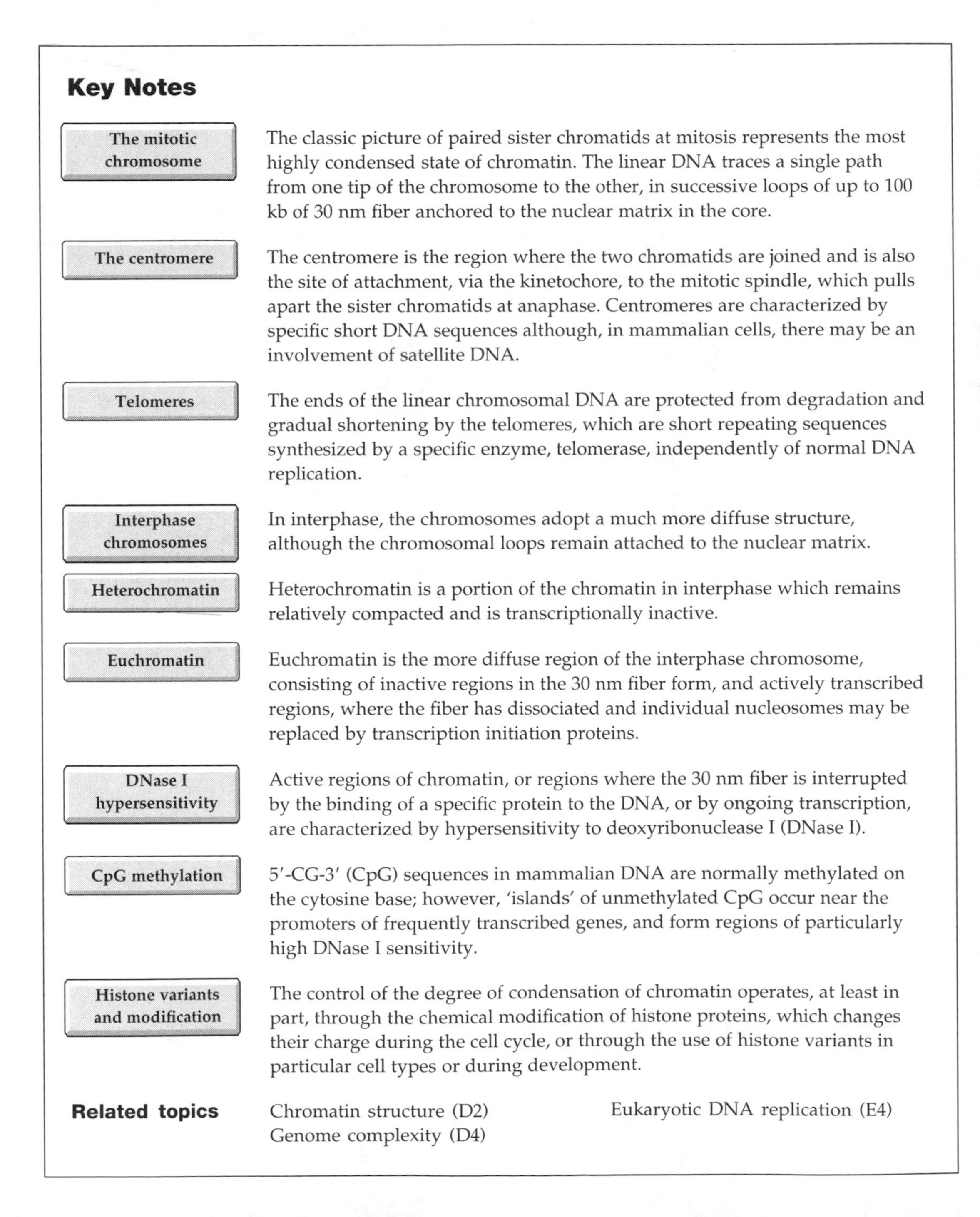

Key Notes

The mitotic chromosome

The classic picture of paired sister chromatids at mitosis represents the most highly condensed state of chromatin. The linear DNA traces a single path from one tip of the chromosome to the other, in successive loops of up to 100 kb of 30 nm fiber anchored to the nuclear matrix in the core.

The centromere

The centromere is the region where the two chromatids are joined and is also the site of attachment, via the kinetochore, to the mitotic spindle, which pulls apart the sister chromatids at anaphase. Centromeres are characterized by specific short DNA sequences although, in mammalian cells, there may be an involvement of satellite DNA.

Telomeres

The ends of the linear chromosomal DNA are protected from degradation and gradual shortening by the telomeres, which are short repeating sequences synthesized by a specific enzyme, telomerase, independently of normal DNA replication.

Interphase chromosomes

In interphase, the chromosomes adopt a much more diffuse structure, although the chromosomal loops remain attached to the nuclear matrix.

Heterochromatin

Heterochromatin is a portion of the chromatin in interphase which remains relatively compacted and is transcriptionally inactive.

Euchromatin

Euchromatin is the more diffuse region of the interphase chromosome, consisting of inactive regions in the 30 nm fiber form, and actively transcribed regions, where the fiber has dissociated and individual nucleosomes may be replaced by transcription initiation proteins.

DNase I hypersensitivity

Active regions of chromatin, or regions where the 30 nm fiber is interrupted by the binding of a specific protein to the DNA, or by ongoing transcription, are characterized by hypersensitivity to deoxyribonuclease I (DNase I).

CpG methylation

5′-CG-3′ (CpG) sequences in mammalian DNA are normally methylated on the cytosine base; however, 'islands' of unmethylated CpG occur near the promoters of frequently transcribed genes, and form regions of particularly high DNase I sensitivity.

Histone variants and modification

The control of the degree of condensation of chromatin operates, at least in part, through the chemical modification of histone proteins, which changes their charge during the cell cycle, or through the use of histone variants in particular cell types or during development.

Related topics

Chromatin structure (D2)
Genome complexity (D4)
Eukaryotic DNA replication (E4)

The mitotic chromosome

The familiar picture of a chromosome (*Fig. 1*) is actually that of its most highly condensed state at **mitosis**. As the daughter chromosomes are pulled apart by the **mitotic spindle** at cell division, the fragile centimeters-long chromosomal DNA would certainly be sheared by the forces generated, were it not in this highly compact state. The structure in *Fig. 1* actually illustrates two identical **sister chromatids**, the products of replication of a single chromosome, joined at their **centromeres**. The tips of the chromosomes are the **telomeres**, which are also the ends of the DNA molecule; the DNA maps in a linear fashion along the length of the chromosome, albeit in a very convoluted path.

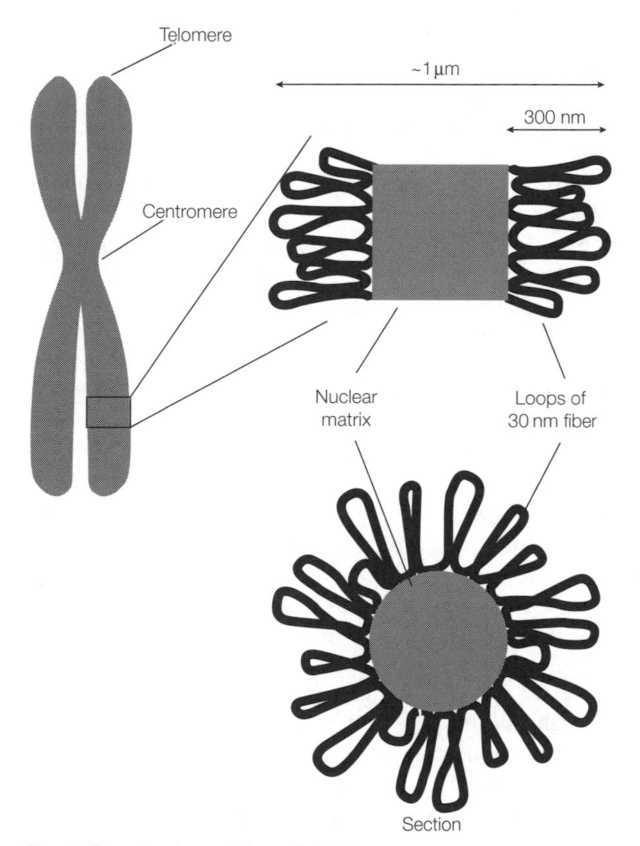

Fig. 1. The structure of the mitotic chromosome.

The structure of a section of a mitotic chromatid is shown in *Fig. 1*. The chromosomal loops (see Topic D2) fan out from a central scaffold or nuclear matrix region consisting of protein. The loops consist of chromatin in the 30 nm fiber form (see Topic D2). One possibility is that consecutive loops may trace a helical path along the length of the chromosome.

The centromere

The **centromere** is the constricted region where the two sister chromatids are joined in the metaphase chromosome. This is the site of assembly of the **kinetochore**, a protein complex which attaches to the **microtubules** (see Topic A4) of the mitotic spindle. The microtubules act to separate the chromatids at **anaphase**. The DNA of the centromere has been shown in yeast to consist merely of a short AT-rich sequence of 88 bp, flanked by two very short conserved regions although, in mammalian cells, centromeres seem to consist of rather longer sequences, and are flanked by a large quantity of repeated DNA, known as **satellite DNA** (see Topic D4).

Telomeres

Telomeres are specialized DNA sequences that form the ends of the linear DNA molecules of the eukaryotic chromosomes. A telomere consists of up to hundreds of copies of a short repeated sequence (5′-TTAGGG-3′ in humans), which is synthesized by the enzyme **telomerase** (an example of a **ribonucleoprotein**; see Topic O1) in a mechanism independent of normal DNA replication. The telomeric DNA forms a special secondary structure, the function of which is to protect the ends of the chromosome proper from degradation. Independent synthesis of the telomere acts to counteract the gradual shortening of the chromosome resulting from the inability of normal replication to copy the very end of a linear DNA molecule (see Topic E4).

Interphase chromosomes

In interphase, the genes on the chromosomes are being transcribed and DNA replication takes place (during S-phase; see Topics E3 and E4). During this time, which is most of the cell cycle, the chromosomes adopt a much more diffuse structure and cannot be visualized individually. It is believed, however, that the chromosomal loops are still present, attached to the nuclear matrix (see Topic D2, *Fig. 3*).

Heterochromatin

Heterochromatin comprises a portion of the chromatin in interphase which remains highly compacted, although not so compacted as at metaphase. It can be visualized under the microscope as dense regions at the periphery of the nucleus, and probably consists of closely packed regions of 30 nm fiber. It has been shown more recently that heterochromatin is transcriptionally inactive. It is believed that much of the heterochromatin may consist of the repeated satellite DNA close to the centromeres of the chromosomes (see Topic D4), although in some cases entire chromosomes can remain as heterochromatin, for example one of the two X chromosomes in female mammals.

Euchromatin

The rest of the chromatin, which is not visible as heterochromatin, is known historically by the catch-all name of **euchromatin**, and is the region where all transcription takes place. Euchromatin is not homogeneous, however, and is comprised of relatively inactive regions, consisting of chromosomal loops compacted in 30 nm fibers, and regions (perhaps 10% of the whole) where genes are actively being transcribed or are destined to be transcribed in that cell type, where the 30 nm fiber has been dissociated to the 'beads on a string' structure (see Topic D2). Parts of these regions may be depleted of nucleosomes altogether, particularly within promoters, to allow the binding of transcription factors and other proteins (*Fig. 2*) (see Topic M5).

DNase I hypersensitivity

The sensitivity of chromatin to the nuclease deoxyribonuclease I (**DNase I**), which cuts the backbone of DNA unless the DNA is protected by bound protein, has been used to map the regions of transcriptionally active chromatin in cells. Short regions of DNase I hypersensitivity are thought to represent regions where the 30 nm fiber is interrupted by the binding of a sequence-specific regulatory protein (*Fig. 2*), so revealing regions of naked DNA which can be attacked readily by DNase I. Longer regions of sensitivity represent sequences where transcription is taking place. These regions vary between different cell types, and correspond to the sites of genes which are expressed specifically in those cells.

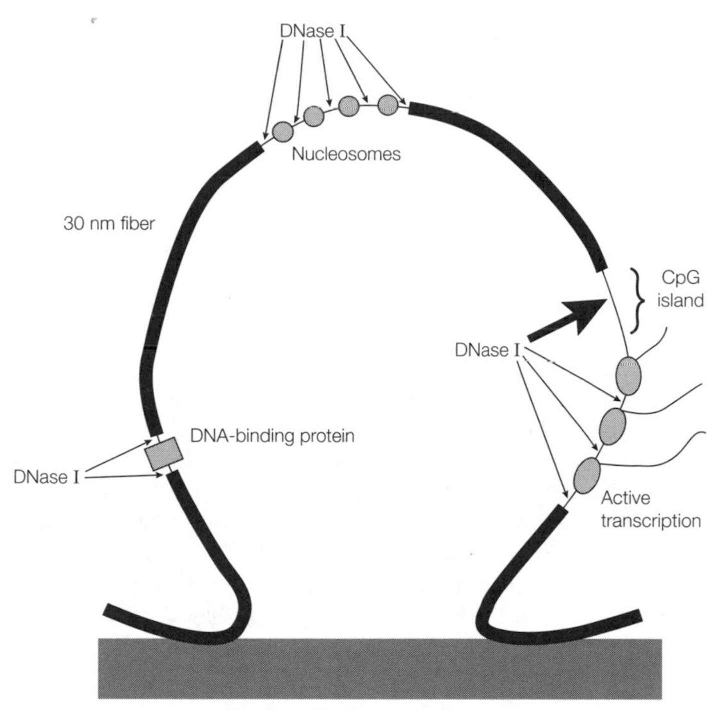

Fig. 2. Euchromatin, showing active and inactive regions (see text for details) and sites of DNase I hypersensitivity indicated by large and small arrows.

CpG methylation

An important chemical modification which may be involved in signaling the appropriate level of chromosomal packing at the sites of expressed genes in mammalian cells is the methylation of C-5 in the cytosine base of 5′-CG-3′ sequences, commonly known as **CpG methylation**. CpG sites are normally methylated in mammalian cells, and are relatively scarce throughout most of the genome. This is because 5-methylcytosine spontaneously deaminates to thymine and, since this error is not always repaired, methylated CpG mutates fairly rapidly to TpG (see Topics F1 and F2). The methylation of CpG is associated with transcriptionally inactive regions of chromatin. However, there exist throughout the genome, '**islands**' of unmethylated CpG, where the proportion of CG dinucleotides is much higher than on average. These islands are commonly around 2000 bp long, and are coincident with regions of particular sensitivity to DNase I (*Fig. 3*). The CpG islands surround the promoter regions of genes which are expressed in almost all cell types, so called **housekeeping genes**, and may be largely free of nucleosomes.

Histone variants and modification

The major mechanisms for the condensing and decondensing of chromatin are believed to operate directly through the histone proteins which carry out the packaging (see Topic D2). Short-term changes in chromosome packing during the cell cycle seem to be modulated by chemical modification of the histone proteins. For example, actively transcribed chromatin is associated with the acetylation of lysine residues in the N-terminal regions (see Topic B2) of the core histones (see Topic D2), whereas the condensation of chromosomes at mitosis is accompanied by the phosphorylation of histone H1. These changes

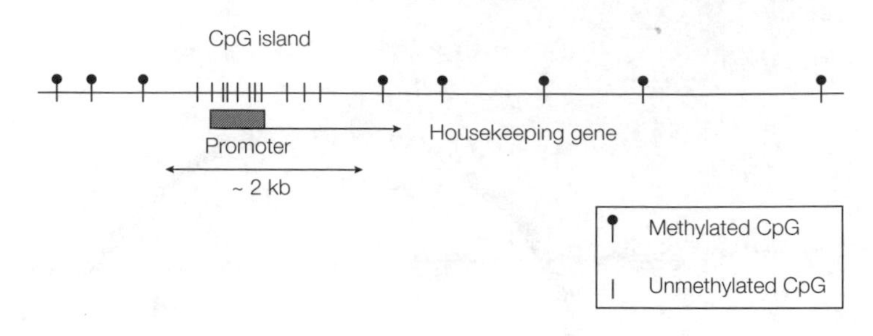

Fig. 3. CpG islands and the promoters of housekeeping genes.

alter the positive charge on the histone proteins, and may affect the stability of the various chromatin conformations, for example the 30 nm fibers or the interactions between them.

Longer term differences in chromatin condensation are associated with changes due to stages in development and different tissue types. These changes are associated with the utilization of alternative histone variants (see Topic D2), which may also act by altering the stability of chromatin conformations. Histone H5 is an extreme example of this effect, replacing H1 in some very inactive chromatin, for example in avian red blood cells (see Topic D2).

D4 GENOME COMPLEXITY

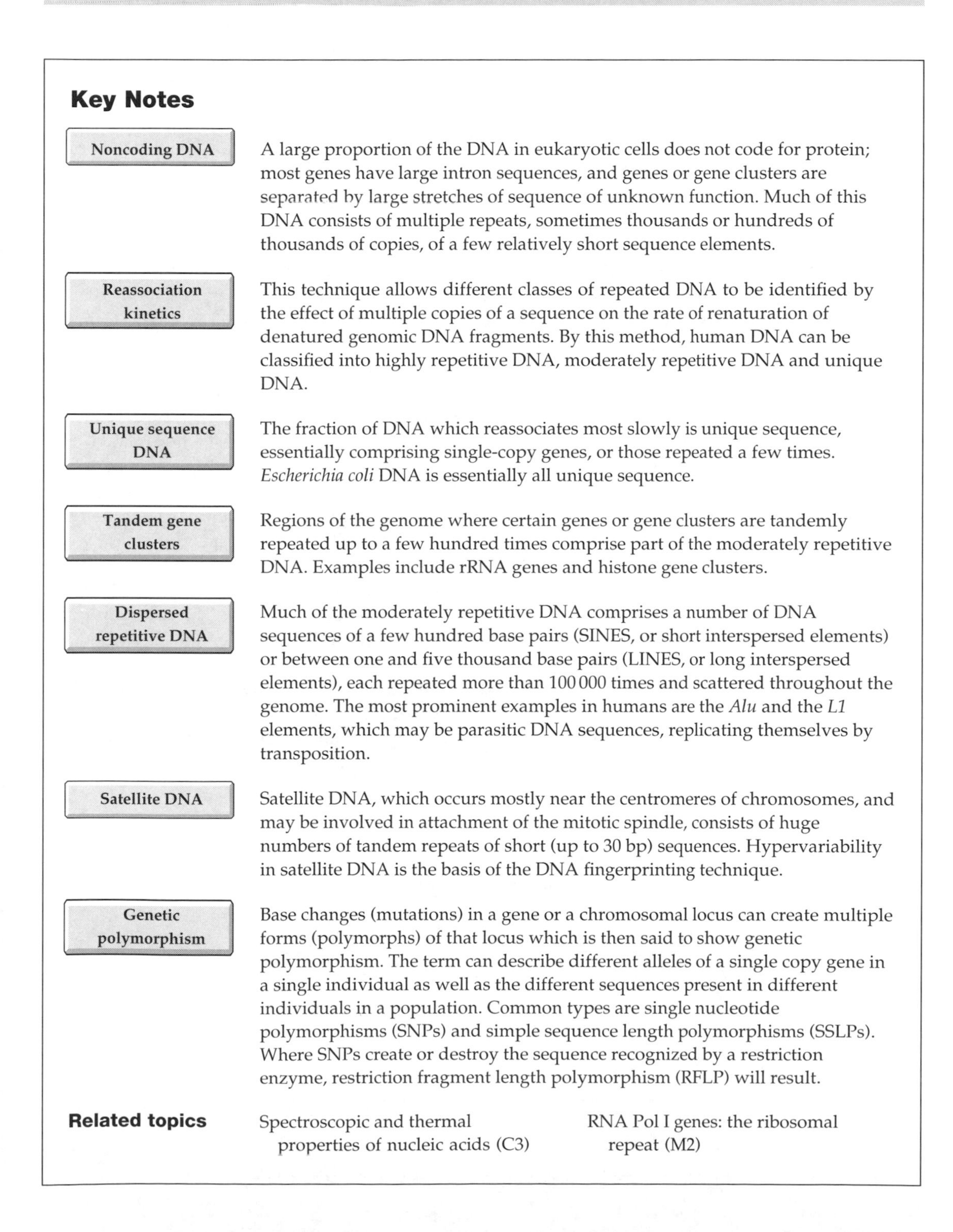

Key Notes

Noncoding DNA

A large proportion of the DNA in eukaryotic cells does not code for protein; most genes have large intron sequences, and genes or gene clusters are separated by large stretches of sequence of unknown function. Much of this DNA consists of multiple repeats, sometimes thousands or hundreds of thousands of copies, of a few relatively short sequence elements.

Reassociation kinetics

This technique allows different classes of repeated DNA to be identified by the effect of multiple copies of a sequence on the rate of renaturation of denatured genomic DNA fragments. By this method, human DNA can be classified into highly repetitive DNA, moderately repetitive DNA and unique DNA.

Unique sequence DNA

The fraction of DNA which reassociates most slowly is unique sequence, essentially comprising single-copy genes, or those repeated a few times. *Escherichia coli* DNA is essentially all unique sequence.

Tandem gene clusters

Regions of the genome where certain genes or gene clusters are tandemly repeated up to a few hundred times comprise part of the moderately repetitive DNA. Examples include rRNA genes and histone gene clusters.

Dispersed repetitive DNA

Much of the moderately repetitive DNA comprises a number of DNA sequences of a few hundred base pairs (SINES, or short interspersed elements) or between one and five thousand base pairs (LINES, or long interspersed elements), each repeated more than 100 000 times and scattered throughout the genome. The most prominent examples in humans are the *Alu* and the *L1* elements, which may be parasitic DNA sequences, replicating themselves by transposition.

Satellite DNA

Satellite DNA, which occurs mostly near the centromeres of chromosomes, and may be involved in attachment of the mitotic spindle, consists of huge numbers of tandem repeats of short (up to 30 bp) sequences. Hypervariability in satellite DNA is the basis of the DNA fingerprinting technique.

Genetic polymorphism

Base changes (mutations) in a gene or a chromosomal locus can create multiple forms (polymorphs) of that locus which is then said to show genetic polymorphism. The term can describe different alleles of a single copy gene in a single individual as well as the different sequences present in different individuals in a population. Common types are single nucleotide polymorphisms (SNPs) and simple sequence length polymorphisms (SSLPs). Where SNPs create or destroy the sequence recognized by a restriction enzyme, restriction fragment length polymorphism (RFLP) will result.

Related topics

Spectroscopic and thermal properties of nucleic acids (C3)

RNA Pol I genes: the ribosomal repeat (M2)

Noncoding DNA

The genomes of complex eukaryotic organisms may contain more than 1000 times as much DNA as prokaryotes such as *E. coli*. It is clear that much of this DNA does not code for protein, since there are not 1000 times as many proteins in humans as in *E. coli*. The coding regions of genes are interrupted by **intron** sequences (see Topic O3) and genes may hence take up many kilobases of sequence but, despite this, genes are by no means contiguous along the genome; they are separated by long stretches of sequence for which the function, if any, is unknown. It has become apparent that much of this **noncoding** DNA consists of multiple repeats of similar or identical copies of a few different types of sequence. These copies may follow one another directly (**tandemly repeated**), for example **satellite DNA**, or they may occur as multiple copies scattered throughout the genome (**interspersed**), such as the ***Alu* elements**.

Reassociation kinetics

Before the advent of large-scale sequencing to investigate the sequence–structure of genomes (see Topic J2), their sequence complexity was studied using measurements of the annealing of denatured DNA (see Topic C3). Genomic DNA fragments of a uniform size (from a few hundred to a few thousand base pairs), normally prepared by shearing or sonication (see Topic C2), are thermally denatured and allowed to re-anneal at a low concentration. The fraction of the DNA which is **reassociated** after a given time (t) will depend on the initial concentration of the DNA, C_0. Crucially, however, single-stranded fragments which have a sequence which is repeated in multiple copies in the genome will be able to anneal with many more alternative complementary strands than fragments with a unique sequence, and will hence reassociate more rapidly. The re-annealing of the strands may be followed spectroscopically (see Topic C3), or more sensitively by separating single-stranded and double-stranded DNA by **hydroxyapatite chromatography**. The fraction of the DNA still in the single-stranded form (f) is plotted against C_0t. The resulting curve represents the **reassociation kinetics** of the DNA sample, and is colloquially known as a **C_0t curve** (pronounced 'cot').

Figure 1 presents idealized C_0t curves for human and *E. coli* genomic DNA. It can be seen that human DNA reassociates in three distinct phases (1–3), corresponding to DNA with decreasing numbers of copies, or increasing

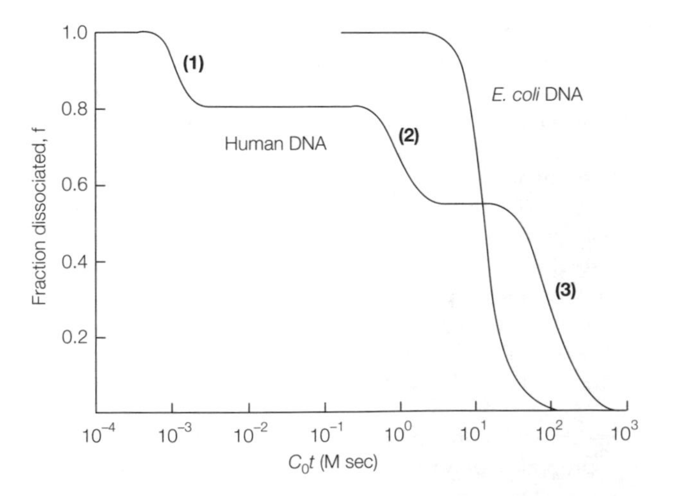

Fig. 1. Idealized reassociation kinetics curves for human and E. coli *DNA.*

'**complexity**'. Phase 1 corresponds to **highly repetitive DNA** (>10^6 copies per genome), phase 2 to **moderately repetitive DNA** (<10^6 copies per genome) and phase 3 to **unique DNA** (one or a few copies per genome). In practice, these divisions are somewhat arbitrary, and the phases are not so distinct. *E. coli* DNA reassociates in a single phase.

Unique sequence DNA

This fraction of genomic DNA is the slowest to reassociate on a C_0t curve, and corresponds to the coding regions of genes which occur in only one or a few copies per haploid genome, and any unique intervening sequence. In the *E. coli* genome, virtually all the DNA has a unique sequence, since it consists predominantly of more or less contiguous (adjacent) single-copy genes. However, since *E. coli* has approximately 1000 times less DNA than a human cell, any given sequence has correspondingly more alternative partners at a given concentration, and its reassociation occurs faster, that is at a lower value of C_0t.

Tandem gene clusters

Moderately repetitive DNA consists of a number of types of repeated sequence. At the lower end of the repeat scale come genes which occur as clusters of multiple repeats. These are genes whose products are required in unusually large quantities. One example is the rRNA-encoding genes (**rDNA**). The gene which encodes the 45S precursor of the 18S, 5.8S and 28S rRNAs, for example, is repeated in **arrays** containing from around 10 to around 10 000 copies depending on the species (see Topic M2, *Fig. 1*). In humans, the 45S gene occurs in arrays on five separate chromosomes, each containing around 40 copies. In interphase, these regions are spatially located together in the **nucleolus**, a dense region of the nucleus, which is a factory for rRNA production and modification (see Topics M2 and O1). A second example of tandem gene clusters is given by the histone genes, whose products are produced in large quantities during S-phase. The five histone genes occur together in a **cluster**, which is directly repeated up to several hundred times in some species.

Dispersed repetitive DNA

Most of the moderately repetitive DNA in many species consists of sequences from a few hundred base pairs (**SINES**, or **short interspersed elements**) to between one and five thousand base pairs (**LINES**, or **long interspersed elements**). Collectively, these are known as **dispersed repetitive DNA** sequences as they are repeated many thousands of times and are scattered throughout the whole genome. The commonest such sequence in humans is the ***Alu* element**, a 300 bp SINE that occurs between 300 000 and 500 000 times. The copies are all 80–90% identical, and most contain the *Alu*I restriction site (see Topic G3); hence the name.

A second dispersed sequence is the ***L1* element**, and together, copies of *Alu* and *L1* make up almost 10% of the human genome, occurring between genes and in introns.

A number of possible functions have been ascribed to these dispersed elements, from origins of replication (see Topic E4) to gene regulation sequences (see Section N). Perhaps most likely, however, is the idea that *Alu* elements and other such families are DNA sequences which can duplicate themselves randomly within the genome by **transposition** (see Topic F4). They may be parasitic, or selfish, DNA with no function, but despite this, they may have had a profound effect on evolution by influencing or disrupting gene sequences.

Satellite DNA

Highly repetitive DNA in eukaryotic genomes consists of very short sequences from 2 bp to 20–30 bp in length, in tandem arrays of many thousands of copies.

Such arrays are known as **satellite DNA**, since they were identified as satellite bands, with a buoyant density different from that of bulk DNA, in CsCl density gradients of chromosomal DNA fragments. Since they consist of repeats of such short sequences, they have nonaverage G+C content (see Topic C2). Satellite DNAs are divided into **minisatellites (variable number tandem repeats, VNTRs)** and **microsatellites (simple tandem repeats, STRs)**, according to the length of the repeating sequence, microsatellites having the shortest repeats. *Figure 2* shows an example of a *Drosophila* satellite DNA sequence, which occurs millions of times in the insect's genome. As with dispersed repetitive sequences, satellite DNA has no function which has been demonstrated conclusively, although it has been suggested that some of these arrays i.e. those which are concentrated near the centromeres of chromosomes and form a large part of heterochromatin, may have a role in the binding of kinetochore components to the centromere (see Topic D3). Minisatellites (VNTRs) are more frequently found near chromosome ends whereas microsatellites (STRs) are more evenly distributed along the chromosomes.

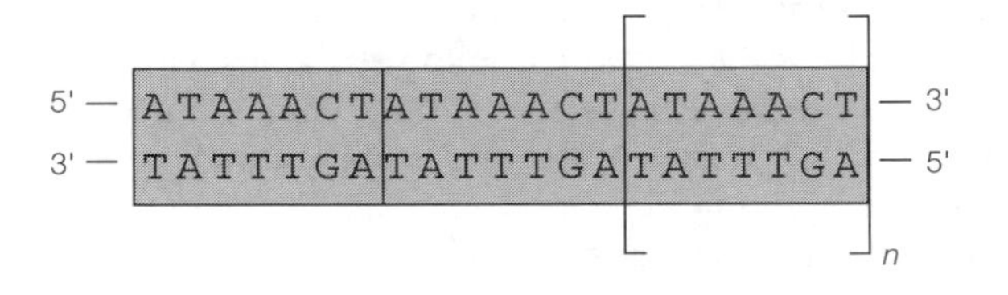

Fig. 2. Drosophila *satellite DNA repeat.*

Minisatellite repeats are the basis of the DNA fingerprinting technique, used to unambiguously identify individuals and their familial relationships. The numbers of repeats in the arrays of some satellite sequences are **hypervariable**, that is they vary significantly between individuals. These variations are examples of genetic polymorphism (see below and Topic F1). The precise lengths of a set of several different minisatellite arrays in the genome, which may be determined by restriction digestion and Southern blotting and hybridization (see Topic J1), are diagnostic of a given individual. That is to say, the probability of another individual (apart from an identical twin, or a clone!) having the same set of array lengths is likely to be vanishingly small. Since roughly half of those arrays will be inherited from each parent, family relationships can be determined by the matching of lengths from different individuals (see Topic J6).

Genetic polymorphism

When the same region, or locus, of a chromosome has two (or more) slightly different DNA sequences in different chromosomes or individuals of the same species, these are described as polymorphs and the locus is said to show (genetic) polymorphism. Genetic polymorphism is caused by mutation (see Topic F1). Within a diploid individual, for example, the two alleles of a particular single copy gene could be different by just one nucleotide and hence the gene can be described as polymorphic. Similarly, in a population of individuals, if other alternative DNA sequences exist for the same gene, there would be many different alleles (in the whole population) encoding that gene. Mutational events (see Topic F1) that cause genetic polymorphism create the **single nucleotide polymorphisms (SNPs)** described above, but they also create length variations in arrays of repeated sequences. Each different length of the repeat is a different form and each is an example of a **simple sequence length polymorphism (SSLPs).**

SNPs can occur within the short sequences that are recognized by restriction enzymes (see Topic G3) and thus the length of a fragment generated by cutting a DNA molecule with a restriction enzyme could be different for each allele. This is known as a **restriction fragment length polymorphism (RFLP).** When a particular RFLP is associated with the allele responsible for a genetic disease, it can be used as a marker in clinical diagnosis. RFLPs can also be used to help create genetic maps of chromosomes and thus help to work out the order of DNA sequences in genome sequencing projects (see Topic J2). It is possible to use a form of gel electrophoresis (see Topic G3) to detect SNPs in DNA fragments of the same length such as restriction fragments (see Topic G3) or PCR products (see Topic J3). To do this the DNA fragments must be denatured so the two separated strands can adopt specific conformations that depend on their nucleotide sequence as they migrate through the gel. This technique is called **single stranded conformational polymorphism (SSCP).**

D5 THE FLOW OF GENETIC INFORMATION

Key Notes

The central dogma

The central dogma is the original proposal that 'DNA makes RNA makes protein', which happen *via* the processes of transcription and translation respectively. This is broadly correct, although a number of examples are known which contradict parts of it. Retroviruses reverse transcribe RNA into DNA, a number of viruses are able to replicate RNA directly into an RNA copy, and a number of organisms can edit a messenger RNA sequence so that the protein coding sequence is not directly specified by DNA sequence.

Prokaryotic gene expression

Transcription of a single gene or an operon starts at the promoter, ends at the terminator and produces a monocistronic or polycistronic messenger RNA. The coding regions of the message are translated by the ribosome from the start codon (close to the ribosome binding site) to the stop codon. Transfer RNAs deliver the appropriate amino acid, according to the genetic code, to the growing protein chain.

Eukaryotic gene expression

In most cases monocistronic messenger RNAs are transcribed from a gene, initiated at a promoter. The resulting pre-messenger RNA is capped at the 5′-end and has a poly(A) tail added to the 3′-end. Introns are removed by splicing before the mature mRNA is exported from the nucleus to be translated by ribosomes in the cytoplasm.

Related topics

Transcription in prokaryotes (Section K)
Transcription in eukaryotes (Section M)
RNA processing and RNPs (Section O)
The genetic code and tRNA (Section P)
Protein synthesis (Section Q)

The central dogma

In the early 1950s, Francis Crick suggested that there was a unidirectional flow of genetic information from DNA through RNA to protein, i.e. 'DNA makes RNA makes protein'. This became known as the **central dogma** of molecular biology, since it was proposed without much evidence for the individual steps. We now know that the broad thrust of the central dogma is correct, although there are a number of modifications we must make to the basic scheme. A more complete diagrammatic version of the genetic information flow is given in *Fig. 1*. The primary route remains from DNA to RNA to protein. In both prokaryotic and eukaryotic cells, DNA is **transcribed** (see Section K) to an RNA molecule (**messenger RNA**), that contains the same sequence information (albeit that Us replace Ts), then that message is **translated** (see Section Q) into a protein sequence according to the **genetic code** (see Topic P1). We can also include

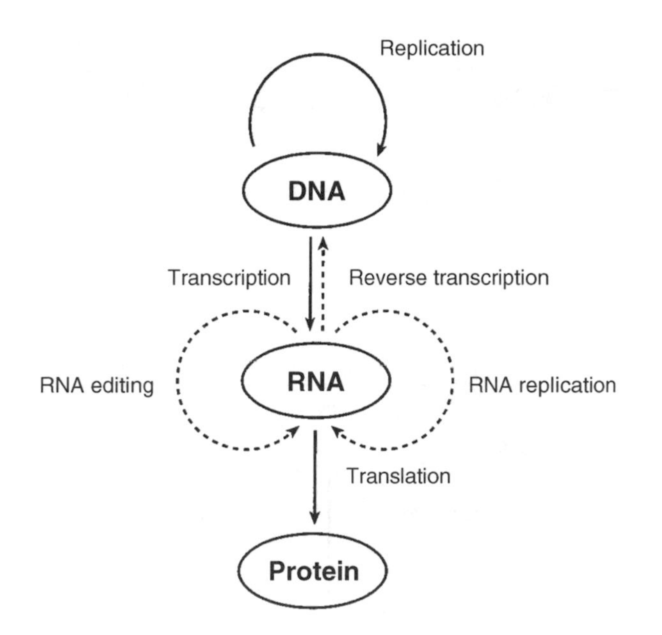

Fig. 1. The flow of genetic information.

DNA **replication** (see Section E) in *Fig. 1*, in which two daughter DNA molecules are formed by duplication of the information in the DNA molecule.

However, a few exceptions to this basic scheme have been identified. A number of classes of virus contain a genome consisting of an RNA molecule (see Topic R4). In the **retroviruses**, which include **HIV**, the causative agent of AIDS, the single-stranded RNA molecule is converted to a double-stranded DNA copy (Ts replace Us), which is then inserted into the genome of the host cell. This process has been termed reverse transcription, and the enzyme responsible, which is encoded in the viral genome, is called **reverse transcriptase**. There are also a number of viruses known whose RNA genome is copied directly into RNA, without the use of DNA as an intermediary (**RNA replication**). Examples include the coronaviruses and hepatitis C virus. As far as is known, there are no examples of a protein being able to specify a specific RNA or DNA sequence, so the translation step of the central dogma does appear to be unidirectional.

In some organisms, examples are known where the sequence of the messenger RNA is altered after it is transcribed from the DNA, in a process known as **RNA editing** (see Topic O4), so that the sequence of the DNA is not faithfully used to encode the product protein.

Prokaryotic gene expression

The similarities and differences in the broad outline of **gene expression** in prokaryotic and eukaryotic cells are illustrated in *Fig. 2* and *Fig. 3*. In prokaryotes (*Fig. 2*), the enzyme **RNA polymerase** (see Topic K2) transcribes a gene from the **promoter**, the sequence in the DNA that specifies the start of transcription (see Topic K3), to the **terminator**, that specifies the end (see Topic K4). The resulting messenger RNA may contain regions coding for one or more proteins. In the latter case, the message is described as **polycistronic**, and the region encoding it in the DNA is called an **operon** (see Topic L1). The coding

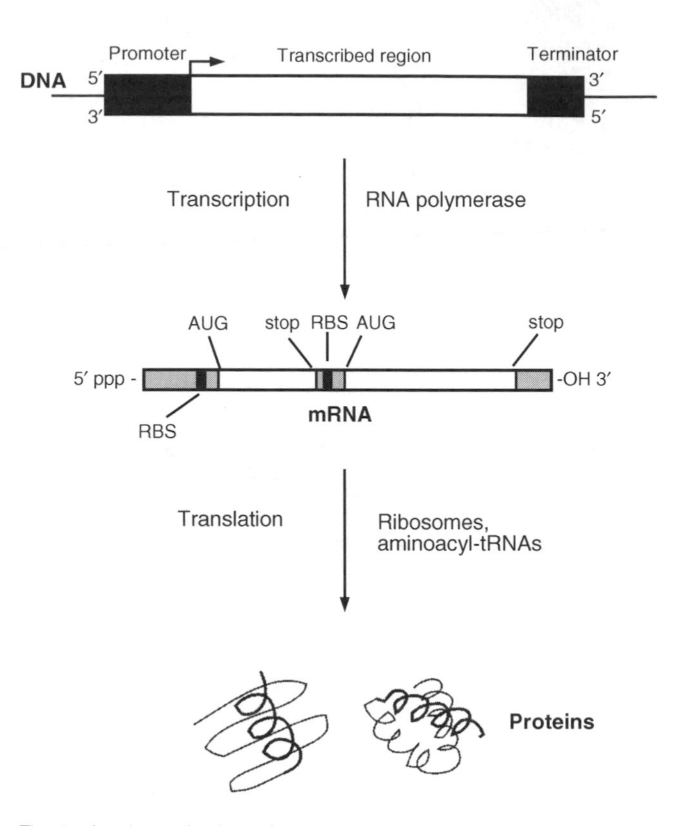

Fig. 2. A schematic view of prokaryotic gene expression.

regions of the messenger RNA are translated into proteins by **ribosomes** (see Topic O1, Section Q), which bind to the messenger RNA at the **ribosome binding site** (RBS) and translate the message into an amino acid sequence from the **start codon** (normally the triplet AUG) to a **stop codon** using **transfer RNAs** (tRNAs; see Topic P2) to deliver the amino acids (as aminoacyl-tRNAs) specified by the genetic code.

Eukaryotic gene expression

In eukaryotes (*Fig. 3*), transcription takes place in the nucleus and translation in the cytoplasm, and the messenger RNAs are consequently longer-lived. Transcribed **pre-messenger RNAs** (pre-mRNAs) are synthesized by one of the three RNA polymerases (**RNA polymerase II**; see Topic M4) and usually only encode one protein (i.e. are **monocistronic**). The RNAs are modified in the nucleus by **capping** at the 5′-end and the addition of a **poly(A) tail** at the 3′-end, increasing their stability (see Topic O3). In most eukaryotic genes, the protein-encoding sections of the gene are interrupted by intervening sequences (known as **introns**) which do not form part of the coding region. Consequently, the introns in the pre-mRNA must be removed to produce continuous protein-coding sequence (the **exons**) in a process known as **splicing**, which is directed by RNA-protein complexes known as **snRNPs** (**small nuclear ribonucleoproteins**; see Topic O3). The mature mRNAs are then exported from the nucleus to be translated into protein on ribosomes in the cytoplasm.

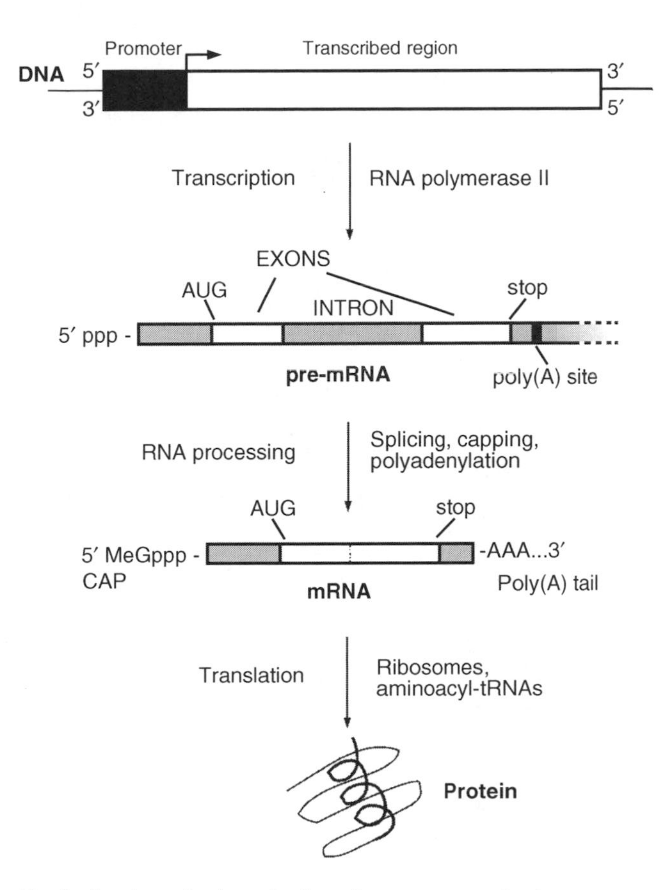

Fig. 3. A schematic view of eukaryotic gene expression.

E1 DNA REPLICATION: AN OVERVIEW

Key Notes

Semi-conservative mechanism

During replication, the strands of the double helix separate and each acts as a template to direct the synthesis of a complementary daughter strand using deoxyribonucleoside 5′-triphosphates as precursors. Thus, each daughter cell receives one of the original parental strands. This mechanism can be demonstrated experimentally by density labeling experiments.

Replicons, origins and termini

Small chromosomes, such as those of bacteria and viruses, replicate as single units called replicons. Replication begins from a unique site, the origin, and proceeds, usually bidirectionally, to the terminus. Large eukaryotic chromosomes contain multiple replicons, each with its own origin, which fuse as they replicate. Origins tend to be AT-rich to make opening easier.

Semi-discontinuous replication

At each replication fork, the leading strand is synthesized as one continuous piece while the lagging strand is made discontinuously as short fragments in the reverse direction. These fragments are 1000–2000 nt long in prokaryotes and 100–200 nt long in eukaryotes. They are joined by DNA ligase. This mechanism exists because DNA can only be synthesized in a 5′→3′ direction.

RNA priming

The leading strand and all lagging strand fragments are primed by synthesis of a short piece of RNA which is then elongated with DNA. The primers are removed and replaced by DNA before ligation. This mechanism helps to maintain high replicational fidelity.

Related topics

Prokaryotic and eukaryotic chromosome structure (Section D)
Bacterial DNA replication (E2)
Eukaryotic DNA replication (E4)
DNA repair (F3)
Bacteriophages and eukaryotic viruses (Section R)

Semi-conservative mechanism

The key to the mechanism of DNA replication is the fact that each strand of the DNA double helix carries the same information – their base sequences are complementary (see Topic C1). Thus, during replication, the two **parental strands** separate and each acts as a **template** to direct the enzyme-catalyzed synthesis of a new complementary **daughter strand** following the normal base-pairing rules (A with T; G with C). The two new double-stranded molecules then pass to the two daughter cells at cell division (*Fig. 1a*). The point at which separation of the strands and synthesis of new DNA takes place is known as the **replication fork**. This **semi-conservative** mechanism was demonstrated experimentally in 1958 by Meselson and Stahl (*Fig. 1b*). *Escherichia coli* cells were grown for several generations in the presence of the stable heavy isotope ^{15}N

so that their DNA became fully **density labeled** (both strands ^{15}N labeled: $^{15}N/^{15}N$). The cells were then transferred to medium containing only normal ^{14}N and, after each cell division, DNA was prepared from a sample of the cells and analyzed on a CsCl gradient using the technique of **equilibrium (isopycnic) density gradient centrifugation**, which separates molecules according to differences in buoyant density (see Topics A2 and C2). After the first cell division, when the DNA had replicated once, it was all of hybrid density, appearing in a position on the gradient half way between fully labeled $^{15}N/^{15}N$ DNA and unlabeled $^{14}N/^{14}N$ DNA. After the second cell generation in ^{14}N, half of the DNA was hybrid density and half fully light ($^{14}N/^{14}N$). After each subsequent generation, the proportion of $^{14}N/^{14}N$ DNA increased but some DNA of hybrid density (i.e. $^{15}N/^{14}N$) persisted. These results could only be interpreted in terms of a semi-conservative model. All DNA molecules replicate in this manner. The template strands are read in the 3′→5′ direction (see Topic C1) while the new strands are synthesized 5′→3′.

The substrates for DNA synthesis are the deoxynucleoside 5′-triphosphates (dNTPs): **dATP**, **dGTP**, **dCTP** and **dTTP**. The reaction mechanism involves nucleophilic attack of the 3′-OH group of the nucleotide at the 3′-end of the growing strand at the α-phosphorus of the dNTP complementary to the next template base, with the elimination of pyrophosphate. The energy for poly-

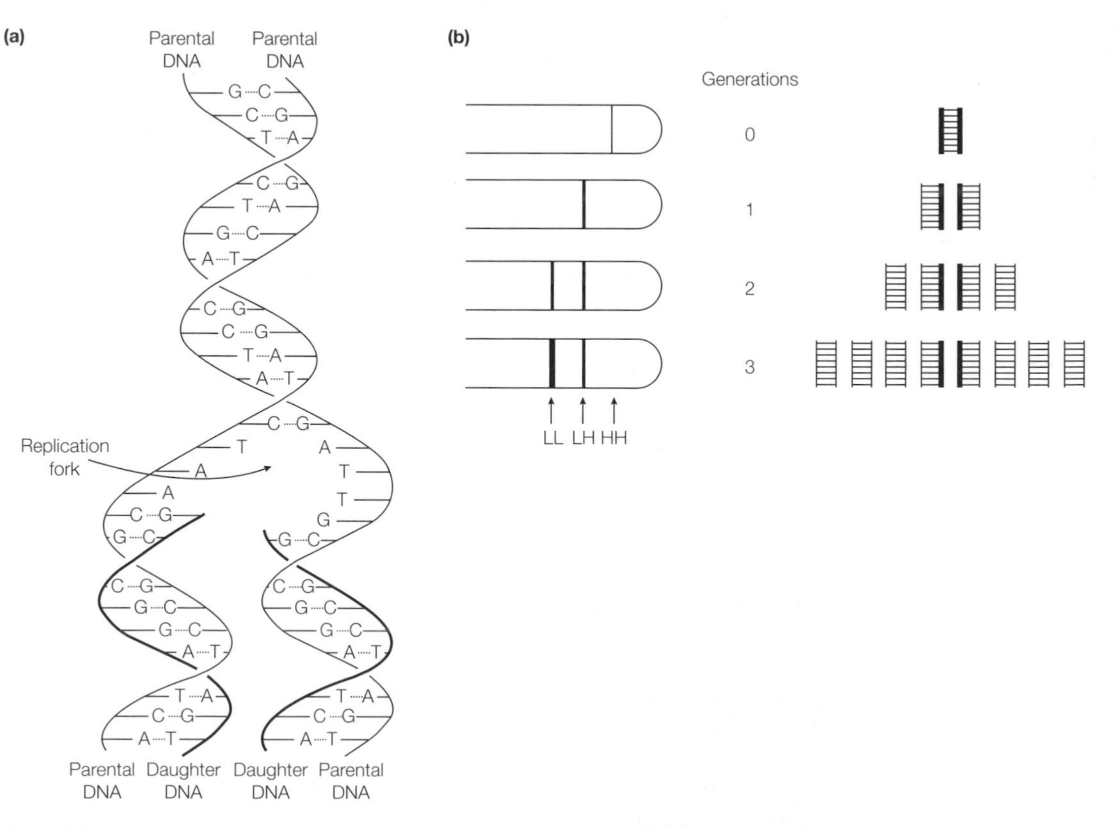

Fig. 1. (a) Semi-conservative replication of DNA at the replication fork. (b) Proof of semi-conservative mechanism. The centrifuge tubes on the left show the positions of fully ^{15}N-labeled (HH), hybrid density (LH) and fully ^{14}N-labeled (LL) DNA in CsCl density gradients after each generation and the diagram on the right shows the corresponding DNA molecules. The original parental ^{15}N molecules are shown as thick lines and all ^{14}N daughter DNA as thin lines.

merization comes from the hydrolysis of the dNTPs and the resulting pyrophosphate.

Replicons, origins and termini

Any piece of DNA which replicates as a single unit is called a **replicon**. The **initiation** of DNA replication within a replicon always occurs at a fixed point known as the **origin** (*Fig. 2a*). Usually, two replication forks proceed **bidirectionally** away from the origin and the strands are copied as they separate until the **terminus** is reached. All prokaryotic chromosomes and many bacteriophage and viral DNA molecules are circular (see Topic D1) and comprise single replicons. Thus, there is a single termination site roughly 180° opposite the unique origin (*Fig. 2a*). Linear viral DNA molecules usually have a single origin, but this is not necessarily in the center of the molecule. In all these cases, the origin is a complex region where the initiation of DNA replication and the control of the growth cycle of the organism are regulated and co-ordinated. In contrast, the long, linear DNA molecules of eukaryotic chromosomes consist of multiple replicons, each with its own origin. A typical mammalian cell has 50 000–100 000 replicons with a size range of 40–200 kb. Where replication forks from adjacent **replication bubbles** meet, they fuse to form the completely replicated DNA (*Fig. 2b*). The sequences of eukaryotic chromosomal origins are much simpler than those of single replicon DNA molecules since the control of eukaryotic DNA replication operates primarily at the onset of S-phase (see Topic E3) rather than at each individual origin. All origins contain AT-rich sequences where the strands initially separate. AT-rich regions are more easily opened than GC-rich regions (see Topic C3).

Semi-discontinuous replication

In semi-conservative replication, both new strands of DNA are synthesized simultaneously at each replication fork. However, the mechanism of DNA replication allows only for synthesis in a 5′→3′ direction. Since the two strands of DNA are **antiparallel**, how is the parental strand that runs 5′→3′ past the replication fork copied? It appears to be made 3′→5′ but this cannot happen. If *E. coli* cells are labeled for just a few seconds with [^{3}H]thymidine, a radioactive precursor of DNA, then a large fraction of the newly synthesized (**nascent**) DNA is found in small fragments 1000–2000 nt long (100–200 nt in eukaryotes) when the DNA is analyzed on alkaline sucrose gradients, which separates single strands of denatured DNA according to size. If the cells are then incubated further with unlabeled thymidine, these **Okazaki fragments** rapidly join into high molecular weight DNA. These results can be explained by a **semi-discontinuous replication** model (*Fig. 3*). One strand, the **leading strand**, is made continuously in a 5′→3′ direction from the origin. The second parental strand is not copied immediately but is displaced as a single strand for a distance of 1000–2000 nt (100–200 nt in eukaryotes). Synthesis of the second, **lagging**, strand is then initiated at the replication fork and proceeds 5′→3′ back towards the origin to form the first Okazaki fragment. As the replication fork progresses, the leading strand continues to be made as one long strand while further lagging strand fragments are made in a discontinuous fashion in the 'reverse' direction as the parental lagging strand is displaced. In fact, the physical direction of both leading and lagging strand synthesis is the same since the lagging strand is looped by 180° at the replication fork. Soon after synthesis, the fragments are joined to make one continuous piece of DNA by the enzyme **DNA ligase** (see Topic E2).

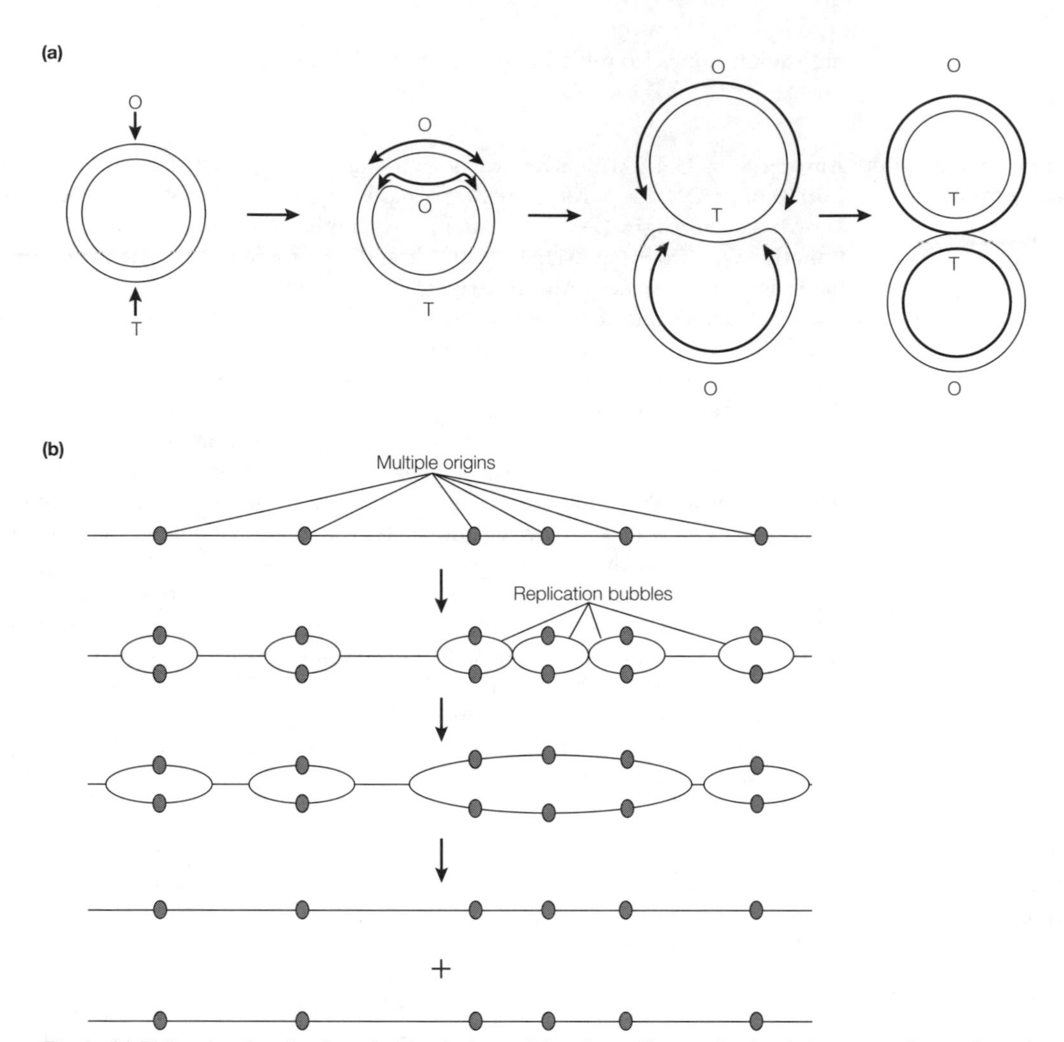

Fig. 2. (a) Bidirectional replication of a circular bacterial replicon. Two replication forks proceed away from the orgin (O) towards the terminus (T). Daughter DNA is shown as a thick line. (b) Multiple eukaryotic replicons. Each origin is marked (●).

RNA priming

Close examination of Okazaki fragments has shown that the first few nucleotides at their 5′-ends are in fact ribonucleotides, as are the first few nucleotides of the leading strand. Hence, DNA synthesis is **primed** by RNA (see Topic E2). These primers are removed and the resulting gaps filled with DNA before the fragments are joined. The reason for initiating each piece of DNA with RNA appears to relate to the need for DNA replication to be of high **fidelity** (see Topic F1).

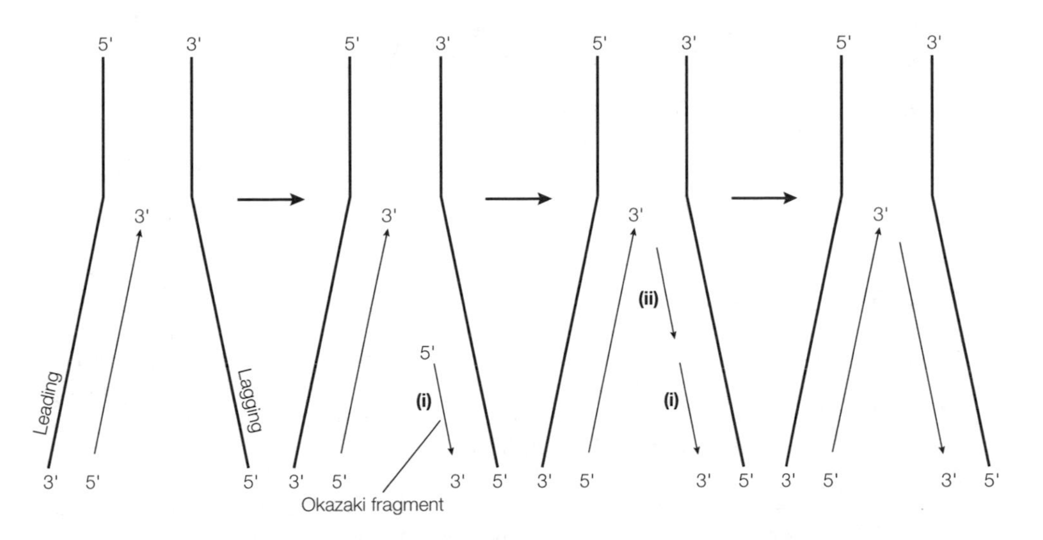

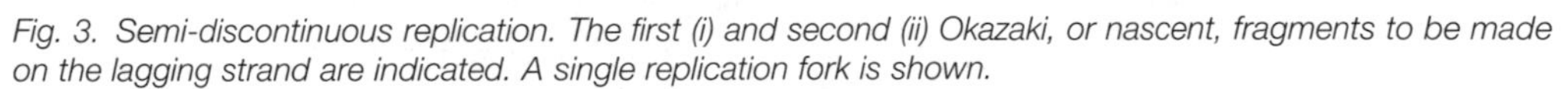

Fig. 3. Semi-discontinuous replication. The first (i) and second (ii) Okazaki, or nascent, fragments to be made on the lagging strand are indicated. A single replication fork is shown.

E2 Bacterial DNA replication

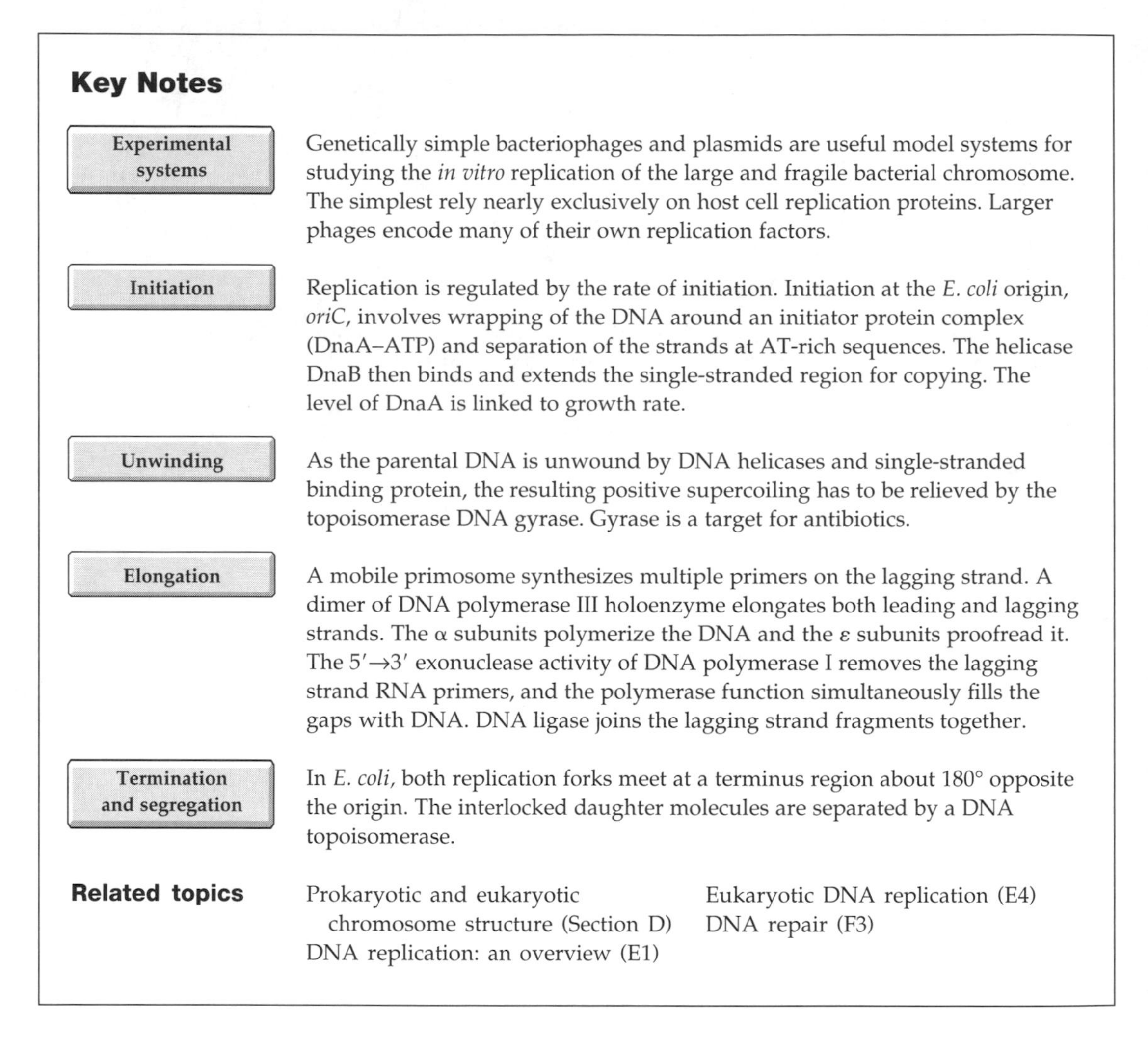

Key Notes

Experimental systems	Genetically simple bacteriophages and plasmids are useful model systems for studying the *in vitro* replication of the large and fragile bacterial chromosome. The simplest rely nearly exclusively on host cell replication proteins. Larger phages encode many of their own replication factors.
Initiation	Replication is regulated by the rate of initiation. Initiation at the *E. coli* origin, *oriC*, involves wrapping of the DNA around an initiator protein complex (DnaA–ATP) and separation of the strands at AT-rich sequences. The helicase DnaB then binds and extends the single-stranded region for copying. The level of DnaA is linked to growth rate.
Unwinding	As the parental DNA is unwound by DNA helicases and single-stranded binding protein, the resulting positive supercoiling has to be relieved by the topoisomerase DNA gyrase. Gyrase is a target for antibiotics.
Elongation	A mobile primosome synthesizes multiple primers on the lagging strand. A dimer of DNA polymerase III holoenzyme elongates both leading and lagging strands. The α subunits polymerize the DNA and the ε subunits proofread it. The 5′→3′ exonuclease activity of DNA polymerase I removes the lagging strand RNA primers, and the polymerase function simultaneously fills the gaps with DNA. DNA ligase joins the lagging strand fragments together.
Termination and segregation	In *E. coli*, both replication forks meet at a terminus region about 180° opposite the origin. The interlocked daughter molecules are separated by a DNA topoisomerase.

Related topics

Prokaryotic and eukaryotic chromosome structure (Section D)
DNA replication: an overview (E1)
Eukaryotic DNA replication (E4)
DNA repair (F3)

Experimental systems

Much of what we know about DNA replication has come from the use of *in vitro* systems consisting of purified DNA and all the proteins and other factors required for its complete replication. However, the chromosome of a prokaryote such as *E. coli* is too large and fragile to be studied in this way. So, smaller and simpler bacteriophage and plasmid DNA molecules have been used extensively as model systems. Of these, the genetically simple phage ϕX174 provides the best model for the replication of the *E. coli* chromosome since it relies almost exclusively on cellular replication factors for its own replication. Its replicative form comprises a supercoiled circle of only 5 kb. Larger phages, for example T7 (40 kb) and T4 (166 kb), encode many of their own replication proteins and employ some unusual mechanisms to complete their replication. Whilst not good model systems, they have been most informative in their own right and show the variety of solutions available for the complex problem of DNA replication.

Initiation

One aspect of bacterial DNA replication for which phages and plasmids are not good models is initiation at the origin. This is because these DNAs are designed to replicate many times within a single cell division cycle whereas replication of the bacterial chromosome is tightly coupled to the growth cycle. The *E. coli* origin is within the genetic locus *oriC* and is bound to the cell membrane. This region has been cloned into plasmids to produce more easily studied **minichromosomes** which behave like the *E. coli* chromosome. *OriC* contains four 9 bp binding sites for the initiator protein DnaA. Synthesis of this protein is coupled to growth rate so that the initiation of replication is also coupled to growth rate. At high cellular growth rates, prokaryotic chromosomes can re-initiate a second round of replication at the two new origins before the first round is completed. In this case, the daughter cells receive chromosomes that are already partly replicated (*Fig. 1*). Once it has attained a critical level, DnaA protein forms a complex of 30–40 molecules, each bound to an ATP molecule, around which the *oriC* DNA becomes wrapped (*Fig. 2*). This process requires that the DNA be negatively supercoiled (see Topic C4). This facilitates **melting** of three 13 bp AT-rich repeat sequences which open to allow binding of DnaB protein. DnaB is a **DNA helicase**. Helicases are enzymes which use the energy of ATP hydrolysis to move into and melt double-stranded DNA (or RNA). The single-stranded bubble created in this way is coated with **single-stranded binding protein** (**Ssb**) to protect it from breakage and to prevent the DNA renaturing. The enzyme **DNA primase** then attaches to the DNA and synthesizes a short **RNA primer** to initiate synthesis of the **leading strand** of the first replication fork. Bidirectional replication then follows.

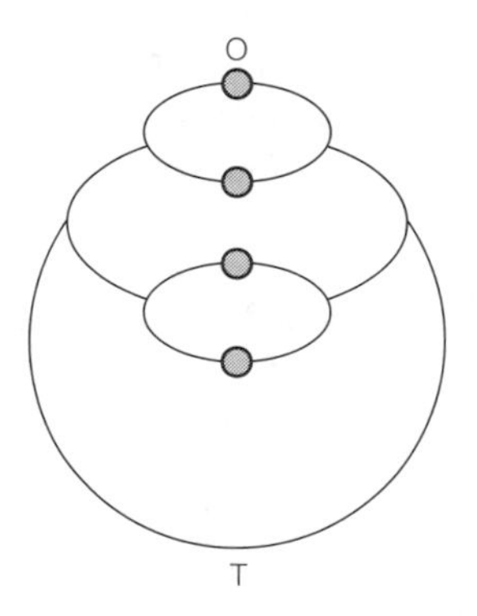

Fig. 1. Re-initiation of bacterial replication at new origins (O) before completion of the first round of replication. T = terminus.

Unwinding

For replication to proceed away from the origin, DNA helicases must travel along the template strands to open the double helix for copying. In addition to DnaB, a second DNA helicase may bind to the other strand to assist unwinding. Binding of Ssb protein further promotes unwinding. In a closed-circular DNA molecule, however, removal of helical turns at the replication fork leads to the introduction of additional turns in the rest of the molecule in the form of

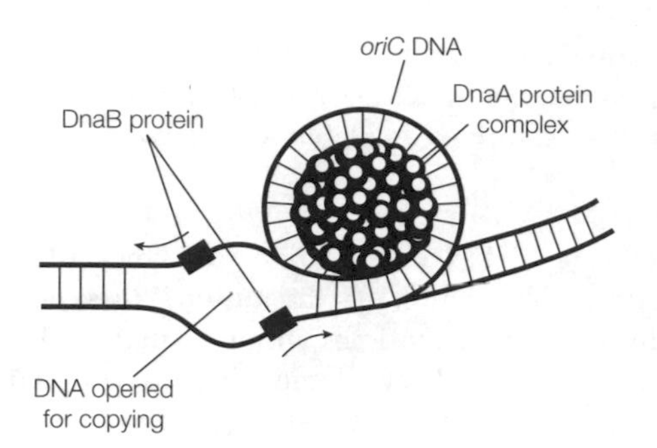

Fig. 2. Opening of origin DNA for copying by wrapping of oriC *round a DnaA protein complex.*

positive supercoiling (see Topic C4). Although the natural negative supercoiling of circular DNA partially compensates for this, it is insufficient to allow continued progression of the replication forks. This positive supercoiling must be relaxed continuously by the introduction of further negative supercoils by a type II **topoisomerase** called **DNA gyrase** (see Topic C4). Inhibitors of DNA gyrase, such as novobiocin and oxolinic acid, are effective inhibitors of bacterial replication and have antibiotic activity.

Elongation

As the newly formed replication fork displaces the parental lagging strand (see Topic E1), a mobile complex called a **primosome**, which includes the DnaB helicase and DNA primase, synthesizes RNA primers every 1000–2000 nt on the lagging strand. Both leading and lagging strand primers are elongated by **DNA polymerase III holoenzyme**. This multisubunit complex is a dimer, one half synthesizing the leading strand and the other the lagging strand. Having two polymerases in a single complex ensures that both strands are synthesized at the same rate. Both halves of the dimer contain an α subunit, the actual polymerase, and an ε subunit, which is a **3′→5′ proofreading exonuclease**. Proofreading helps to maintain high replicational fidelity (see Topic F1). The β subunits clamp the polymerase to the DNA. The remaining subunits in each half are different and may allow the holoenzyme to synthesize short and long stretches of DNA on the lagging and leading strands, respectively. Once the lagging strand primers have been elongated by DNA polymerase III, they are removed and the gaps filled by **DNA polymerase I**. This enzyme has 5′→3′ polymerase, 5′→3′ exonuclease and 3′→5′ proofreading exonuclease activities on a single polypeptide chain. The 5′→3′ exonuclease removes the primers while the polymerase function simultaneously fills the gaps with DNA by elongating the 3′-end of the adjacent Okazaki fragment (*Fig. 3*). The final phosphodiester bond between the fragments is made by **DNA ligase**. The enzyme from *E. coli* uses the co-factor NAD^+ as an unusual energy source. The NAD^+ is cleaved to nicotinamide mononucleotide (NMN) and adenosine monophosphate (AMP), and the AMP is covalently attached to the 5′-end of one fragment via a 5′–5′ linkage, The energy available from the hydrolysis of the pyrophosphate linkage of NAD^+ is thus maintained and subsequently used to link this 5′-end to the 3′-OH of the adjacent fragment with the release of AMP. *In vivo*, the DNA polymerase III holoenzyme dimer, the primosome and the DNA helicases are believed to be physically associated in a large complex called a **replisome** which synthesizes DNA at a rate of 900 bp per sec.

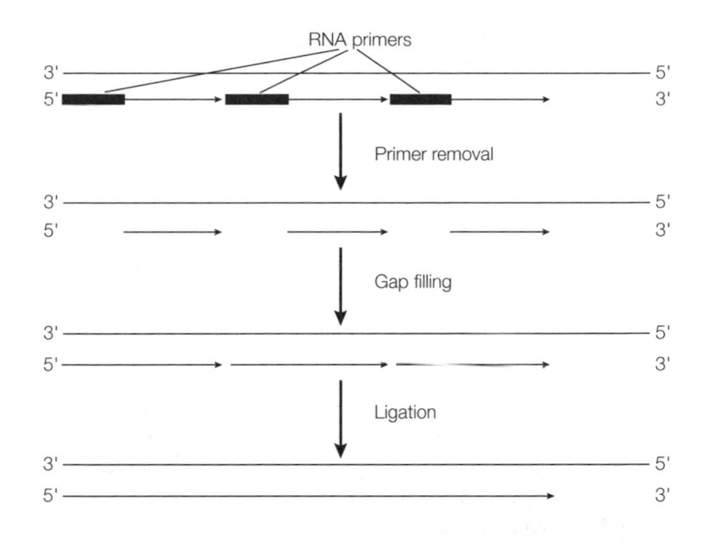

Fig. 3. Removal of RNA primers from lagging strand fragments, and subsequent gap filling and ligation. Primer removal and gap filling by DNA polymerase I occur simultaneously.

Termination and segregation

The two replication forks meet approximately 180° opposite *oriC*. Around this region are several terminator sites which arrest the movement of the forks by binding the *tus* gene product, an inhibitor of the DnaB helicase. Hence, if one fork is delayed for some reason, they will still meet within the terminus. Once replication is completed, the two daughter circles remain interlinked (catenated). They are unlinked by **topoisomerase IV**, a type II DNA topoisomerase (see Topic C4). They can then be segregated into the two daughter cells by movement apart of their membrane attachment sites (see Topic D1).

E3 THE CELL CYCLE

Key Notes

The cell cycle

The cell cycle involves DNA replication followed by cell division to produce two daughter cells from one parent. It is an ordered process, which is controlled by the cell cycle machinery.

Cell cycle phases

There are four main phases, G1, S, G2 and M phase. S phase is the DNA synthesis phase, whereas M phase, or mitosis, is the cell division phase. These two phases are separated by two gap phases G1 and G2. M phase can be divided into three further phases, prophase, metaphase and anaphase. G1, S and G2 comprise the interphase. Cells can enter a nonproliferative phase from G1, which is called G0 phase or quiescence.

Checkpoints and their regulation

The cell cycle is regulated in response to the cell's environment and to avoid the proliferation of damaged cells. Checkpoints are stages at which the cell cycle may be halted if the circumstances are not right for cell division. Principal checkpoints occur at the end of the G1 and G2 gap phases. At the R point in G1 phase, cells starved of mitogens withdraw from the cell cycle into the resting G0 phase.

Cyclins and cyclin-dependent kinases

The cell cycle is controlled through protein phosphorylation, which is catalysed by multiple protein kinase complexes. These complexes consist of cyclins, the regulatory subunits, and cyclin-dependent kinases (CDKs), the catalytic subunits. Different cyclin-CDK complexes control different phases of the cell cycle. In turn, their activity is regulated through transcriptional control of their synthesis, alteration of their enzyme activity by inhibitor proteins, and by regulation of their proteolytic destruction.

Regulation by E2F and Rb

G1 progression is controlled by activation of the transcription factor E2F, which regulates expression of genes required for later phases of the cell cycle. E2F is inhibited in early G1 by the binding of hypophosphorylated Rb. Phosphorylation of Rb by G1 cyclin-CDK complexes frees E2F, which can then activate transcription.

Cell cycle activation, inhibition and cancer

The CIP and INK4 classes of proteins halt cell cycle progression by inhibition of the activity of cyclin-CDK complexes. The G1 to S phase transition is regulated by proto-oncogenes and by tumor suppressor proteins. B cell tumors are associated with over-expression of the cyclin D1 gene. Rb, a critical regulator of G1 to S progression, and the INK4 p16 protein are tumor suppressor proteins. The CIP protein p21waf1/cip1 is activated by the tumor suppressor p53 in response to DNA damage.

Related topics

Eukaryotic chromosome structure (D3)
Eukaryotic DNA replication (E4)
RNA Pol II genes: promoters and enhancers (M4)
Examples of transcriptional regulation (N2)
Categories of oncogenes (S2)
Tumor suppressor genes (S3)
Apoptosis (S4)

The cell cycle

The two main stages of cell division are replication of cellular DNA and the division of the cell into two daughter cells. The repeated process of cell division, where daughter cells continue to divide to form further generations of cells can be thought of as a repeated cycle of DNA replication followed by cell division. This is known as the **cell cycle**.

Cell division is an ordered and regulated process. It could be catastrophic if a cell attempted to divide without first replicating its DNA, or alternatively if it attempted to replicate another copy of its DNA before cell division had occurred. Thus, we can consider cell proliferation as a cyclic process, co-ordinated by the cell cycle machinery to ensure the correct ordering of the cellular processes.

Cell cycle phases

Since the processes of DNA replication and cell division occur at distinct and regulated time intervals, the cell cycle can be conceptually divided into four phases *(Fig. 1)*.

G1
The longest phase (called a **gap phase**) during which the cells are preparing for replication;

S
Or **DNA synthesis phase** during which the DNA is replicated and a complete copy of each of the chromosomes is made (see Topic E4);

G2
A short gap phase, which occurs after S phase and before mitosis;

M
Mitosis, during which the new chromosomes are segregated equally between the two daughter cells as the cell divides.

Mitosis can be subdivided into **prophase,** during which the chromosomes condense; **metaphase,** during which the sister chromatids attached at the centromere (see Topic D3) become aligned in the centre of the cell; and finally **anaphase,** where the sister chromatids separate and move to opposite poles or spindles and segregate into the daughter cells.

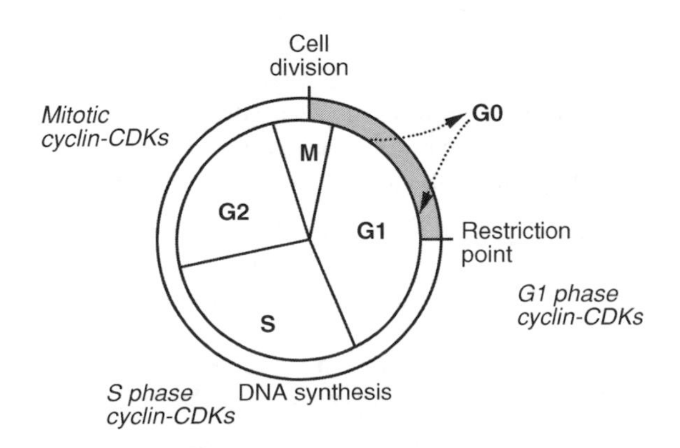

Fig. 1. Schematic diagram of the phases of the cell cycle. The regulatory complexes which are important in each phase are shown in italics.

Together, G1, S and G2 comprise the **interphase**. After mitosis, proliferating cells will enter the G1 phase of the next cell cycle. Cells can also exit the cell cycle after mitosis and enter a non-proliferative resting state, **G0,** which is also called **quiescence.**

Checkpoints and their regulation

The initiation of a cell division cycle requires the presence of extracellular growth factors, or **mitogens**. In the absence of mitogens, cells withdraw from the cell cycle in G1 and enter the G0 resting phase (*Fig. 1*). The point in G1 at which information regarding the environment of the cell is assessed, and the cell decides whether to enter another division cycle is called the **restriction point** (or **R point**). Cells starved of mitogens before reaching the R point re-enter G0 and fail to undergo cell division. Cells that are starved of mitogens after passing through the R point continue through the cell cycle to complete cell division before entering G0. In most cell types, the R point occurs a few hours after mitosis. The R point is clearly of crucial importance in understanding the commitment of cells to undergoing a cell division cycle. The interval in G1 between mitosis and the R point is the period in which multiple signals coincide and interact to determine the fate of the cell.

Those parts of the cell cycle, such as the R point, where the process may be stopped are known as **checkpoints**. Checkpoints operate during the gap phases. These ensure that the cell is competent to undergo another round of DNA replication (at the R point in G1 phase) and that replication of the DNA has been successfully completed before cell division (G2 phase checkpoint).

Cyclins and cyclin-dependent kinases

A major mechanism for control of cell cycle progression is by regulation of protein phosphorylation. This is controlled by specific protein kinases made up of a regulatory subunit and a catalytic subunit. The regulatory subunits are called **cyclins** and the catalytic subunits are called **cyclin-dependent kinases (CDKs)**. The CDKs have no catalytic activity unless they are associated with a cyclin and each can associate with more than one type of cyclin. The CDK and the cyclin present in a specific CDK-cyclin complex jointly determine which target proteins are phosphorylated by the protein kinase.

There are three different classes of cyclin-CDK complexes, which are associated with either the G1, S or M phases of the cell cycle. The **G1 CDK complexes** prepare the cell for S phase by activating transcription factors that cause expression of enzymes required for DNA synthesis and the genes encoding S phase CDK complexes. **The S phase CDK complexes** stimulate the onset of organized DNA synthesis. The machinery ensures that each chromosome is replicated only once. The **mitotic CDK complexes** induce chromosome condensation and ordered chromosome separation into the two daughter cells.

The activity of the CDK complexes is regulated in three ways:

(i) By control of the transcription of the CDK complex subunits.
(ii) By inhibitors that reduce the activity of the CDK complexes. For example, the mitotic CDK complexes are synthesized in S and G2 phase, but their activity is repressed until DNA synthesis is complete.
(iii) By organized proteolysis of the CDK complexes at a defined stage in the cell cycle where they are no longer required.

Regulation by E2F and Rb

Cell cycle progression through G1 and into S phase is in part regulated by activation (and in some cases inhibition) of gene transcription, whereas progression through the later cell cycle phases appears to be regulated primarily by post-transcriptional mechanisms. Passage through the key G1 R point critically depends on the activation of a transcription factor, **E2F.** E2F stimulates the transcription and expression of genes encoding proteins required for DNA replication and deoxyribonucleotide synthesis as well as for cyclins and CDKs required in later cell cycle phases. The activity of E2F is inhibited by the binding of the protein **Rb** (the **retinoblastoma tumor suppressor protein**, see Section S3) and related proteins. When Rb is hypophosphorylated (under-phosphorylated) E2F activity is inhibited. The phosphorylation of Rb by cyclin-CDK complexes during middle and late G1 phase frees E2F so that it can activate transcription (*Fig. 2*).

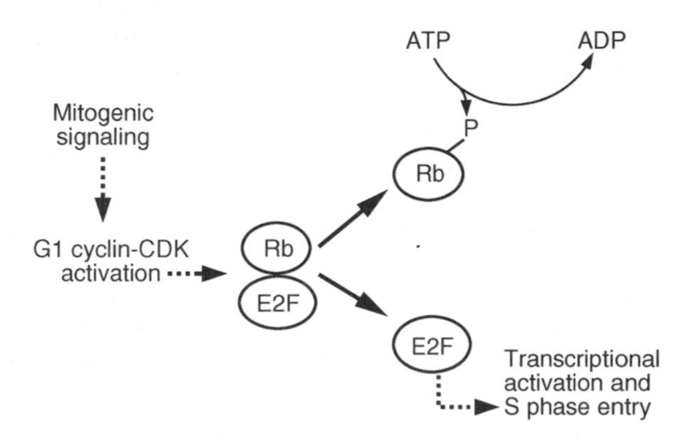

Fig. 2. Schematic diagram showing the regulation of E2F by Rb and G1 cyclin-CDK complexes.

Cell cycle activation, inhibition and cancer

Small inhibitor proteins can delay cell cycle progression by repression of the activity of cyclin-CDK complexes. There are two classes of these inhibitors, CIP proteins and INK4 proteins. For example, during skeletal muscle cell differentiation, the differentiating cells are induced to withdraw from the cell cycle via activation of one of the CIP protein genes, encoding the p21waf1/cip1 protein. This is controlled by the master regulator transcription factor MyoD (see Topic N2).

There is a fundamental link between regulation of the cell cycle and cancer. This is supported by the observation that proto-oncogenes (see Topic S2) and tumor suppressor genes (see Topic S3) regulate the passage of cells from G1 to S phase. One of the G1 cyclins, cyclin D1, is often over-expressed in tumors of immunoglobulin producing B cells. This is caused by a chromosomal translocation that brings the immunoglobulin gene enhancer (see Topic M4) to a chromosomal position adjacent to the cyclin D1 gene. The INK4 p16 protein has been shown to have all the characteristics of a tumor suppressor. Importantly, the products of the two most significant tumor suppressor genes in human cancer, Rb and p53 (see Topic S3), are both critically involved in cell cycle regulation. Rb is involved in direct regulation of E2F activity (see above), and p53 is involved in the inhibition of the cell cycle in response to DNA damage. This occurs through p53-induced transcriptional activation of the *p21waf1/cip1* gene in response to DNA damage.

E4 Eukaryotic DNA replication

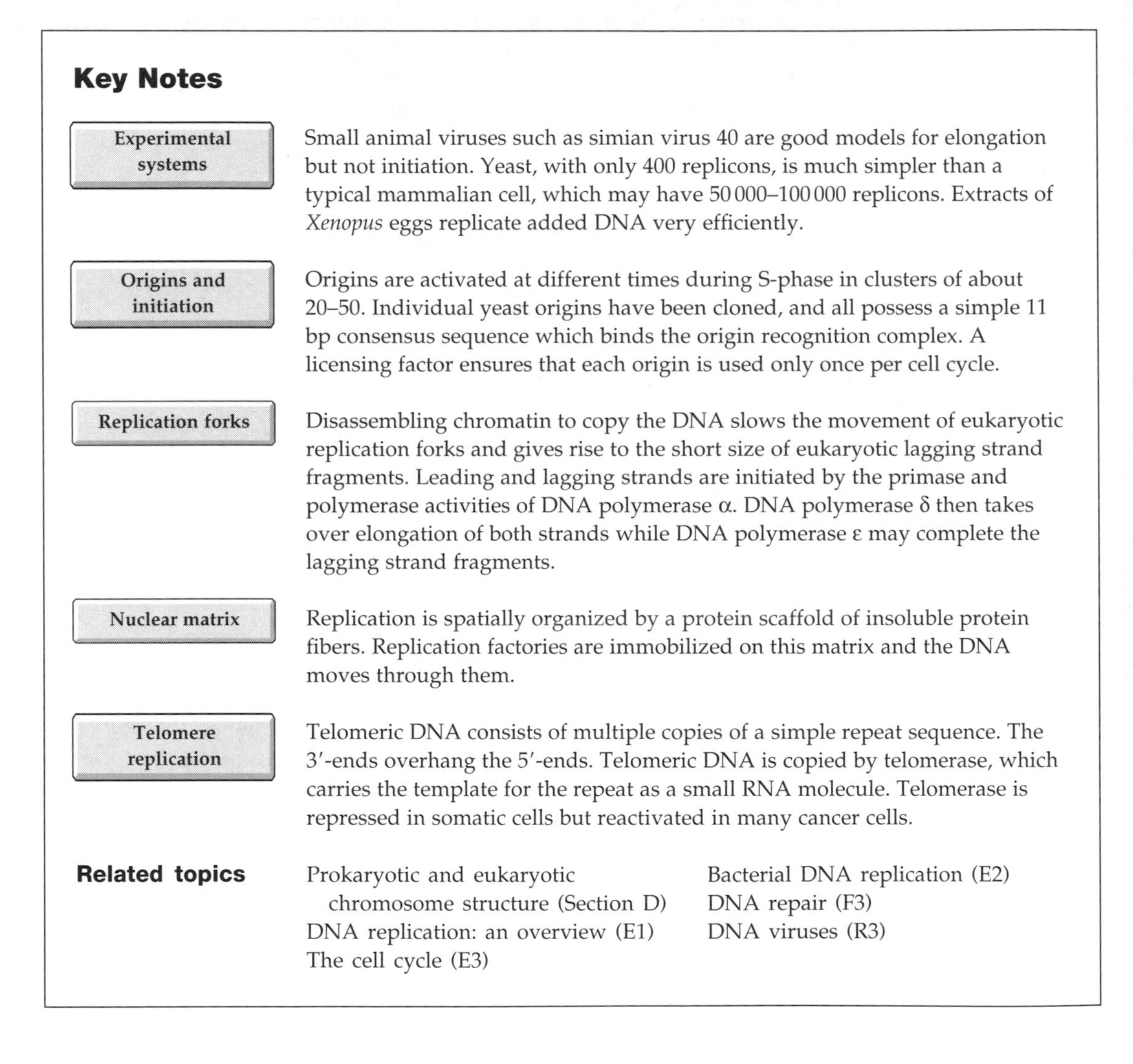

Key Notes

Experimental systems	Small animal viruses such as simian virus 40 are good models for elongation but not initiation. Yeast, with only 400 replicons, is much simpler than a typical mammalian cell, which may have 50 000–100 000 replicons. Extracts of *Xenopus* eggs replicate added DNA very efficiently.
Origins and initiation	Origins are activated at different times during S-phase in clusters of about 20–50. Individual yeast origins have been cloned, and all possess a simple 11 bp consensus sequence which binds the origin recognition complex. A licensing factor ensures that each origin is used only once per cell cycle.
Replication forks	Disassembling chromatin to copy the DNA slows the movement of eukaryotic replication forks and gives rise to the short size of eukaryotic lagging strand fragments. Leading and lagging strands are initiated by the primase and polymerase activities of DNA polymerase α. DNA polymerase δ then takes over elongation of both strands while DNA polymerase ε may complete the lagging strand fragments.
Nuclear matrix	Replication is spatially organized by a protein scaffold of insoluble protein fibers. Replication factories are immobilized on this matrix and the DNA moves through them.
Telomere replication	Telomeric DNA consists of multiple copies of a simple repeat sequence. The 3′-ends overhang the 5′-ends. Telomeric DNA is copied by telomerase, which carries the template for the repeat as a small RNA molecule. Telomerase is repressed in somatic cells but reactivated in many cancer cells.

Related topics

Prokaryotic and eukaryotic chromosome structure (Section D)
DNA replication: an overview (E1)
The cell cycle (E3)
Bacterial DNA replication (E2)
DNA repair (F3)
DNA viruses (R3)

Experimental systems

Because of the complexities of chromatin structure (see Topic D2), eukaryotic replication forks move at around 50 bp per sec. At this rate, it would take about 30 days to copy a typical mammalian chromosomal DNA molecule of 10^5 kb using two replication forks, hence the need for multiple replicons – typically 50 000–100 000 in a mammalian cell. To study all these simultaneously would be a daunting prospect. Fortunately, the yeast *Saccharomyces cerevisiae* has a much smaller genome (14 000 kb in 16 chromosomes) and only 400 replicons. The complete genome sequence is known. Simpler still are viruses such as simian virus 40 (SV40) (see Topic R3). SV40 DNA is a 5 kb double-stranded circle which forms nucleosomes when it enters the cell nucleus. It provides an excellent model for a mammalian replication fork. Another useful system is a cell-free extract prepared from the eggs of the African clawed frog, *Xenopus laevis*. Because of its high concentration of replication proteins (sufficient for several cell doublings *in vivo*), this extract can support the extensive replication of added DNA or even whole nuclei.

Origins and initiation

In eukaryotes, **clusters (tandem arrays)** of about 20–50 replicons initiate simultaneously at defined times throughout S-phase. Those which replicate early in S-phase comprise predominantly **euchromatin** (which includes transcriptionally active DNA) while those activated late in S-phase are mainly within **heterochromatin** (see Topic D3). Centromeric and telomeric DNA replicates last. This pattern reflects the accessibility of the different chromatin structures to initiation factors. Individual yeast replication origins have been cloned into prokaryotic plasmids. Since they allow these plasmids to replicate in yeast (a eukaryote) they are termed **autonomously replicating sequences (ARS**s). The minimum length of DNA that will support replication is only 11 bp and has the consensus sequence [A/T]TTTAT[A/G]TTT[A/T], though additional copies of this sequence are required for optimal efficiency. This sequence is bound by the **origin recognition complex (ORC)** which, when activated by CDKs, permits opening of the DNA for copying. Defined origin sequences have not been isolated from mammalian cells, and it is believed that initiation of each replicon may occur at random within an **initiation zone** which may be several kilobases in length and which may be part of the dispersed repetitive DNA fraction (see Topic D4). In contrast to prokaryotes, eukaryotic replicons can only initiate once per cell cycle. A protein (**licensing factor**) which is absolutely required for initiation and inactivated after use can only gain access to the nucleus when the nuclear envelope dissolves at mitosis, thus preventing premature re-initiation.

Replication forks

Before copying, the DNA must be unwound from the nucleosomes at the replication forks (*Fig. 1*). This slows the movement of the forks to about 50 bp per sec. After the fork has passed, new nucleosomes are assembled from a mixture of old and newly synthesized histones. As in prokaryotes, one or more DNA helicases and a single-stranded binding protein, **replication protein A (RP-A)**, are required to separate the strands. Three different DNA polymerases are involved in elongation. The leading strands and each lagging strand fragment are initiated with RNA by a primase activity that is an integral part of **DNA polymerase α**. This polymerase continues elongation with DNA but is quickly replaced by **DNA polymerase-δ (delta)** on the leading strand and probably also on the lagging strand. The role of **DNA polymerase-ε (epsilon)** is less clear, but it may simply complete the lagging strand fragments after primer removal. Both these enzymes have associated proofreading activity. The ability to synthesize long DNA is conferred on DNA polymerase δ by **proliferating cell nuclear antigen (PCNA)**, the functional equivalent of the β subunit of *E. coli* DNA polymerase III holoenzyme. The small size of eukaryotic lagging strand fragments (e.g. 135 bp in SV40) appears to reflect the amount of DNA unwound from each nucleosome as the fork progresses (*Fig. 1*). In addition to doubling the DNA, the histone content of the cell is also doubled during S-phase.

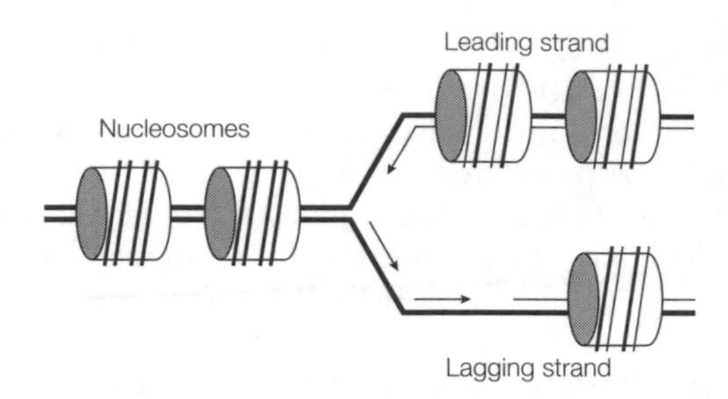

Fig. 1. Nucleosomes at a eukaryotic replication fork.

Nuclear matrix

The nuclear matrix is a scaffold of insoluble protein fibers which acts as an organizational framework for nuclear processes, including DNA replication (see Topic D3). Huge **replication factories** containing all the enzymes and DNA associated with the replication forks of all replicons within a cluster are immobilized on the matrix, and the DNA moves through these sites as it replicates. These factories can be visualized in the microscope by pulse-labeling the replicating DNA with the thymidine analog, **bromodeoxyuridine** (**BUdR**), and visualizing the labeled DNA by immunofluorescence using an antibody that recognizes BUdR (*Fig. 2*).

Telomere replication

The ends of linear chromosomes cannot be fully replicated by semi-discontinuous replication as there is no DNA to elongate to replace the RNA removed from the 5′-end of the lagging strand. Thus, genetic information could be lost from the DNA. To overcome this, the ends of eukaryotic chromosomes (**telomeres**, see Topic D3) consist of hundreds of copies of a simple, noninformational repeat sequence (e.g. TTAGGG in humans) with the 3′-end overhanging the 5′-end (*Fig. 3*). The enzyme **telomerase** contains a short RNA molecule, part of whose sequence is complementary to this repeat. This RNA acts as a template for the addition of these repeats to the 3′-overhang by repeated cycles of

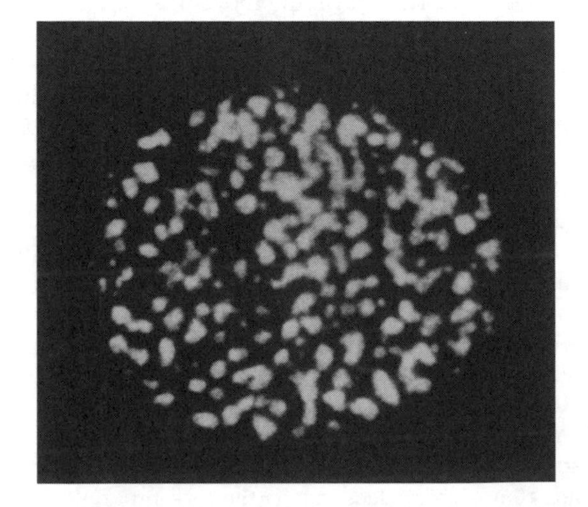

Fig. 2. Replication factories in a eukaryotic nucleus. Reproduced from H. Nakamura et al. *(1986)* Exp. Cell Res. ***165****, 291–297 with permission from Academic Press.*

3'-A A T C C C A A T C C C-5'
5'-T T A G G G T T A G G G (T T A G G G)$_n$ T T A G G G-3'

Fig. 3. Sequence of human telomeric DNA (n = several hundred).

elongation (polymerization) and translocation. The complementary strand is then synthesized by normal lagging strand synthesis, leaving a 3′-overhang. Interestingly, telomerase activity is repressed in the somatic cells of multicellular organisms, resulting in a gradual shortening of the chromosomes with each cell generation. As this shortening reaches informational DNA, the cells senesce and die. This phenomenon may be important in cell aging. Furthermore, the unlimited proliferative capacity of many cancer cells is associated with reactivation of telomerase activity.

F1 MUTAGENESIS

Key Notes

Mutation

Mutations are heritable permanent changes in the base sequence of DNA. Point mutations may be transitions (e.g. G·C→A·T) or transversions (e.g. G·C→T·A). Deletions and insertions involve the loss or addition of bases and can cause frameshifts in reading the genetic code. Silent mutations have no phenotypic effect, while missense and nonsense mutations change the amino acid sequence of the encoded protein.

Replication fidelity

The high accuracy of DNA replication (one error per 10^{10} bases incorporated) depends on a combination of proper base pairing of template strand and incoming nucleotide in the active site of the DNA polymerase, proofreading of the incorporated base by 3′→5′ exonuclease and mismatch repair.

Physical mutagens

Ionizing (e.g. X- and γ-rays) and nonionizing (e.g. UV) radiation produce a variety of DNA lesions. Pyrimidine dimers are the commonest product of UV irradiation.

Chemical mutagens

Base analogs can mispair during DNA replication to cause mutations. Nitrous acid deaminates cytosine and adenine. Alkylating and arylating agents generate a variety of adducts that can block transcription and replication and cause mutations by direct or, more commonly, indirect mutagenesis. Most chemical mutagens are carcinogenic.

Direct mutagenesis

If a base analog or modified base whose base pairing properties are different from the parent base is not removed by a DNA repair mechanism before passage of a replication fork, then an incorrect base will be incorporated. A second round of replication fixes the mutation permanently in the DNA.

Indirect mutagenesis

Most lesions in DNA are repaired by error-free direct reversal or excision repair mechanisms before passage of a replication fork. If this is not possible, an error-prone form of translesion DNA synthesis may take place involving specialized DNA polymerases and one or more incorrect bases become incorporated opposite the lesion.

Related topics

Nucleic acid structure (C1)
DNA replication (Section E)
DNA damage (F2)
DNA repair (F3)

Mutation

Mutations are permanent, heritable alterations in the base sequence of the DNA. They arise either through spontaneous errors in DNA replication or meiotic recombination (see Topic F4) or as a consequence of the damaging effects of physical or chemical agents on the DNA. The simplest mutation is a **point mutation** – a single base change. This can be either a **transition**, in which one purine (or pyrimidine) is replaced by the other, or a **transversion**, where a purine replaces a pyrimidine or vice versa. The phenotypic effects of such a mutation can be various. If it is in a noncoding or nonregulatory piece of DNA or in the third position of a codon, which often has no effect on the amino acid incorporated into a protein (see Topic P1), then it may be **silent**. If it results in an altered amino acid in a gene product then it is a **missense** mutation whose effect can vary from none to lethality, depending on the amino acid affected. Mutations which generate new stop codons are **nonsense** mutations and give rise to truncated protein products. Insertions or deletions involve the addition or loss of one or more bases. These can produce **frameshift** mutations in genes, where the translated protein sequence to the C-terminal side of the mutation is completely changed. Mutations that affect the processes of cell growth and cell death can result in tumorigenesis (see Section S). The accumulation of many silent and other nonlethal mutations in populations produces **genetic polymorphisms** – acceptable variations in the 'normal' DNA and protein sequences (see Topic D4).

Replication fidelity

The error rate of DNA replication is much lower than that of transcription because of the need to preserve the meaning of the genetic message from one generation to the next. For example, the **spontaneous mutation** rate in *E. coli* is about one error per 10^{10} bases incorporated during replication. This is due primarily to the presence of the minor tautomeric forms of the bases which have altered base pairing properties (see Topic C2). The error rate is minimized by a variety of mechanisms. DNA polymerases will only incorporate an incoming nucleotide if it forms the correct Watson–Crick base pair with the template nucleotide in its active site. The occasional error (*Fig. 1a*) is detected by the 3′→5′ **proofreading exonuclease** associated with the polymerase (see Topic E2). This removes the incorrect nucleotide from the 3′-end before any further incorporation, allowing the polymerase then to insert the correct base (*Fig. 1b*). In order for the proofreading exonuclease to work properly, it must be able to distinguish a correct base pair from an incorrect one. The increased mobility of 'unanchored' base pairs at the very 5′-end of newly initiated lagging strand fragments of DNA (see Topic E1) means that they can never appear correct and so cannot be proofread. Hence, the first few nucleotides are **ribo**nucleotides (RNA) so that they subsequently can be identified as low fidelity material and replaced with DNA elongated (and proofread) from the adjacent fragment. Errors which escape proofreading are corrected by a **mismatch repair** mechanism (see Topic F3).

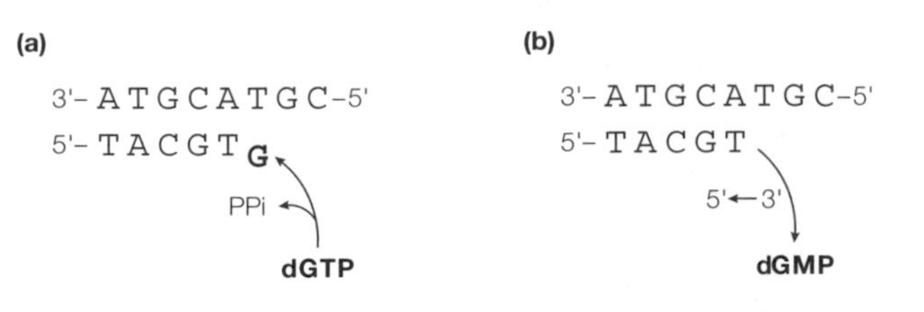

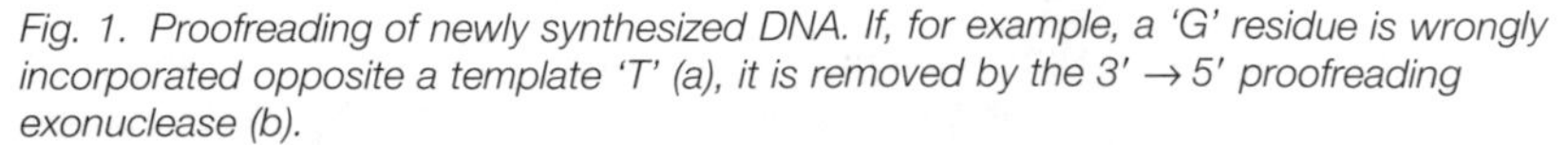
Fig. 1. Proofreading of newly synthesized DNA. If, for example, a 'G' residue is wrongly incorporated opposite a template 'T' (a), it is removed by the 3′ → 5′ proofreading exonuclease (b).

Physical mutagens

Absorption of high-energy ionizing radiation such as X-rays and γ-rays causes the target molecules to lose electrons. These electrons can cause extensive chemical alterations to DNA, including strand breaks and base and sugar destruction. Nonionizing radiation causes molecular vibrations or promotion of electrons to higher energy levels within the target molecules. This can lead to the formation of new chemical bonds. The most important form causing DNA damage is UV light which produces **pyrimidine dimers** (see Topic F2) from adjacent pyrimidine bases.

Chemical mutagens

Base analogs are derivatives of the normal bases with altered base pairing properties and can cause **direct mutagenesis**. A wide range of other natural and synthetic organic and inorganic chemicals can react with DNA and alter its properties. **Nitrous acid** deaminates cytosine to produce uracil (*Fig. 2*), which base-pairs with adenine and causes G·C→A·T transitions upon subsequent replication. Deamination of adenine to the guanine analog hypoxanthine results in A·T→G·C transitions. **Alkylating agents**, such as **methylmethane sulfonate (MMS)** and **ethylnitrosourea (ENU)**, and **arylating agents** produce **lesions** (see Topic F2) that usually have to be repaired to prevent serious disruption to the processes of transcription and replication. Processing of these lesions by the cell may give rise to mutations by **indirect mutagenesis**. Intercalators (see Topic C4) generate insertion and deletion mutations. Most chemical mutagens are **carcinogens** and cause cancer.

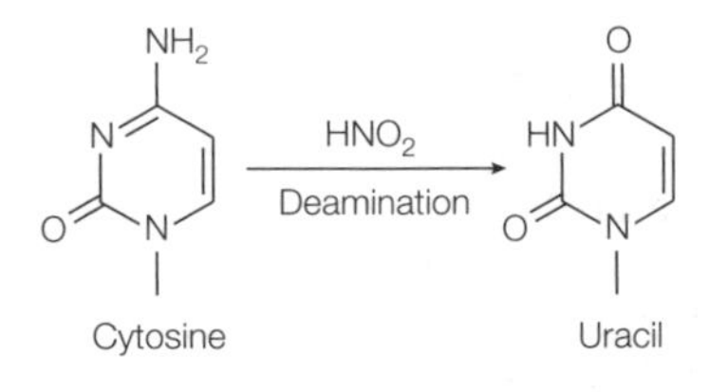

Fig. 2. Deamination of cytosine to uracil by nitrous acid.

Direct mutagenesis

Direct mutagenesis results from the presence of a stable, unrepaired base with altered base pairing properties in the DNA. All that is required for this lesion (nonpermanent) to be fixed as a mutation (permanent and heritable) is DNA replication. For example, the keto tautomer of the thymine analog 5-bromouracil base-pairs with adenine as expected, but the frequently occurring enol form (due to electronegativity of Br) pairs with guanine (*Fig. 3a*). After one round of replication, a guanine may be inserted opposite the 5-bromouracil. After a second round of replication, a cytosine will be incorporated opposite this guanine and so the mutation is **fixed** (*Fig. 3b*). The net effect is an A·T→G·C transition. 8-Oxo-dGTP, a naturally occurring derivative of dGTP produced by intracellular oxidation (see Topic F2), can base-pair with A and, if incorporated into DNA, gives rise to A·T→C·G transversions.

Indirect mutagenesis

The great majority of lesions introduced by chemical and physical mutagens are substrates for one or more of the DNA repair mechanisms that attempt to restore the original structure to the DNA in an error-free fashion before the damage is encountered by a replication fork (see Topic F3). Occasionally this is not possible; for example, when the DNA is damaged immediately ahead of an advancing fork producing a lesion unsuitable for **recombination repair** (see

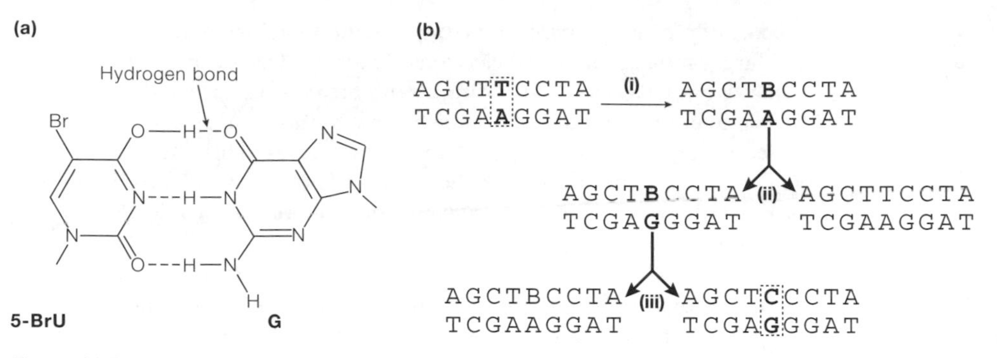

Fig. 3. (a) Base pairing of the enol form of 5-bromouracil with guanine; (b) direct mutagenesis by 5-bromouracil (B): (i) the 'T' analog 'B' can be readily incorporated in place of 'T'; (ii) after one round of replication 'G' may be incorporated opposite the 'B'; (iii) after a second round, a 'C' will be incorporated opposite the 'G'.

Topic F4). This could lead to a potentially lethal gap in the daughter strand so in such cases the cell resorts to **translesion DNA synthesis**. This involves the temporary replacement of the normal replicative apparatus by one of several specialized DNA polymerase activities which insert bases opposite the unrepaired lesion. After this, normal replication resumes. Thus, the lesion is not removed or repaired but tolerated and so the integrity of the replicated DNA is maintained. The template strand can then be repaired before the next round of replication.

In some cases, this translesion DNA synthesis may be accurate and insert the correct bases; however, the specific base pairing properties of damaged bases are often lost and so some of the translesion DNA polymerases insert wrong bases opposite such lesions simply to ensure chromosomal integrity. This is known as **indirect mutagenesis**. Indirect mutagenesis may be **targetted**, if the mutation occurs only at the site of the lesion, or **untargetted**, if mutations are also generated elsewhere in the genome. In prokaryotes, translesion DNA synthesis is part of the **SOS response** to DNA damage and is sometimes called **error-prone repair**, though it is not strictly a repair mechanism. The SOS response involves the induction of many genes whose functions are to increase survival after DNA damage. The damage-inducible RecA, UmuC, UmuD and DinB proteins are essential for mutation induction by indirect mutagenesis. Together, UmuC and a modified form of UmuD (UmuD′) form a protein complex ($UmuD'_2C$), recently termed DNA polymerase V, that can insert a base opposite a **non-coding** lesion such as an **apurinic** site (see Topic F2) leading to targetted mutagenesis. Induction of DinB, another translesion polymerase called DNA polymerase IV, leads to untargetted mutagenesis. Untargetted mutagenesis may be a useful and deliberate response to a hostile, DNA-damaging environment in order to increase the mutation rate with the hope of producing a more resistant strain that could be selected for survival. In eukaryotes, translesion DNA polymerases include DNA polymerase-ζ (zeta), which is error-prone, and DNA polymerase-η (eta), which may be predominantly error-free.

F2 DNA DAMAGE

Key Notes

DNA lesions

The chemical reactivity of DNA with exogenous chemicals or radiation can give rise to changes in its chemical or physical structure. These may block replication or transcription and so be lethal, or they may generate mutations through direct or indirect mutagenesis. The chemical instability of DNA can generate spontaneous lesions such as deamination and depurination.

Oxidative damage

Reactive oxygen species such as superoxide and hydroxyl radicals produce a variety of lesions including 8-oxoguanine and 5-formyluracil. Such damage occurs spontaneously but is increased by some exogenous agents including γ-rays.

Alkylation

Electrophilic alkylating agents such as methylmethane sulfonate and ethylnitrosourea can modify nucleotides in a variety of positions. Most lesions are indirectly mutagenic, but O^6-alkylguanine is directly mutagenic.

Bulky adducts

Bulky lesions such as pyrimidine dimers and arylating agent adducts distort the double helix and cause localized denaturation. This disrupts the normal functioning of the DNA.

Related topics

Nucleic acid structure (C1)
Mutagenesis (F1)
DNA repair (F3)
Apoptosis (S4)

DNA lesions

A **lesion** is an alteration to the normal chemical or physical structure of the DNA. Some of the nitrogen and carbon atoms in the heterocyclic ring systems of the bases and some of the exocyclic functional groups (i.e. the keto and amino groups of the bases) are chemically quite reactive. Many exogenous agents, such as chemicals and radiation, can cause changes to these positions. The altered chemistry of the bases may lead to loss of base pairing or altered base pairing (e.g. an altered A may base-pair with C instead of T). If such a lesion was allowed to remain in the DNA, a mutation could become fixed in the DNA by **direct** or **indirect mutagenesis** (see Topic F1). Alternatively, the chemical change may produce a physical distortion in the DNA which blocks replication and/or transcription, causing cell death. Thus, DNA lesions may be **mutagenic** and/or **lethal**. Some lesions are **spontaneous** and occur because of the inherent chemical reactivity of the DNA and the presence of normal, reactive chemical species within the cell. For example, the base cytosine undergoes spontaneous hydrolytic **deamination** to give uracil (see Topic F1, *Fig. 2*). If left unrepaired, the resulting uracil would form a base pair with adenine during subsequent replication, giving rise to a point mutation (see Topic F1). In fact, the generation of uracil in DNA in this way is the probable reason why DNA contains thymine instead of uracil. Any uracil found in DNA is removed by an enzyme called uracil DNA glycosylase and is replaced by cytosine (see Topic F3). 5-Methylcytosine, a modified base found in small amounts in DNA (see Topic D3), deaminates to thymine, a normal base. This is much more difficult to detect.

Depurination is another spontaneous hydrolytic reaction that involves cleavage of the **N-glycosylic** bond between *N*-9 of the purine bases A and G and *C*-1′ of the deoxyribose sugar and hence loss of purine bases from the DNA. The sugar–phosphate backbone of the DNA remains intact. The resulting **apurinic site** is a **noncoding** lesion, as information encoded in the purine bases is lost. Depurination occurs at the rate of 10 000 purines lost per human cell per hour at 37°C. Though less frequent, **depyrimidination** can also occur.

Oxidative damage

This occurs under normal conditions due to the presence of reactive oxygen species (ROS) in all aerobic cells, for example superoxide, hydrogen peroxide and, most importantly, the hydroxyl radical (•OH). This radical can attack DNA at a number of points, producing a range of oxidation products with altered properties, for example 8-oxoguanine, 2-oxoadenine and 5-formyluracil (*Fig. 1*). The levels of these can be increased by hydroxyl radicals from the radiolysis of water caused by ionizing radiation.

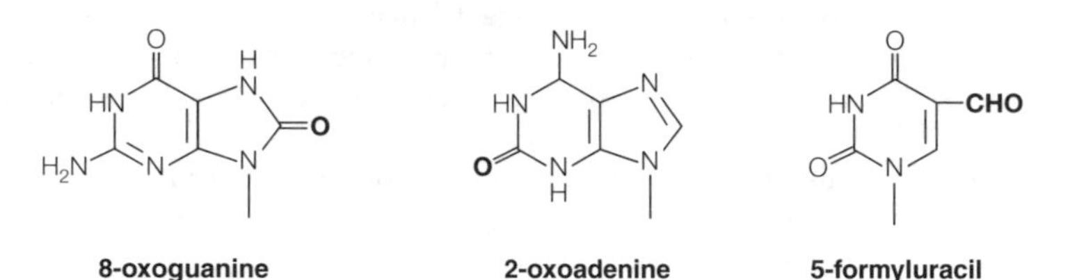

Fig. 1. Examples of oxidized bases produced in DNA by reactive oxygen species.

Alkylation

Alkylating agents are electrophilic chemicals which readily add alkyl (e.g. methyl) groups to various positions on nucleic acids distinct from those methylated by normal methylating enzymes (see Topics C1 and G3). Common examples are **methylmethane sulfonate (MMS)** and **ethylnitrosourea (ENU)** (*Fig. 2a*). Typical examples of methylated bases are 7-methylguanine, 3-methyladenine, 3-methylguanine and O^6-methylguanine (*Fig. 2b*). Some of these lesions are potentially lethal as they can interfere with the unwinding of DNA during

(a)

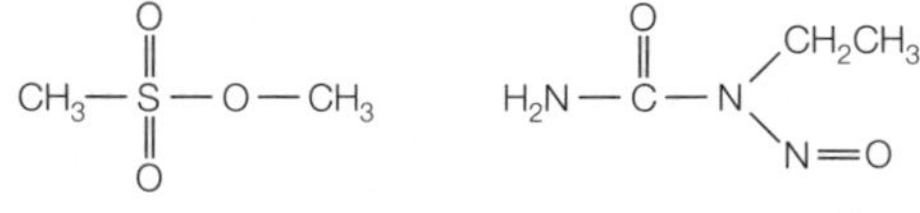
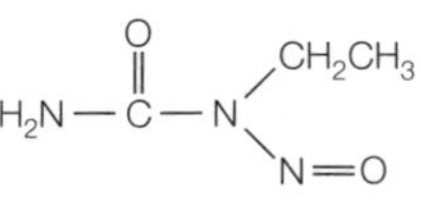

(b)

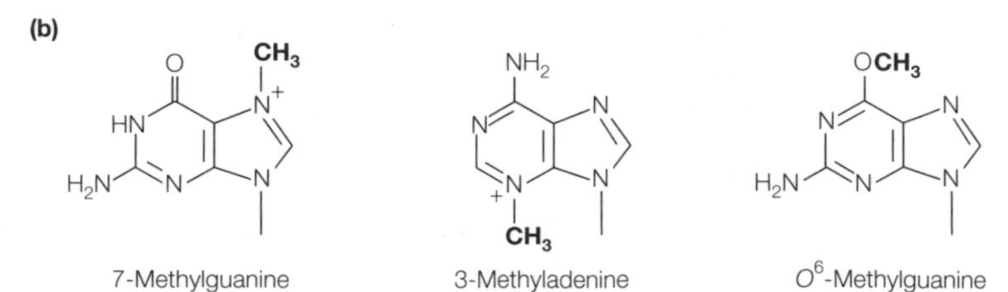

Fig. 2. Examples of (a) alkylating agents; (b) alkylated bases.

replication and transcription. Most are also indirectly mutagenic; however, O^6-methylguanine is a directly mutagenic lesion as it can base-pair with thymine during replication.

Bulky adducts

Cyclobutane pyrimidine dimers are formed by ultraviolet light from adjacent pyrimidines on one strand by cyclization of the double-bonded C5 and C6 carbon atoms of each base to give a cyclobutane ring (*Fig. 3a*). The resulting loss of base pairing with the opposite strand causes localized denaturation of the DNA producing a **bulky lesion** which would disrupt replication and transcription. Another type of pyrimidine dimer, the **6,4-photoproduct**, results from the formation of a bond between C6 of one pyrimidine base and C4 of the adjacent base (see ring carbon numbers in *Fig. 3a*). When the coal tar carcinogen **benzo[*a*]pyrene** is metabolized by cytochrome P-450 in the liver one of its metabolites (a diol epoxide) can covalently attach to the 2-amino group of guanine residues (*Fig. 3b*). Many other aromatic **arylating agents** form covalent **adducts** with DNA. The liver carcinogen aflatoxin B_1 also covalently binds to DNA.

(a)

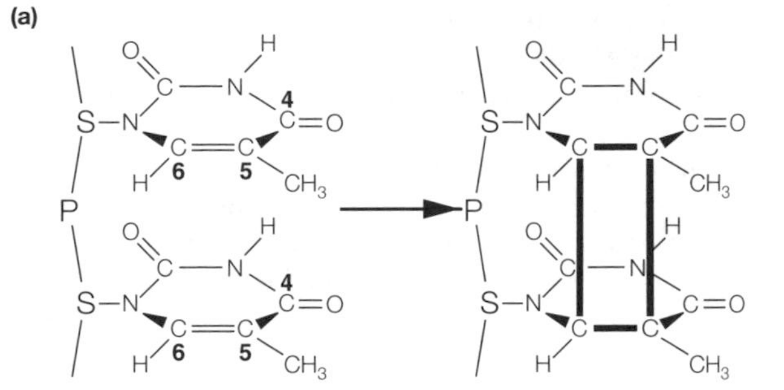

(b)

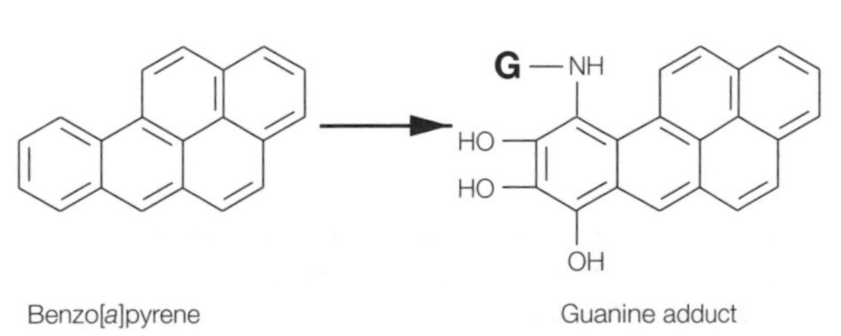

Fig. 3. Formation of (a) cyclobutane thymine dimer from adjacent thymine residues. S = sugar, P = phosphate; (b) guanine adduct of benzo[a]pyrene diol epoxide.

F3 DNA REPAIR

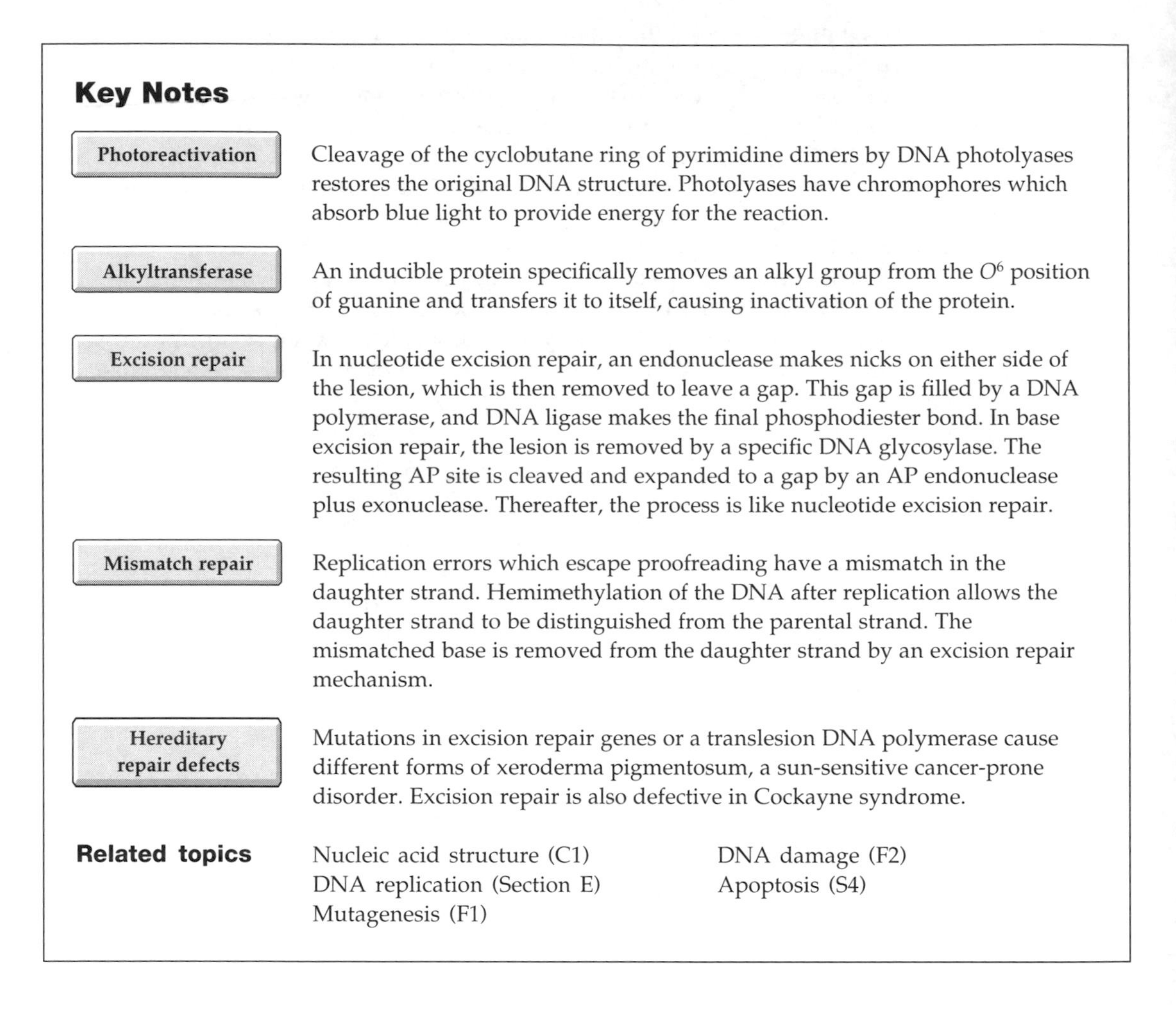

Key Notes

Photoreactivation	Cleavage of the cyclobutane ring of pyrimidine dimers by DNA photolyases restores the original DNA structure. Photolyases have chromophores which absorb blue light to provide energy for the reaction.
Alkyltransferase	An inducible protein specifically removes an alkyl group from the O^6 position of guanine and transfers it to itself, causing inactivation of the protein.
Excision repair	In nucleotide excision repair, an endonuclease makes nicks on either side of the lesion, which is then removed to leave a gap. This gap is filled by a DNA polymerase, and DNA ligase makes the final phosphodiester bond. In base excision repair, the lesion is removed by a specific DNA glycosylase. The resulting AP site is cleaved and expanded to a gap by an AP endonuclease plus exonuclease. Thereafter, the process is like nucleotide excision repair.
Mismatch repair	Replication errors which escape proofreading have a mismatch in the daughter strand. Hemimethylation of the DNA after replication allows the daughter strand to be distinguished from the parental strand. The mismatched base is removed from the daughter strand by an excision repair mechanism.
Hereditary repair defects	Mutations in excision repair genes or a translesion DNA polymerase cause different forms of xeroderma pigmentosum, a sun-sensitive cancer-prone disorder. Excision repair is also defective in Cockayne syndrome.

Related topics

Nucleic acid structure (C1)
DNA replication (Section E)
Mutagenesis (F1)
DNA damage (F2)
Apoptosis (S4)

Photoreactivation

Cyclobutane pyrimidine dimers can be monomerized again by **DNA photolyases** (**photoreactivating enzymes**) in the presence of visible light. These enzymes have prosthetic groups (see Topic B2) which absorb blue light and transfer the energy to the cyclobutane ring which is then cleaved. The *E. coli* photolyase has two chromophores, N^5,N^{10}-methenyltetrahydrofolate and reduced flavin adenine dinucleotide (FADH). Photoreactivation is specific for pyrimidine dimers. It is an example of **direct reversal** of a lesion and is error-free.

Alkyltransferase

Another example of error-free direct reversal forms part of the **adaptive response** to alkylating agents. This response is induced in *E. coli* by low levels of alkylating agents and gives increased protection against the lethal and mutagenic effects of subsequent high doses. Mutagenic protection is afforded by an **alkyltransferase** which removes the alkyl group from the directly mutagenic O^6-alkylguanine (which can base-pair with thymine, see Topic F2). Curiously, the alkyl group is transferred to the protein itself and inactivates it. Thus, each alkyltransferase can only be used once. This protein is also present in mammalian cells. Protection against lethality involves induction of a DNA glycosylase which removes other alkylated bases through **base excision repair**.

Excision repair

This ubiquitous mechanism operates on a wide variety of lesions and is essentially error-free. There are two forms, **nucleotide excision repair (NER)** and **base excision repair (BER)**. In NER, an endonuclease cleaves the DNA a precise number of bases on either side of the lesion (*Fig. 1a*, step 1) and an oligonucleotide containing the lesion is removed leaving a gap (*Fig. 1a*, step 2). For example, in *E. coli*, the UvrABC endonuclease removes pyrimidine dimers and other bulky lesions by recognizing the distortion these produce in the double helix. In BER, modified bases are recognized by relatively specific **DNA glycosylases** which cleave the *N*-glycosylic bond between the altered base and the sugar (see Topic C1), leaving an **apurinic** or **apyrimidinic (AP)** site (*Fig. 1b*, step 1a). AP sites are also produced by spontaneous base loss (see Topic F2). An **AP endonuclease** then cleaves the DNA at this site and a gap may be created by further exonuclease activity (*Fig. 1b*, steps 1b and 2). The gap is generally larger in NER and can be as small as one nucleotide in BER. From this point on, both forms of excision repair are essentially the same. In *E. coli*, the gap is filled by DNA polymerase I (*Fig. 1*, step 3) and the final phosphodiester bond made by DNA ligase (*Fig. 1*, step 4), much as in the final stages of processing of the lagging strand fragments during DNA replication (see Topic E1). In eukaryotes, gap filling in BER involves predominantly **DNA polymerase** β whereas the longer gaps generated in NER are filled by **DNA polymerases** δ or ε. In eukaryotic NER, recognition and excision of DNA damage is a complex process involving at least 18 polypeptide factors, including the transcription factor TFIIH (see Topic M5). Excision is coupled to transcription so that transcribed (genetically active) regions of the DNA are repaired more rapidly than non-transcribed DNA. This helps to limit the production of defective gene products.

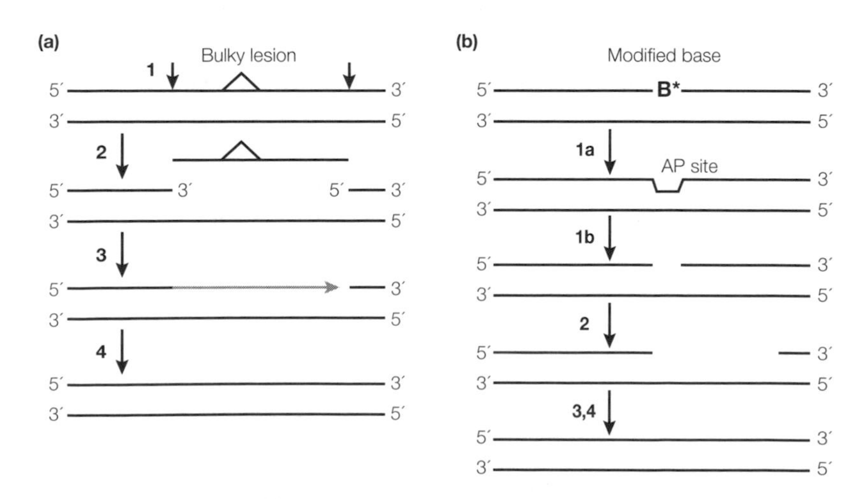

Fig. 1. (a) Nucleotide excision repair; (b) base excision repair. Numbered stages are described in the text.

Mismatch repair

This is a specialized form of excision repair that deals with any base mispairs produced during replication and that have escaped proofreading (see Topic F1). In a replicational mispair, the wrong base is in the daughter strand. This system must, therefore, have a way of distinguishing the parental and daughter strands after the replication fork has passed to ensure that the mismatched base is removed only from the daughter strand. In prokaryotes, certain adenine residues are normally methylated in the sequence GATC on both strands (see Topic C1). Methylation of daughter strands lags several minutes behind replication. Thus, newly replicated DNA is **hemimethylated** — the parental strands are methylated but the daughter strands are not, so they can be readily distinguished. The mismatched base pair (e.g. GT or CA) is recognized and bound by a complex of the MutS and MutL proteins which then associates with the MutH endonuclease, which specifically nicks the daughter strand at a nearby GATC site. This nick initiates excision of a region containing the wrong base. The discriminatory mechanism in eukaryotes is not known, but mismatch repair is clearly important in maintaining the overall error rate of DNA replication and, therefore, the spontaneous mutation rate: **hereditary nonpolyposis carcinoma of the colon** is caused by mutational loss of one of the human mismatch repair enzymes. Mismatch repair may also correct errors that arise from sequence misalignments during **meiotic recombination** in eukaryotes (see Topic F4).

Hereditary repair defects

Xeroderma pigmentosum (**XP**) is an autosomal recessive disorder characterized phenotypically by extreme sensitivity to sunlight and a high incidence of skin tumors. XP sufferers are defective in the NER of bulky DNA damage, including that caused by ultraviolet light. Defects in at least seven different genes can cause XP, indicating the complexity of excision repair in mammalian cells. **Xeroderma pigmentosum variant** (**XP-V**) is clinically very similar to classical XP but cells from XP-V individuals can carry out normal NER. In this case, the defect is in the gene encoding the translesion DNA polymerase η, which normally inserts dAMP in an error-free fashion opposite the thymine residues in a cyclobutane thymine dimer (see Topics F1 and F2). XP-V cells may, therefore, have to rely more heavily on alternative error-prone modes of translesion DNA synthesis to maintain DNA integrity after radiation damage. Sufferers of **Cockayne syndrome** are also sun-sensitive and defective in transcription-coupled excision repair, but are not cancer-prone.

F4 RECOMBINATION

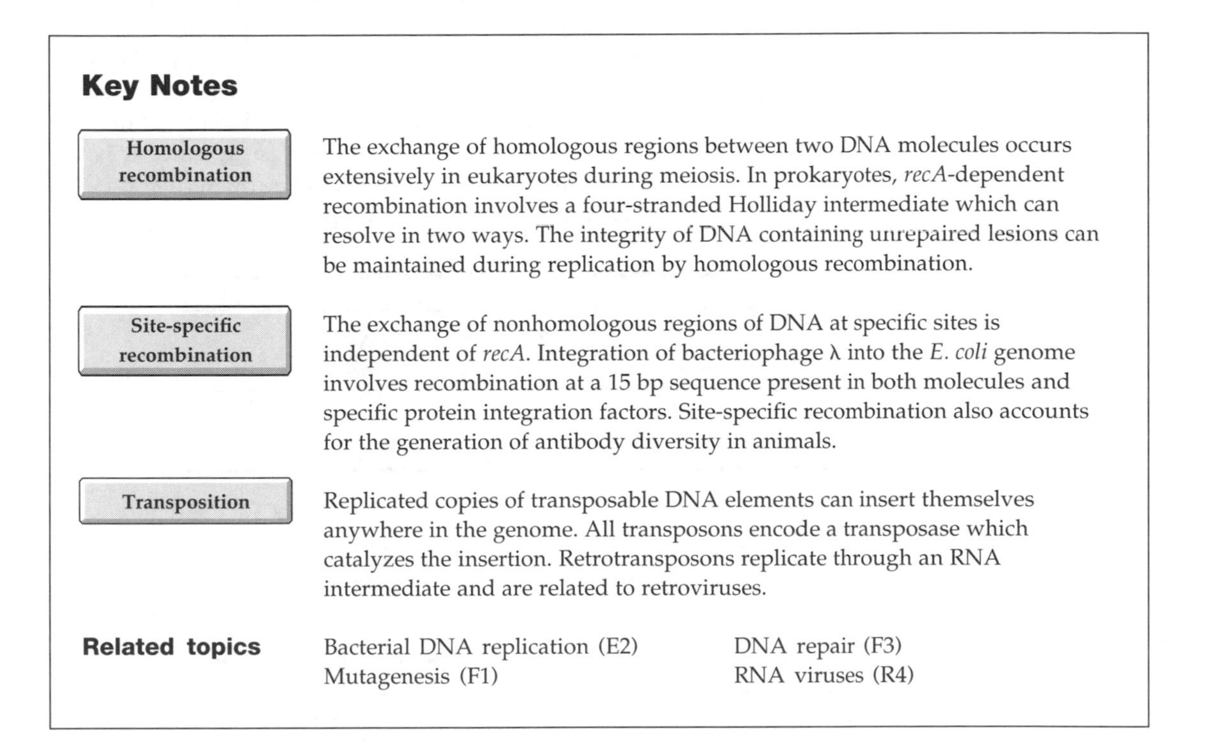

Key Notes

Homologous recombination

The exchange of homologous regions between two DNA molecules occurs extensively in eukaryotes during meiosis. In prokaryotes, *recA*-dependent recombination involves a four-stranded Holliday intermediate which can resolve in two ways. The integrity of DNA containing unrepaired lesions can be maintained during replication by homologous recombination.

Site-specific recombination

The exchange of nonhomologous regions of DNA at specific sites is independent of *recA*. Integration of bacteriophage λ into the *E. coli* genome involves recombination at a 15 bp sequence present in both molecules and specific protein integration factors. Site-specific recombination also accounts for the generation of antibody diversity in animals.

Transposition

Replicated copies of transposable DNA elements can insert themselves anywhere in the genome. All transposons encode a transposase which catalyzes the insertion. Retrotransposons replicate through an RNA intermediate and are related to retroviruses.

Related topics

Bacterial DNA replication (E2)
Mutagenesis (F1)
DNA repair (F3)
RNA viruses (R4)

Homologous recombination

Also known as **general recombination**, this process involves the exchange of homologous regions between two DNA molecules. In diploid eukaryotes, this commonly occurs during **meiosis** when the homologous duplicated chromosomes line up in parallel in metaphase I and the nonsister chromatids exchange equivalent sections by **crossing over**. After meiosis is complete, the resulting haploid gametes contain information derived from both maternally and paternally inherited chromosomes, thus ensuring that an individual will inherit genes from all four grandparents. Haploid bacteria also perform recombination, for example between the replicated portions of a partially duplicated DNA or between the chromosomal DNA and acquired 'foreign' DNA such as plasmids or phages. In *E. coli*, the two homologous DNA **duplexes** (double helices) align with each other (*Fig. 1a*) and nicks are made in a pair of **sister** strands (ones with identical sequence) near a specific sequence called *chi* (GCTGGTGG), which occurs approximately every 4 kb, by a nuclease associated with the **RecBCD** protein complex (*Fig. 1b*). The ssDNA carrying the 5′-ends of the nicks becomes coated in **RecA** protein to form RecA–ssDNA filaments. These cross over and search the opposite DNA duplex for the corresponding sequence (**invasion**, *Fig. 1c*), after which the nicks are sealed and a four-branched **Holliday** structure is formed (*Fig. 1d*). This structure is dynamic and the cross-over point can move a considerable distance in either direction (**branch migration**, *Fig. 1e*). The Holliday intermediate can be resolved into two DNA duplexes in one of two ways. If the two invading strands are cut, the resulting **recombinants** are similar to the original molecules apart from the exchange of a section in the middle (*Fig. 1f*). If the noninvading strands are cut, the products have one half from one parental duplex and one half from the other, with a hybrid (**heteroduplex**) section in between (*Fig. 1g*). Homologous recombination is also important for DNA repair. When a replication fork encounters an unrepaired, noncoding lesion it can skip the damaged section of DNA and re-initiate on the other side, leaving a **daughter strand gap**. This gap can be filled by replacing it with the corresponding section from the parental sister strand by recombination. The resulting gap in the parental sister strand can be filled easily since it is not opposite a lesion. The original lesion can be removed later by normal excision repair. This mechanism has also been called **post-replication repair**.

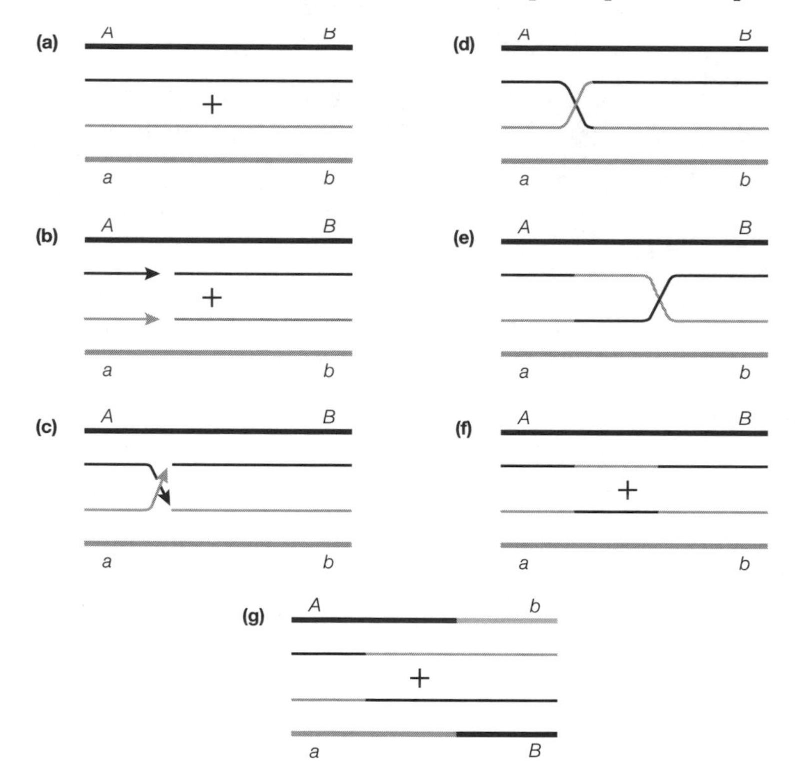

Fig. 1. The Holliday model for homologous recombination between two DNA duplexes.

Site-specific recombination

This involves the exchange of nonhomologous but specific pieces of DNA and is mediated by proteins that recognize specific DNA sequences. It does not require RecA or ssDNA. Bacteriophage λ (see Topic R2) has the ability to insert its genome into a specific site on the *E. coli* chromosome. The attachment site on the λ DNA shares an identical 15 bp sequence with the attachment site on the *E. coli* DNA. A λ-encoded **integrase (Int)** makes staggered cuts in these sequences with 7 bp overhangs (cf. restriction enzymes) and, in combination with the bacterially encoded **integration host factor (IHF)**, promotes recombination between these sites and insertion of the λ DNA into the host chromosome. This form of integrated λ is called a **prophage** and is stably inherited with each cell division until the λ-encoded **excisionase** is activated and the process reversed. In eukaryotes, site-specific recombination is responsible for the generation of antibody diversity. Immunoglobulins are composed of two heavy (H) chains and two light (L) chains of various types, both of which contain regions of constant and variable amino acid sequence. The sequences of the variable regions of these chains in germ cells are encoded by three gene segments: V, D and J. There are a total of 250 V, 15 D and five J genes for H chains and 250 V and four J for L chains. Recombination between these segments during differentiation of antibody-producing cells can produce an enormous number (> 10^8) of different H and L gene sequences and hence antibody specificities. Each V, D and J segment is associated with short recognition sequences for the recombination proteins. V genes are followed by a 39 bp sequence, D genes are flanked by two 28 bp sequences while J genes are preceded by the 39 bp sequence. Since recombination can only occur between one 39 bp and one 28 bp sequence, this ensures that only VDJ combinations can be produced.

Transposition

Transposons or **transposable elements** are small DNA sequences that can move to virtually any position in a cell's genome. Transposition has also been called **illegitimate recombination** because it requires no homology between sequences nor is it site-specific. Consequently, it is relatively inefficient. The simplest transposons are the *E. coli* **IS elements** or **insertion sequences**, of which there may be several copies in the genome. These are 1–2 kb in length and comprise a **transposase** gene flanked by short (~20 bp) **inverted terminal repeats** (identical sequences but with opposite orientation). The transposase makes a staggered cut in the chromosomal DNA and, in a replicative process, a copy of the transposon inserts at the target site (*Fig. 2*). The gaps are filled and sealed by DNA polymerase I and DNA ligase, resulting in a duplication of the target site and formation of a new **direct** repeat sequence. The gene into which the transposon inserts is usually inactivated, and genes between two copies of a transposon can be deleted by recombination between them. Inversions and other rearrangements of host DNA sequences can also occur. **Transposon mutagenesis** is a useful way of creating mutants. In addition to a transposase, the **Tn** transposon series carry other genes, including one for a **β-lactamase**, which

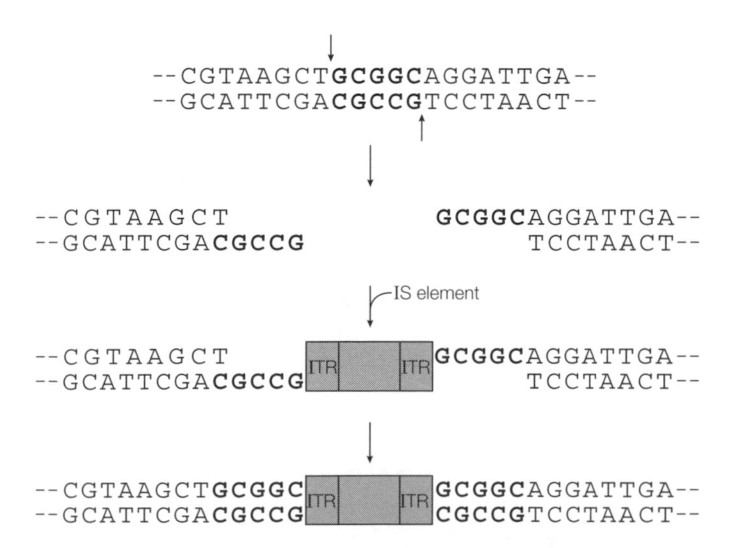

Fig. 2. Transposition of an insertion sequence (IS) element into a host DNA with duplication of the target site (shown in bold). ITR, inverted terminal repeat sequence.

confers penicillin resistance on the organism. The spread of antibiotic resistance among bacterial populations is a consequence of the transposition of resistance genes into plasmids, which replicate readily within the bacteria, and the ability of transposons to cross prokaryotic species barriers.

Many eukaryotic transposons have a structure similar to retroviral genomes (see Topic R4). The 6 kb yeast **Ty element** (with about 30 full-length copies per genome) encodes proteins similar to retroviral reverse transcriptases and integrases and is flanked by long terminal repeat sequences (**LTR**s). It replicates and moves to other genomic sites by transcription into RNA and subsequent reverse transcription of the RNA into a DNA duplex copy which then inserts elsewhere. The ***copia*** element from *Drosophila* (50 copies of a 5 kb sequence) is similar in structure. Such elements are called **retrotransposons** and it is believed that retroviruses are retrotransposons that have acquired the ability to exist outside the cell and pass to other cells. The dispersed repetitive sequences found in higher eukaryotic DNA (e.g. LINES and SINES) probably spread through the genome by transposition (see Topic D4). LINES are similar to retroviral genomes while the *Alu* element, a SINE, bears a strong resemblance to the **7SL RNA** component of the **signal recognition particle**, a complex involved in the secretion of newly synthesized polypeptides through the endoplasmic reticulum (see Topic Q4), and probably originated as a reverse transcript of this RNA.

G1 DNA CLONING: AN OVERVIEW

Key Notes

DNA cloning — DNA cloning facilitates the isolation and manipulation of fragments of an organism's genome by replicating them independently as part of an autonomous vector.

Hosts and vectors — Most of the routine manipulations involved in gene cloning use *Escherichia coli* as the host organism. Plasmids and bacteriophages may be used as cloning vectors in *E. coli*. Vectors based on plasmids, viruses and whole chromosomes have been used to carry foreign genes into other prokaryotic and eukaryotic organisms.

Subcloning — Subcloning is the simple transfer of a cloned fragment of DNA from one vector to another; it serves to illustrate many of the routine techniques involved in gene cloning.

DNA libraries — DNA libraries, consisting of sets of random cloned fragments of either genomic or cDNA, each in a separate vector molecule, are used in the isolation of unknown genes.

Screening libraries — Libraries are screened for the presence of a gene sequence by hybridization with a sequence derived from its protein product or a related gene, or through the screening of the protein products of the cloned fragments.

Analysis of a clone — Once identified, a cloned gene may be analyzed by restriction mapping, and ultimately by DNA sequencing, before being used in any of the diverse applications of DNA cloning.

Related topics

Cloning vectors (Section H)
Gene libraries and screening (Section I)
Analysis and uses of cloned DNA (Section J)

DNA cloning

Classically, detailed molecular analysis of proteins or other constituents of most organisms was rendered difficult or impossible by their scarcity and the consequent difficulty of their purification in large quantities. One approach is to isolate the gene(s) responsible for the expression of a protein or the formation of a product. However, every organism's genome is large and complex (see Section D), and any sequence of interest usually occurs only once or twice per cell. Hence, standard chemical or biochemical methods cannot be used to isolate a specific region of the genome for study, particularly as the required sequence of DNA is chemically identical to all the others. The solution to this dilemma is to place a relatively short fragment of a genome, which might contain the gene or other sequence of interest, in an autonomously replicating piece of DNA, known as a **vector**, forming **recombinant DNA**, which can be replicated independently of the original genome, and normally in another host species altogether. Propagation of the host organism containing the recombinant DNA forms a set of genetically identical organisms, or a **clone**. This process is hence known as **DNA cloning**.

Amongst the exploding numbers of applications of DNA cloning, often collected together under the term **genetic engineering**, are the following:

- DNA sequencing, and hence the derivation of protein sequence (see Topic J2).
- Isolation and analysis of gene promoters and other control sequences (see Topic J4).
- Investigation of protein/enzyme/RNA function by large-scale production of normal and altered forms (see Topic J5).
- Identification of mutations, for example gene defects leading to disease (see Topic J6).
- Biotechnology; the large-scale commercial production of proteins and other molecules of biological importance, for example human insulin and growth hormone (see Topic J6).
- Engineering animals and plants, and gene therapy (see Topic J6).
- Engineering proteins to alter their properties (see Topic J6).

Hosts and vectors

The initial isolation and analysis of DNA fragments is almost always carried out using the bacterium *E. coli* as the **host organism**, although the yeast *Saccharomyces cerevisiae* is being used to manipulate very large fragments of the human genome (see Topic H3). A wide variety of natural replicons have the properties required to allow them to act as **cloning vectors**. Vectors must normally be capable of being replicated and isolated independently of the host's genome, although some are designed to incorporate DNA into the host genome for longer term expression of cloned genes. Vectors also incorporate a **selectable marker**, a gene which allows host cells containing the vector to be **selected** from amongst those which do not, usually by conferring resistance to a toxin (see Topic G2), or enabling their survival under certain growth conditions (see Topic H3).

The first *E. coli* vectors were extrachromosomal (separate from the chromosome) circular **plasmids** (see Topic G2), and a number of **bacteriophages** (viruses infecting bacteria; see Topic R2) have also been used in *E. coli*. **Phage λ** can be used to clone fragments larger than plasmid vectors, and **phage M13** allows cloned DNA to be isolated in single-stranded form (see Topic H2). More specialist vectors have been engineered to use aspects of plasmids and bacteriophages, such as the plasmid–bacteriophage λ hybrids known as **cosmids** (see Topic H3). Very large genomic fragments from humans and other species have been cloned in *E. coli* as **bacterial artificial chromosomes** (**BACs**) and *S. cerevisiae* as **yeast artificial chromosomes** (**YACs**; see Topic H3).

Plasmid and phage vectors have been used to express genes in a range of bacteria other than *E. coli*, and some phages may be used to incorporate DNA into the host genome, for example phage λ (see Topic H2). Plasmid vectors have been developed for use in yeast (**yeast episomal plasmids**), while in plants, a bacterial plasmid (*Agrobacterium tumefaciens* **Ti plasmid**) can be used to integrate DNA into the genome. In other eukaryotic cells in culture, vectors have often been based on viruses that naturally infect the required species, either by maintaining their DNA extrachromosomally or by integration into the host genome (examples include **SV40, baculovirus, retroviruses**; see Topic H4).

Subcloning

The simplest kind of cloning experiment, which exemplifies many of the basic techniques of DNA cloning, is the transfer of a fragment of cloned DNA from one vector to another, a process known as **subcloning**. This might be used to investigate a short region of a large cloned fragment in more detail, or to transfer a gene to a vector designed to express it in a particular species, for example. In the case of plasmid vectors in *E. coli*, the most common situation, the process may be divided into the following steps, which are considered in greater detail in Topics G2–G5:

- **Isolation** of plasmid DNA containing the cloned sequence of interest (see Topic G2).
- **Digestion** (cutting) of the plasmid into discrete fragments with **restriction endonucleases** (see Topic G3).
- **Separation** of the fragments by **agarose gel electrophoresis** (see Topic G3).
- **Purification** of the desired **target** fragment (see Topic G3).
- **Ligation** (joining) of the fragment into a new plasmid vector, to form a new recombinant molecule (see Topic G4).
- Transfer of the ligated plasmid into an *E. coli* strain (**transformation**) (see Topic G4)
- **Selection** of transformed bacteria (see Topic G4).
- **Analysis** of recombinant plasmids (see Topic G4).

DNA libraries

There are two main sources from which DNA is derived for cloning experiments designed to identify an unknown gene – bulk genomic DNA from the species of interest and, in the case of eukaryotes, bulk mRNA from a cell or tissue where the gene is known to be expressed. They are used in the formation of **genomic libraries** and **cDNA libraries** respectively (see Topics I1 and I2).

DNA libraries are sets of DNA clones (a clone is a genetically distinct individual or set of identical individuals), each of which has been derived from the insertion of a different fragment into a vector followed by propagation in the host. Genomic libraries are prepared from random fragments of genomic DNA. However, genomic libraries may be an inefficient method of finding a gene, particularly in large eukaryotic genomes, where much of the DNA is noncoding (see Topic D4). The alternative is to use as the source of the library the mRNA from a cell or tissue which is known to express the gene. DNA copies (cDNA) are synthesized from the mRNA by **reverse transcription** and are then inserted into a vector to form a cDNA library. cDNA libraries are efficient for cloning a gene sequence, but yield only the coding region, and not the surrounding genomic sequences.

Screening libraries

Since it is not apparent which clone in a library contains the gene of interest, a method for screening for its presence is required. This is often based on the use of a radioactively or fluorescently labeled DNA **probe** which is complementary or partially complementary to a region of the gene sequence, and which can be used to detect it by **hybridization** (see Topic C3). The probe sequence might be

Table 1. Enzymes used in DNA cloning

Enzyme	Use
Alkaline phosphatase	Removes phosphate from 5′-ends of double- or single-stranded DNA or RNA (see Topic G4).
DNA ligase (from phage T4)	Joins sugar–phosphate backbones of dsDNA with a 5′-phosphate and a 3′-OH in an ATP-dependent reaction. Requires that the ends of the DNA be compatible, i.e. blunt with blunt, or complementary cohesive ends (see Topics E2 and G4).
DNA polymerase I	Synthesizes DNA complementary to a DNA template in a 5′ to 3′ direction, beginning with a primer with a free 3′-OH (see Topic E2). The Klenow fragment is a truncated version of DNA polymerase I which lacks the 5′ to 3′ exonuclease activity.
Exonuclease III	Exonucleases cleave from the ends of linear DNA. Exonuclease III digests dsDNA from the 3′-end only (see Topic J4).
Mung bean nuclease	Digests single-stranded nucleic acids, but will leave intact any region which is double helical (see Topic J4).
Nuclease S1	As mung bean nuclease. However, the enzyme will also cleave a strand opposite a nick on the complementary strand (see Topic J4).
Polynucleotide kinase	Adds phosphate to 5′-OH end of double- or single-stranded DNA or RNA in an ATP-dependent reaction. If [γ-^{32}P]ATP is used, then the DNA will become radioactively labeled (see Topic I3).
Restriction enzymes	Cut both strands of dsDNA within a (normally symmetrical) recognition sequence. Hydrolyze sugar–phosphate backbone to give a 5′-phosphate on one side and a 3′-OH on the other. Yield blunt or 'sticky' ends (5′- or 3′-overhang) (see Topic G3).
Reverse transcriptase	RNA-dependent DNA polymerase. Synthesizes DNA complementary to an RNA template in a 5′ to 3′ direction, beginning with a primer with a free 3′-OH. Requires dNTPs (see Topic I2).
RNase A	Nuclease which digests RNA, but not DNA (see Topic G2).
RNase H	Nuclease which digests the RNA strand of an RNA–DNA heteroduplex (see Topic I2).
T7, T3 and SP6 RNA polymerases	Specific RNA polymerases encoded by the respective bacteriophages. Each enzyme recognizes only the promoters from its own phage DNA, and can be used specifically to transcribe DNA downstream of such a promoter (see Topics H1 and I3).
Taq DNA polymerase	DNA polymerase derived from a thermostable bacterium (*Thermus aquaticus*). Operates at 72°C and is reasonably stable above 90°C. Used in PCR (see Topic J3).
Terminal transferase	Adds a number of nucleotides to the 3′-end of linear single- or double-stranded DNA or RNA. If only GTP is used, for example, then only Gs will be added (see Topic I2).

an oligonucleotide derived from the sequence of the protein product of the gene, if it is available, or from a related gene from another species (see Topic I3). An increasingly important method for the generation of probes is the **polymerase chain reaction** (**PCR**; see Topic J3). Other screening methods rely on the expression of the coding sequences of the clones in the library, and identification of the protein product from its activity, or with a specific antibody, for example (see Topic I3).

Analysis of a clone

Once a clone containing a target gene is identified, the structure of the cloned fragment may be investigated further using **restriction mapping**, the analysis of the fragmentation of the DNA with restriction enzymes (see Topic J1), or ultimately by the **sequencing** of the entire fragment (see Topic J2). The sequence can then be analyzed by comparison with other known sequences from databases, and the complete sequence of the protein product determined (see Topic J2). The sequence is then available for manipulation in any of the applications of cloning described above.

Many enzymes are used *in vitro* in DNA cloning and analysis. The properties of the common enzymes are given in *Table 1*, along with the section where their use is discussed in more detail.

G2 Preparation of plasmid DNA

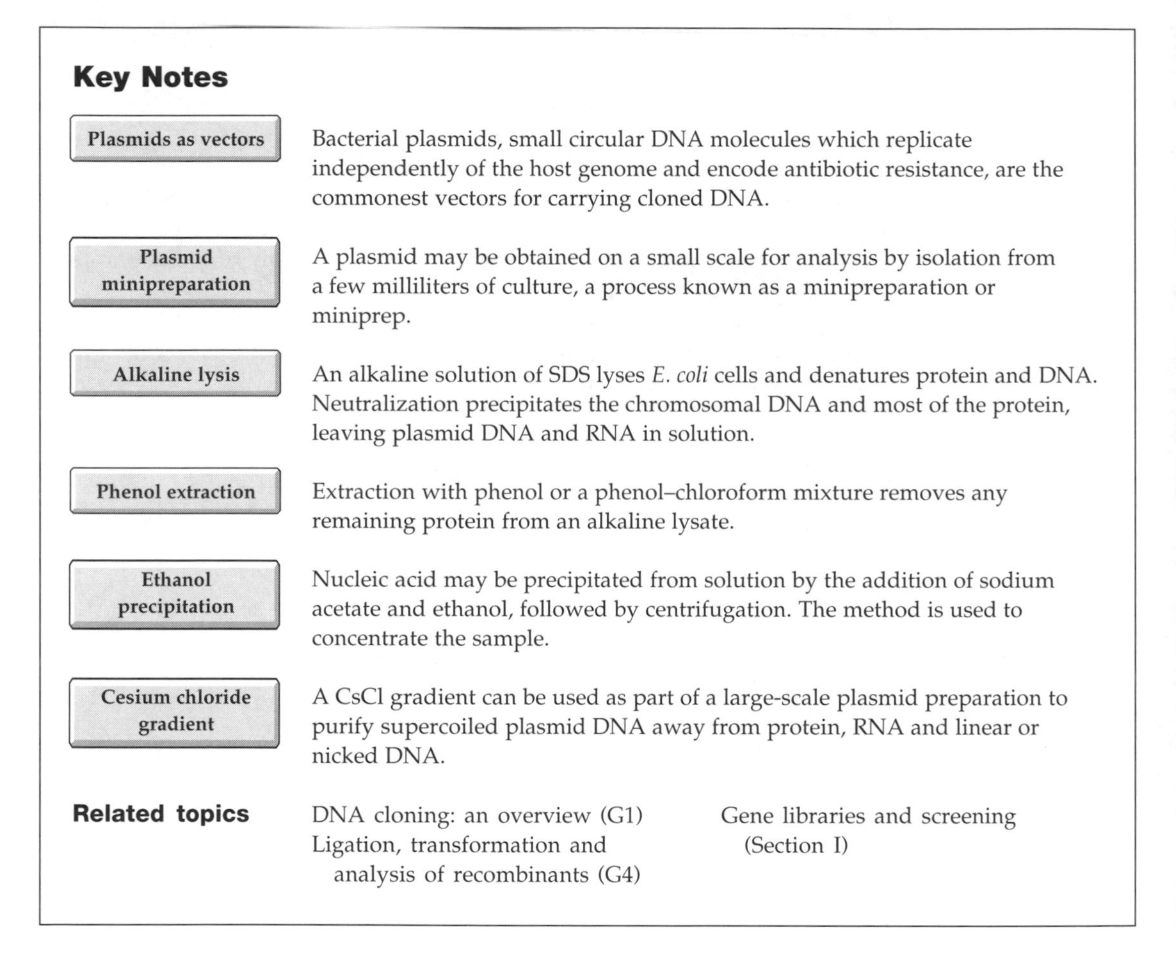

Key Notes

Plasmids as vectors	Bacterial plasmids, small circular DNA molecules which replicate independently of the host genome and encode antibiotic resistance, are the commonest vectors for carrying cloned DNA.
Plasmid minipreparation	A plasmid may be obtained on a small scale for analysis by isolation from a few milliliters of culture, a process known as a minipreparation or miniprep.
Alkaline lysis	An alkaline solution of SDS lyses *E. coli* cells and denatures protein and DNA. Neutralization precipitates the chromosomal DNA and most of the protein, leaving plasmid DNA and RNA in solution.
Phenol extraction	Extraction with phenol or a phenol–chloroform mixture removes any remaining protein from an alkaline lysate.
Ethanol precipitation	Nucleic acid may be precipitated from solution by the addition of sodium acetate and ethanol, followed by centrifugation. The method is used to concentrate the sample.
Cesium chloride gradient	A CsCl gradient can be used as part of a large-scale plasmid preparation to purify supercoiled plasmid DNA away from protein, RNA and linear or nicked DNA.

Related topics

DNA cloning: an overview (G1)
Ligation, transformation and analysis of recombinants (G4)
Gene libraries and screening (Section I)

Plasmids as vectors

The first cloning vectors to be used, in the mid 1970s, were naturally occurring bacterial **plasmids**, originally from *E. coli*. Plasmids are small, **extrachromosomal** circular molecules, from 2 to around 200 kb in size, which exist in multiple copies (up to a few hundred) within the host *E. coli* cell. They contain an **origin of replication** (*ori*; see Topic E1), which enables them to be replicated independently, although this normally relies on polymerases and other components of the host cell's machinery. They usually carry a few genes, one of which may confer **resistance** to antibacterial substances. The most widely known resistance gene is the *bla*, or amp^r gene, encoding the enzyme β-lactamase, which degrades penicillin antibiotics such as **ampicillin**. Another is the *tetA* gene, which encodes a transmembrane pump able to remove the antibiotic tetracycline from the cell.

Plasmid minipreparation

The first step in a subcloning procedure, for example the transfer of a gene from one plasmid to another (see Topic G1), is the isolation of plasmid DNA. Since plasmids are so much smaller than *E. coli* chromosomal DNA (see Topic D1), they can be separated from the latter by physico-chemical methods, such as alkaline lysis (see below). It is normally possible to isolate sufficient plasmid DNA for initial manipulation from a few milliliters of bacterial culture. Such an isolation is normally known as a **minipreparation** or **miniprep**. A sample of an *E. coli* strain harboring the required plasmid is inoculated into a few milliliters of culture broth. After growth to stationary phase (overnight), the suspension is centrifuged to yield a cell pellet.

Alkaline lysis

The most commonly used method for the purification of plasmid DNA away from chromosomal DNA and most of the other cell constituents is called **alkaline lysis** (*Fig. 1*). The cell pellet is resuspended in a buffer solution which may optionally contain lysozyme to digest the cell wall of the bacteria. The cell **lysis** solution, which contains the detergent sodium dodecyl sulfate (SDS) in an alkaline sodium hydroxide solution, is then added. The SDS disrupts the cell membrane (see Topic A1) and denatures the proteins; the alkaline conditions denature the DNA and begin the hydrolysis of RNA (see Topic C2). The preparation is then **neutralized** with a concentrated solution of potassium acetate (KOAc) at pH 5. This has the effect of precipitating the denatured proteins, along with the chromosomal DNA and most of the detergent (potassium dodecyl sulfate is insoluble in water). The sample is centrifuged again, and the resulting supernatant (the **lysate**) now contains plasmid DNA, which, being small and closed-circular, is easily renatured after the alkali treatment, along with a lot of small RNA molecules and some protein.

Phenol extraction

There are now innumerable proprietary methods for the isolation of pure plasmid DNA from the lysate above, many of which involve the selective binding of DNA to a resin or membrane and the washing away of protein and RNA. The classical method, which is slower but perfectly effective, involves the extraction of the lysate with **phenol** or a **phenol–chloroform** mixture. The phenol is immiscible with the aqueous layer but, when mixed vigorously with it, then allowed to separate, denatures the remaining proteins, which form a precipitate at the interface between the layers (*Fig. 1*).

Ethanol precipitation

The DNA and RNA remaining in the aqueous layer are now concentrated by **ethanol precipitation**. This is a general procedure, which may be used with any nucleic acid solution. If sodium acetate is added to the solution until the Na^+ concentration is more than 0.3 M, the DNA and/or RNA may be precipitated by the addition of 2–3 volumes of ethanol. Centrifugation will pellet the nucleic acid, which may then be resuspended in a smaller volume, or in a buffer with new constituents, etc. In the case of a minipreparation, ethanol is added directly to the phenol-extracted lysate, and the pellet is taken up in Tris–EDTA solution, the normal solution for the storage of DNA. This solution contains Tris-hydrochloride to buffer the solution (usually pH 8) and a low concentration of EDTA, which chelates any Mg^{2+} ions in the solution, protecting the DNA against degradation by nucleases, most of which require magnesium. **Ribonuclease A (RNase A)** may also be added to the solution to digest away any remaining RNA contamination. This enzyme digests RNA but leaves DNA untouched and does not require Mg^{2+}.

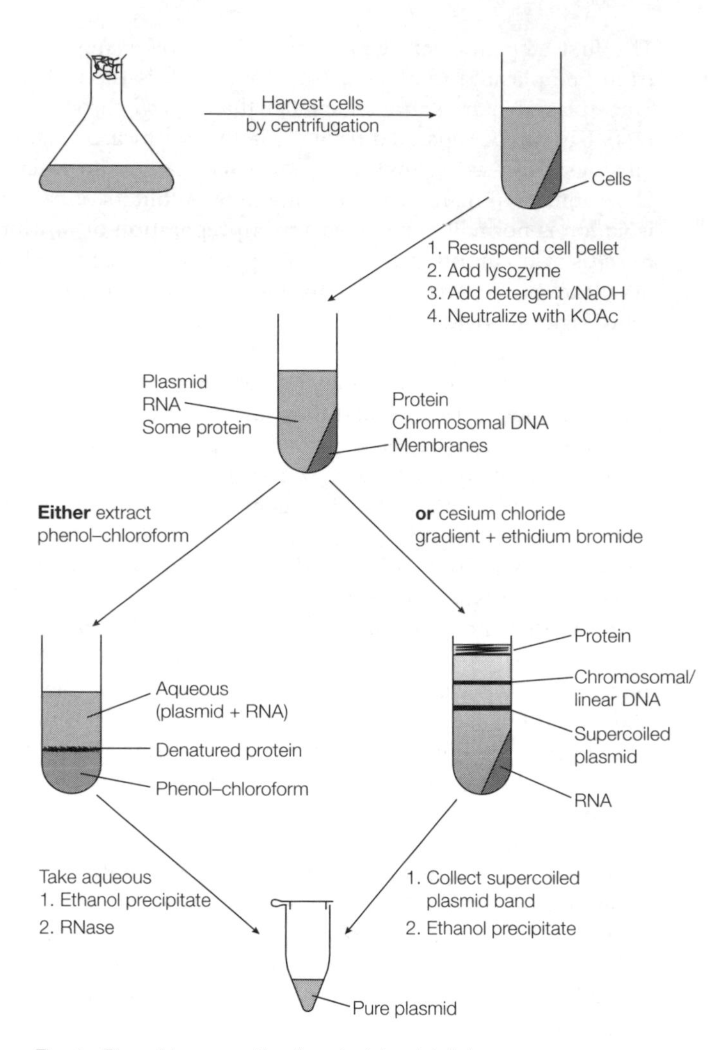

Fig. 1. Plasmid preparation (see text for details).

Cesium chloride gradient

The alkaline lysis method may also be used on a larger scale to prepare up to milligram quantities of plasmid DNA. The production of plasmids on a large scale may be required for stocks of common cloning vectors, or for those who wish to use plasmids on a large scale as substrates for enzymic reactions. In this case, **CsCl density gradient centrifugation** (see Topic C2) may be used as a final purification step. This is somewhat laborious, but is the best method for the production of very pure supercoiled plasmid DNA. If a crude lysate is fractionated on a CsCl gradient (see Topic C2) in the presence of ethidium bromide (see Topic C4), the various constituents may all be separated (*Fig. 1*). Supercoiled DNA binds less ethidium bromide than linear or nicked DNA (see Topic C4) and, therefore, has a higher density (it is less unwound). Hence supercoiled plasmid may be isolated in large quantities away from protein, RNA and contaminating chromosomal DNA in one step. The solution containing the plasmid band is removed from the centrifuge tube and may then be concentrated using ethanol precipitation as described above after removal of ethidium bromide.

G3 Restriction enzymes and electrophoresis

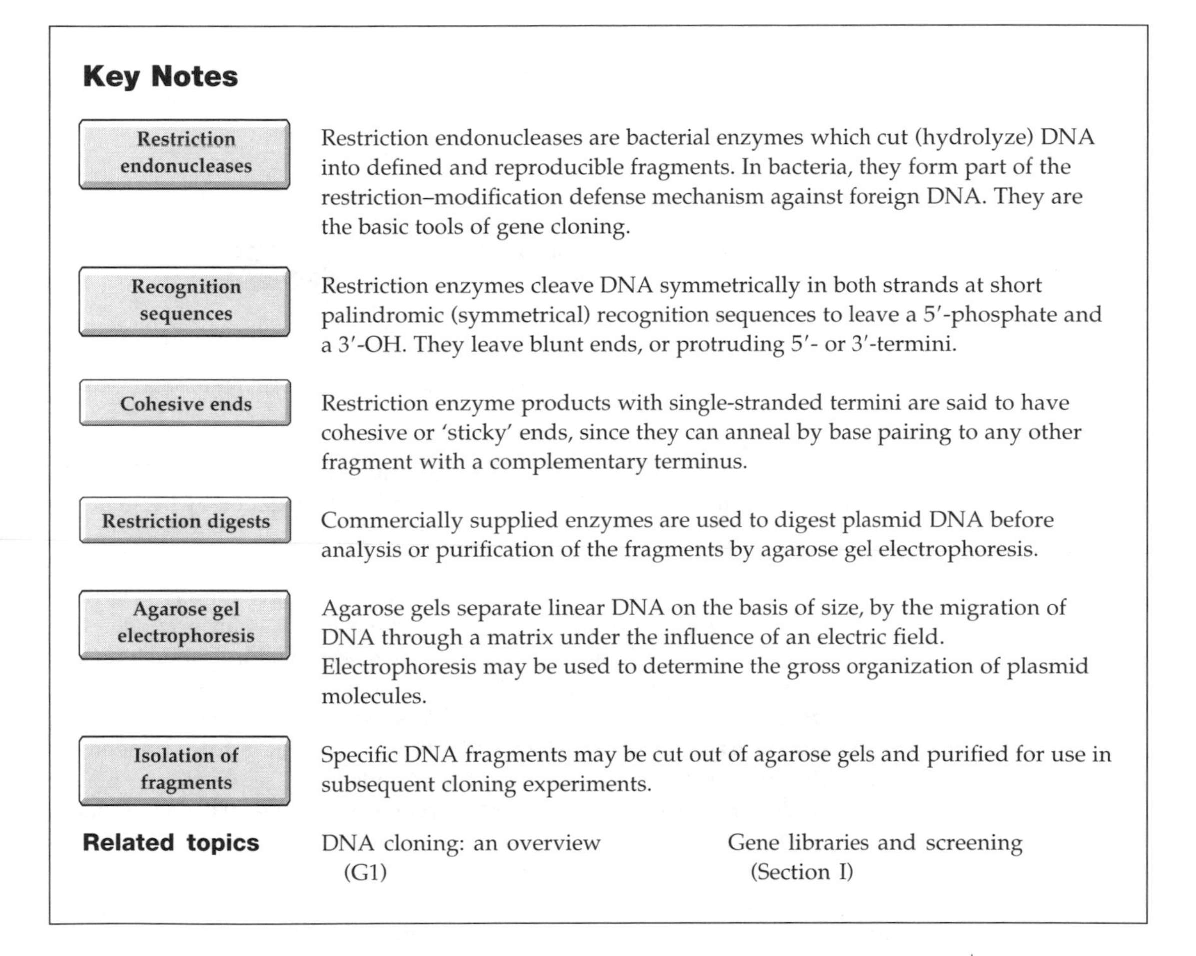

Key Notes

Restriction endonucleases — Restriction endonucleases are bacterial enzymes which cut (hydrolyze) DNA into defined and reproducible fragments. In bacteria, they form part of the restriction–modification defense mechanism against foreign DNA. They are the basic tools of gene cloning.

Recognition sequences — Restriction enzymes cleave DNA symmetrically in both strands at short palindromic (symmetrical) recognition sequences to leave a 5′-phosphate and a 3′-OH. They leave blunt ends, or protruding 5′- or 3′-termini.

Cohesive ends — Restriction enzyme products with single-stranded termini are said to have cohesive or 'sticky' ends, since they can anneal by base pairing to any other fragment with a complementary terminus.

Restriction digests — Commercially supplied enzymes are used to digest plasmid DNA before analysis or purification of the fragments by agarose gel electrophoresis.

Agarose gel electrophoresis — Agarose gels separate linear DNA on the basis of size, by the migration of DNA through a matrix under the influence of an electric field. Electrophoresis may be used to determine the gross organization of plasmid molecules.

Isolation of fragments — Specific DNA fragments may be cut out of agarose gels and purified for use in subsequent cloning experiments.

Related topics

DNA cloning: an overview (G1)

Gene libraries and screening (Section I)

Restriction endonucleases

To incorporate fragments of foreign DNA into a plasmid vector, methods for the cutting and rejoining of dsDNA are required. The identification and manipulation of **restriction endonucleases** in the 1960s and early 1970s was the key discovery which allowed the cloning of DNA to become a reality. Restriction–modification systems occur in many bacterial species, and constitute a defense mechanism against the introduction of foreign DNA into the cell. They consist of two components; the first is a restriction endonuclease, which recognizes a short, symmetrical DNA sequence (*Fig. 1*), and cuts (**hydrolyzes**) the DNA backbone in each strand at a specific site within that sequence. Foreign DNA will hence be degraded to relatively short fragments. The second component of the system is a **methylase**, which adds a methyl group to a C or A

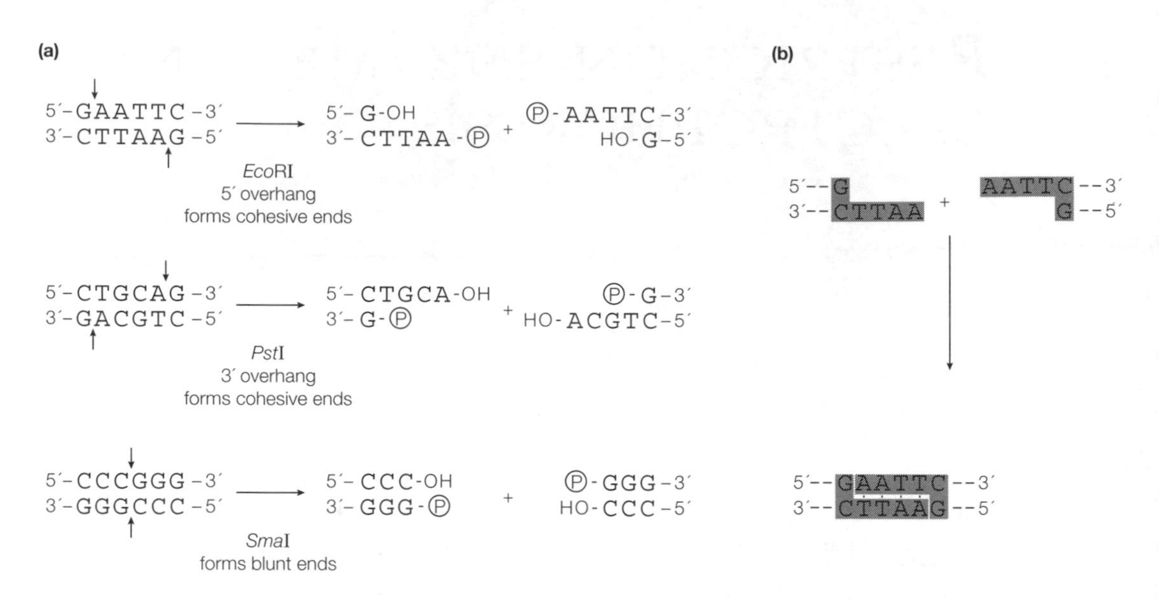

Fig. 1. (a) The action of restriction endonucleases at their recognition sequences; (b) the annealing of cohesive ends.

base within the same recognition sequences in the cellular DNA (see Topic C1). This modification renders the host DNA resistant to degradation by the endonuclease.

Recognition sequences

The action of restriction endonucleases (**restriction enzymes** for short) is illustrated in *Fig. 1a* including the archetypal enzyme *Eco*RI as an example. This enzyme, which acts as a dimer, will only recognize a 6 bp **palindromic** sequence (the sequence is the same, reading 5′→3′, on each strand). The product of the cutting reaction at this site on a linear DNA is two double-stranded fragments (**restriction fragments**), each with an identical protruding single-stranded 5′-end with a phosphate group attached. The 3′-ends have free hydroxyl groups. A 6 bp recognition sequence will occur on average every $4^6 = 4096$ bp in random sequence DNA; hence, a very large DNA molecule will be cut into specific fragments averaging 4 kb by such an enzyme. Hundreds of restriction enzymes are now known, and a large number are commercially available. They recognize sites ranging in size from 4 to 8 bp or more, and may give products with protruding 5′- or 3′-tails or blunt ends. The newly formed 5′-ends always retain the phosphate groups. Two further examples are illustrated in *Fig. 1*. The extremely high specificity of restriction enzymes for their sites of action allows large DNA molecules and vectors to be cut reproducibly into defined fragments.

Cohesive ends

Those products of restriction enzyme digestion with protruding ends have a further property; these ends are known as **cohesive**, or **'sticky'** ends, since they can bind to any other end with the same overhanging sequence, by base pairing (annealing) of the single-stranded tails. Hence, for example, any fragment formed by an *Eco*RI cut can anneal to any other fragment formed in the same way (*Fig. 1b*), and may subsequently be joined covalently by ligation (see Topic E4). In fact, in some cases, DNA ends formed by enzymes with different recognition sequences may be compatible, provided the single-stranded tails can base-pair together.

Restriction digests

Digestion of plasmid or genomic DNA (see Topic I1) is carried out with restriction enzymes for analytical or preparative purposes, using commercial enzymes and buffer solutions. All restriction enzymes require Mg^{2+}, usually at a concentration of up to 10 mM, but different enzymes require different pHs, NaCl concentrations or other solution constituents for optimum activity. The buffer solution required for a particular enzyme is supplied with it as a concentrate. The digestion of a sample plasmid with two different restriction enzymes, *Bam*HI and *Eco*RI, is illustrated in *Fig. 2*.

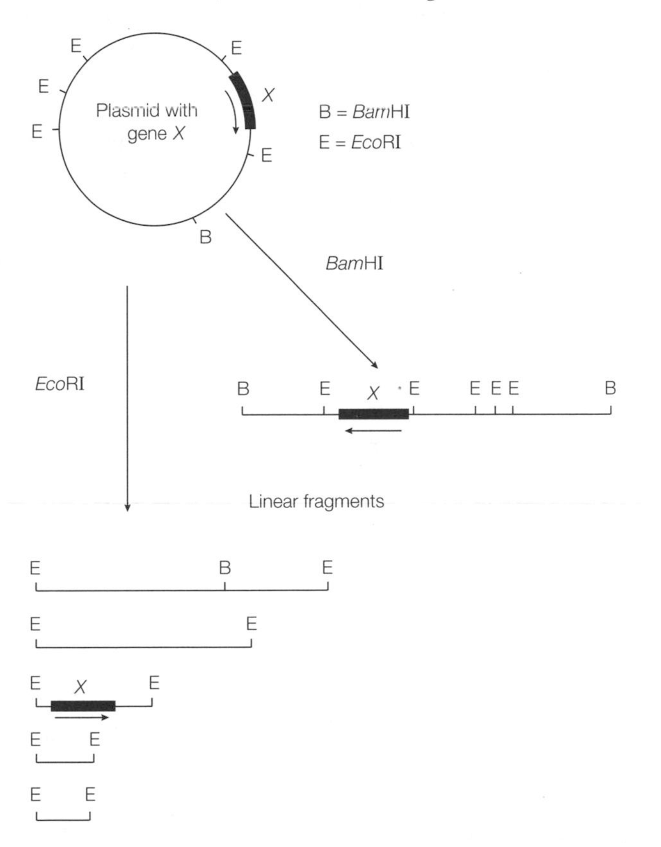

Fig. 2. The digestion of a plasmid with two different restriction enzymes.

The digestion of a few hundred nanograms (<1 μg) of plasmid DNA is sufficient for analysis by agarose gel electrophoresis; preparative purposes may require a few micrograms. The former amount corresponds to a few percent of a miniprep sample, as described in Topic G2. The DNA is incubated with the enzyme and the appropriate buffer at the optimum temperature (usually 37°C), in a volume of perhaps 20 μl. A dye mixture is then added to the solution, and the sample is loaded on to an agarose gel.

Agarose gel electrophoresis

Agarose is a polysaccharide derived from seaweed, which forms a solid gel when dissolved in aqueous solution at concentrations between 0.5 and 2% (w/v). Agarose used for electrophoresis is a more purified form of the agar used to make bacterial culture plates.

When an electric field is applied to an agarose gel in the presence of a buffer solution which will conduct electricity, DNA fragments move through the gel towards the positive electrode (DNA is highly negatively charged; see Topic C1) at a rate which is dependent on their size and shape (*Fig. 3*). Small linear fragments move more quickly than large ones, which are retarded by entanglement with the network of agarose fibers forming the gel. Hence, this process of **electrophoresis** may be used to separate mixtures of DNA fragments on the basis of size. Different concentrations of gel [1%, 1.5% (w/v), etc.] will allow the optimal resolution of fragments in different size ranges. The DNA samples are placed in **wells** in the gel surface (*Fig.* 3), the power supply is switched on and the DNA is allowed to migrate through the gel in separate **lanes** or **tracks**. The added dye also migrates, and is used to follow the progress of electrophoresis. The DNA is stained by the inclusion of **ethidium bromide** (see Topic C4) in the gel, or by soaking the gel in a solution of ethidium bromide after electrophoresis. The DNA shows up as an orange band on illumination by UV light.

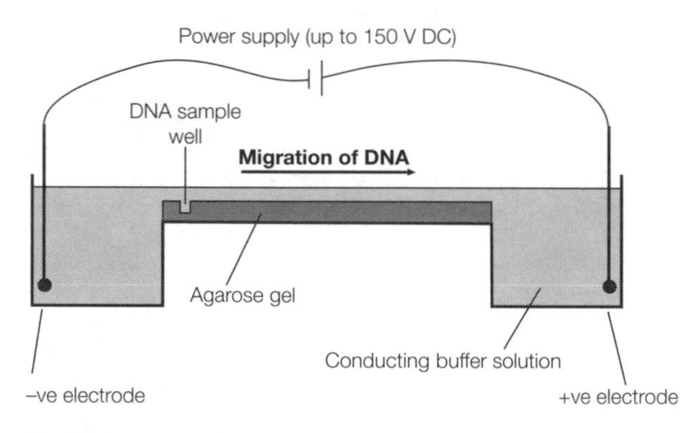

Fig. 3. Agarose gel apparatus.

Figure 4a illustrates the result of gel electrophoresis of the fragments formed by the digestions in *Fig.* 2. The plasmid has been run on the gel without digestion (track U), and after digestion with *Bam*HI (track B) and *Eco*RI (track E). A set of linear marker DNA fragments of known sizes (tracks M) have been run alongside the samples at two different concentrations; the sizes are marked on the figure. A number of points may be noted.

- Undigested plasmid DNA (track U) run on an agarose gel commonly consists of two bands. The lower, more mobile, band consists of negatively supercoiled plasmid DNA isolated intact from the cell. This has a high mobility, because of its compact conformation (see Topic C4). The upper band is open-circular, or nicked DNA, formed from supercoiled DNA by breakage of one strand; this has an opened-out circular conformation and lower mobility.
- The lanes containing the digested DNA clearly reveal a single fragment (track B) and five fragments (track E), whose sizes can be estimated by comparison with the marker tracks (M). The intensities of the bands in track E are proportional to the sizes of the fragments, since a small fragment has less mass of DNA at a given molar concentration. This is also true of the markers, since in this case they are formed by digestion of the 48.5 kb linear DNA of bacteriophage λ (see Topic H2). The amount of DNA present in tracks U, B

and E is not equal; the quantities have been optimized to show all the fragments clearly.

- A more accurate determination of the sizes of the linear fragments can be made by plotting a calibration curve of the log of the size of the known fragments in track M against the distance migrated by each fragment. This plot (*Fig. 4b*) is a fairly straight line, often with a deviation at large fragment sizes. This may be used to derive the size of an unknown linear fragment on the same gel from its mobility, by reading off the log(size) as shown. It is not possible to derive the sizes of undigested circular plasmids by the same method, since the relative mobility of circular and linear DNA on a gel depends on the conditions (temperature, electric field, etc.).

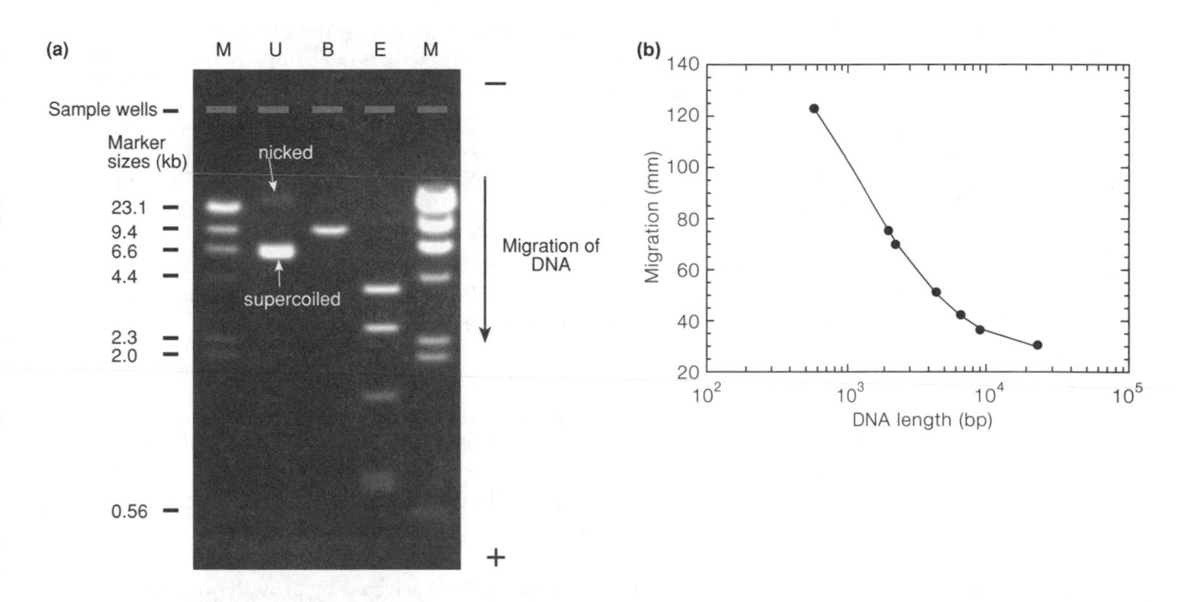

Fig. 4. (a) An agarose gel of DNA restriction fragments (see text for details); (b) a calibration curve of migration distance against fragment size.

Isolation of fragments

Agarose gels may also be used preparatively to isolate specific fragments for use in subsequent ligation and other cloning experiments. Fragments are excised from the gel, and treated by one of a number of procedures to purify the DNA away from the contaminating agarose and ethidium bromide stain. If we assume that the *Eco*RI fragment containing the gene *X* (*Fig. 2*) is the target DNA for a subcloning experiment (see Topic G1), then the third largest fragment in track E of *Fig. 4a* could be purified from the gel ready for ligation into a new vector (see Topic G4).

G4 LIGATION, TRANSFORMATION AND ANALYSIS OF RECOMBINANTS

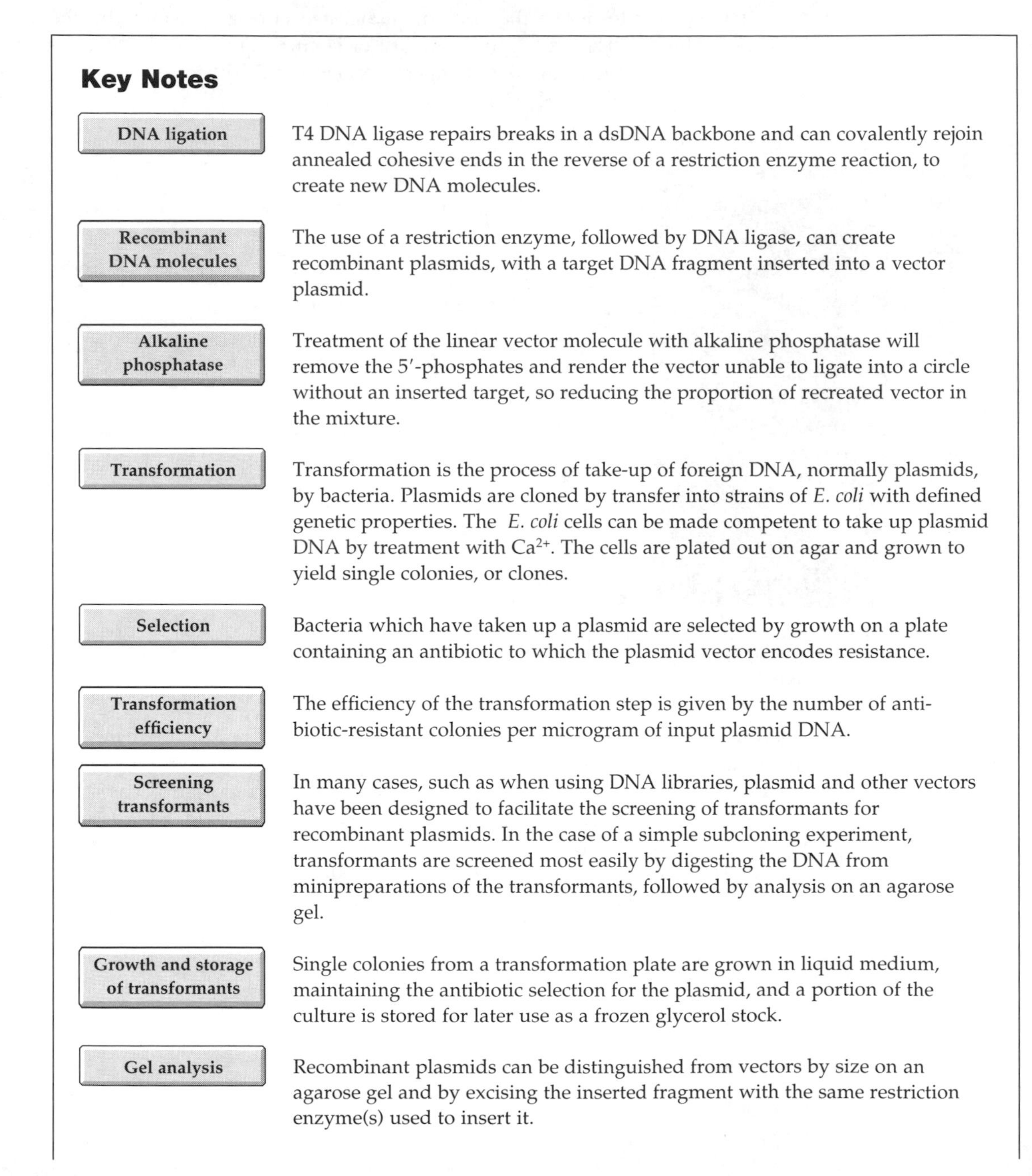

Key Notes

DNA ligation

T4 DNA ligase repairs breaks in a dsDNA backbone and can covalently rejoin annealed cohesive ends in the reverse of a restriction enzyme reaction, to create new DNA molecules.

Recombinant DNA molecules

The use of a restriction enzyme, followed by DNA ligase, can create recombinant plasmids, with a target DNA fragment inserted into a vector plasmid.

Alkaline phosphatase

Treatment of the linear vector molecule with alkaline phosphatase will remove the 5′-phosphates and render the vector unable to ligate into a circle without an inserted target, so reducing the proportion of recreated vector in the mixture.

Transformation

Transformation is the process of take-up of foreign DNA, normally plasmids, by bacteria. Plasmids are cloned by transfer into strains of *E. coli* with defined genetic properties. The *E. coli* cells can be made competent to take up plasmid DNA by treatment with Ca^{2+}. The cells are plated out on agar and grown to yield single colonies, or clones.

Selection

Bacteria which have taken up a plasmid are selected by growth on a plate containing an antibiotic to which the plasmid vector encodes resistance.

Transformation efficiency

The efficiency of the transformation step is given by the number of antibiotic-resistant colonies per microgram of input plasmid DNA.

Screening transformants

In many cases, such as when using DNA libraries, plasmid and other vectors have been designed to facilitate the screening of transformants for recombinant plasmids. In the case of a simple subcloning experiment, transformants are screened most easily by digesting the DNA from minipreparations of the transformants, followed by analysis on an agarose gel.

Growth and storage of transformants

Single colonies from a transformation plate are grown in liquid medium, maintaining the antibiotic selection for the plasmid, and a portion of the culture is stored for later use as a frozen glycerol stock.

Gel analysis

Recombinant plasmids can be distinguished from vectors by size on an agarose gel and by excising the inserted fragment with the same restriction enzyme(s) used to insert it.

Fragment orientation

The orientation of the insert in the vector may be determined using an agarose gel by digestion of the plasmid with a restriction enzyme known to cut asymmetrically within the insert sequence.

Related topics

DNA cloning: an overview (G1)
Cloning vectors (Section H)

DNA ligation

To insert a target DNA fragment into a vector, a method for the covalent joining of DNA molecules is essential. **DNA ligase** enzymes perform just this function; they will repair (**ligate**) a break in one strand of a dsDNA molecule (see Topic E2), provided the 5′-end has a phosphate group. They require an adenylating agent to activate the phosphate group for attack by the 3′-OH; the *E. coli* enzyme uses NAD^+, and the more commonly used enzyme from bacteriophage T4 uses ATP. Ligases are efficient at sealing the broken phosphodiester bonds in an annealed pair of cohesive ends (see Topic G2), essentially the reverse of a restriction enzyme reaction, and T4 ligase can even ligate one blunt end to another, albeit with rather lower efficiency.

Recombinant DNA molecules

We can now envisage an experiment in which a DNA fragment containing a gene (*X*) of interest (the target DNA) is inserted into a plasmid vector (*Fig. 1*). The target DNA may be a single fragment isolated from an agarose gel (see Topic G3), or a mixture of many fragments from, for example, genomic DNA (see Topic I1). If the target has been prepared by digestion with *Eco*RI, then the fragment can be ligated with vector DNA cut with the same enzyme (*Fig. 1*) In practice, the vector should have only one site for cleavage with the relevant enzyme, since otherwise, the correct product could only be formed by the ligation of three or more fragments, which would be very inefficient. There are many possible products from this ligation reaction, and the outcome will depend on the relative concentrations of the fragments as well as the conditions, but the products of interest will be circular molecules with the target fragment inserted at the *Eco*RI site of the vector molecule (with either orientation), to form a **recombinant** molecule (*Fig. 1*). The recreation of the original vector plasmid, by circularization of the linear vector alone, is a competing side reaction which can make the identification of recombinant products problematic. One solution is to prepare both the target and the vector using a pair of distinct restriction enzymes, such that they have noncompatible cohesive ends at either end. The likelihood of ligating the vector into a circle is then much reduced.

Alkaline phosphatase

If it is inconvenient to use two restriction enzymes, then the linear vector fragment may be treated with the enzyme **alkaline phosphatase** after restriction enzyme digestion. Alkaline phosphatase removes phosphate groups from the 5′-ends of DNA molecules. The linear vector will hence be unable to ligate into a circle, since no phosphates are available for the ligation reaction (*Fig. 2*). A ligation with a target DNA insert can still proceed, since one phosphate is present to ligate one strand at each cut site (*Fig. 2*). The remaining nicks in the other strands will be repaired by cellular mechanisms after transformation (see following).

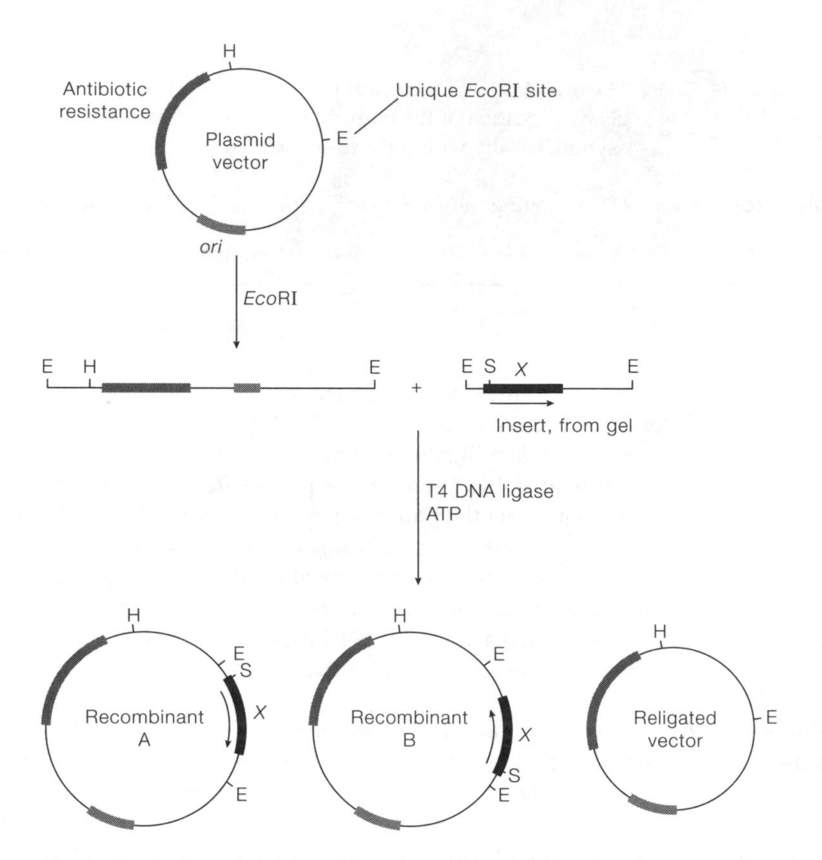

Fig. 1. The ligation of vector and target to yield recombinant and nonrecombinant products.

Transformation

The components of the mixture of recombinant and other plasmid molecules formed by ligation (*Fig. 1*) must now be isolated from one another and replicated (**cloned**) by transfer into a **host organism**. By far the most common hosts for simple cloning experiments are strains of *E. coli* which have specific genetic properties. One obvious requirement, for example, is that they must not express a restriction–modification system (see Topic G3).

It was discovered that *E. coli* cells treated with solutions containing Ca^{2+} ions were rendered susceptible to take up exogenous DNA, a process known as **transformation**. Cells pre-treated with Ca^{2+} (and sometimes more exotic metal ions such as Rb^{+} and Mn^{2+}), in order to render them able to take up DNA, are known as **competent cells**.

In transformation of *E. coli*, a solution of a plasmid molecule, or a mixture of molecules formed in a ligation reaction, is combined with a suspension of competent cells for a period, to allow the DNA to be taken up. The precise mechanism of the transfer of DNA into the cells is obscure. The mixture is then **heat-shocked** at 42°C for 1–2 min. This induces enzymes involved in the repair of DNA and other cellular components, which allow the cells to recover from the unusual conditions of the transformation process, and increases the efficiency. The cells are then incubated in a growth medium and finally spread on an agar plate and incubated until single colonies of bacteria grow (*Fig. 3*). All the cells within a colony originate from division of a single individual. Thus,

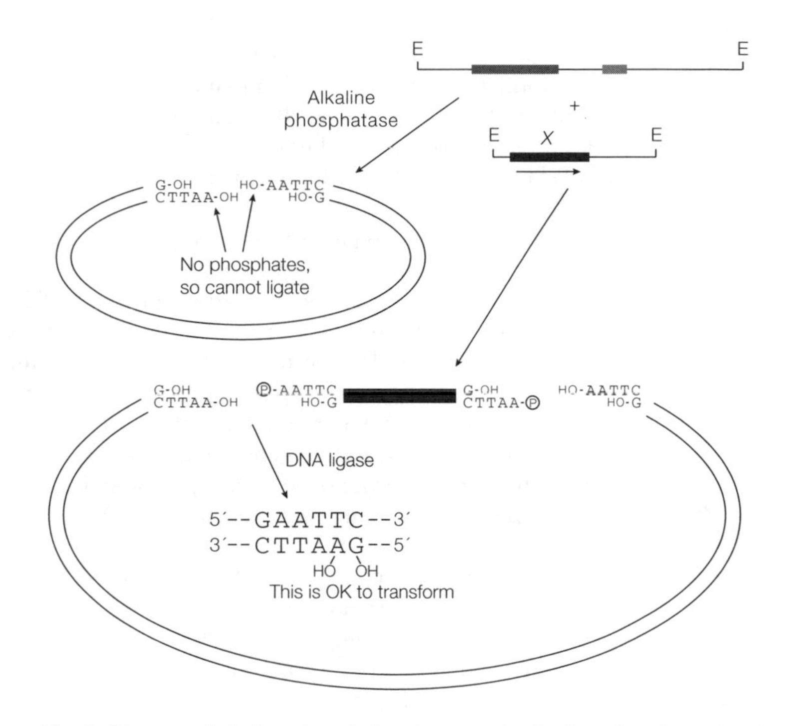

Fig. 2. The use of alkaline phosphatase to prevent religation of vector molecules.

all the cells will have the same genotype, barring spontaneous mutations (see Topic F1), including the presence of any plasmid introduced in the transformation step (in other words, they will be **clones**).

Selection

If all the competent cells present in a transformation reaction were allowed to grow on an agar plate, then many thousands or millions of colonies would result. Furthermore, transformation is an inefficient process; most of the resultant colonies would not contain a plasmid molecule and it would not be obvious which did. A method for the **selection** of clones containing a plasmid is required. This is almost always provided by the presence of an antibiotic resistance gene on the plasmid vector, for example the β-lactamase gene (*amp*^{r}) conferring resistance to ampicillin (see Topic G2). If the transformed cells are grown on plates containing ampicillin, only those cells which are expressing β-lactamase due to the presence of a transformed plasmid will survive and

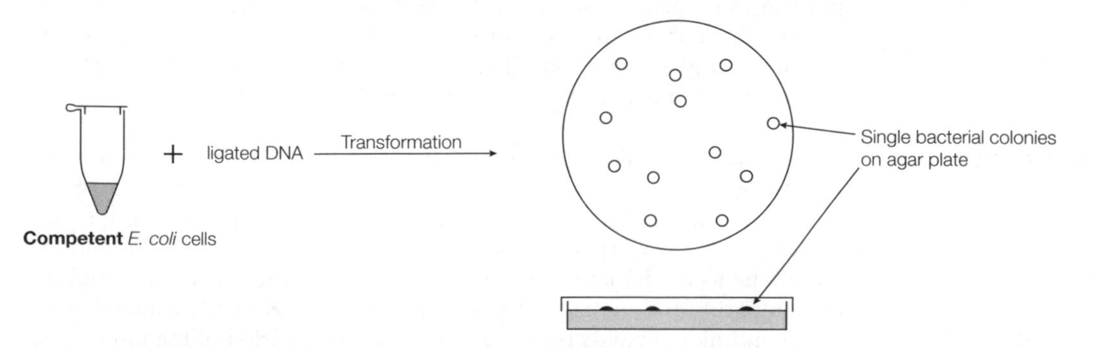

Fig. 3. The formation of single colonies after transformation of E. coli.

grow. We can therefore be sure that colonies formed on an ampicillin plate after transformation have grown from single cells which contained a plasmid with an intact β-lactamase gene. If a ligation mixture had been used for the transformation, we would not know at this stage which clones contain recombinant plasmids with a target fragment incorporated (*Fig. 1*).

Transformation efficiency

The quality of a given preparation of competent cells may be measured by determining the **transformation efficiency**, defined as the number of colonies formed (on a selective plate) per microgram of input DNA, where that DNA is a pure plasmid, most commonly the vector to be used in a cloning experiment.

Transformation efficiencies can range from 10^3 per μg for crude transformation protocols, which would only be appropriate for transferring an intact plasmid to a new host strain, to more than 10^9 per μg for very carefully prepared competent cells to be used for the generation of libraries (see Topic I2). A transformation efficiency of around 10^5 per μg would be adequate for a simple cloning experiment of the kind outlined here.

Screening transformants

Once a set of transformant clones has been produced in a cloning experiment, the first requirement is to know which clones contain a recombinant plasmid, with inserted target fragment. Plasmids have been designed to facilitate this process, and are described in Topic H1. In many cases, such as the screening of a DNA library (see Topic I3), it will then be necessary to identify the clone of interest from amongst thousands or even hundreds of thousands of others. This can be the most time-consuming part of the process, and it is discussed in Topic I3. In the case of a simple subcloning experiment, the design of the experiment can maximize the production of recombinant clones, for example by alkaline phosphatase treatment of the vector. In this case, the normal method of screening is to prepare the plasmid DNA from a number of clones and analyze it by agarose gel electrophoresis.

Growth and storage of transformants

Single colonies from a transformation plate are transferred to culture broth and grown overnight to stationary phase. The broth must include the antibiotic used to select the transformants on the original plate, to maintain the selection for the presence of the plasmid. Some plasmids may be lost from their host strains during prolonged growth without selection, since the plasmid-bearing bacteria may be out-competed by those which accidentally lose the plasmid, enabling them to replicate with less energy cost. The plasmids are then prepared from the cultures by the minipreparation technique (see Topic G2). It is normal practice to prepare a stock of each culture at this stage, by freezing a portion of the culture in the presence of glycerol, to protect the cells from ice crystal formation (a **glycerol stock**). The stock will enable the same strain/plasmid to be grown and prepared again if and when it is required.

Gel analysis

Recombinant plasmids can usually be simply distinguished from recreated vectors by the relative sizes of the plasmids, and further by the pattern of restriction digests. *Figure 4* shows a hypothetical gel representing the analysis of the plasmids in *Fig. 1*. Tracks corresponding to the vector plasmid and to recombinants are indicated. The larger size of the recombinant plasmid is seen by comparing the undigested plasmid samples (tracks U), containing supercoiled and nicked bands (see Topic G3), and the excision of the insert from the recombinant is seen in the *Eco*RI digest (track E).

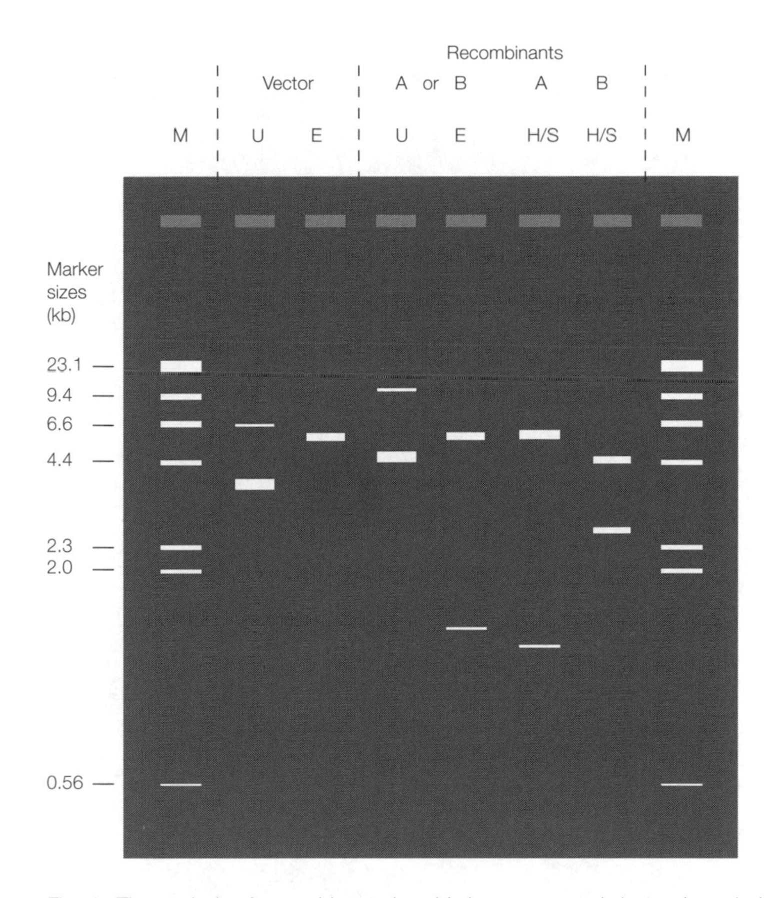

Fig. 4. The analysis of recombinant plasmids by agarose gel electrophoresis (see text for details).

Fragment orientation

If a ligation reaction has been carried out using a vector and target prepared with a single restriction enzyme (see Topic G3), then the insert can be ligated into the vector in either orientation. This might be important if, for example, the target insert contained the coding region of a gene which was to be placed downstream of a promoter in the vector. The orientation of the fragment can be determined using a restriction digest with an enzyme which is known to cut asymmetrically within the insert sequence, together with one which cuts at some specified site in the vector. This is illustrated in *Figs 1* and *4*, using a **double digest** with the enzyme *Sal*I (S), which cuts in the insert sequence, and *Hin*dIII (H), which cuts once in the vector. The patterns expected from the two orientations, A and B, of the inserted fragment (*Fig. 1*) are illustrated in *Fig. 4* (tracks H/S).

H1 DESIGN OF PLASMID VECTORS

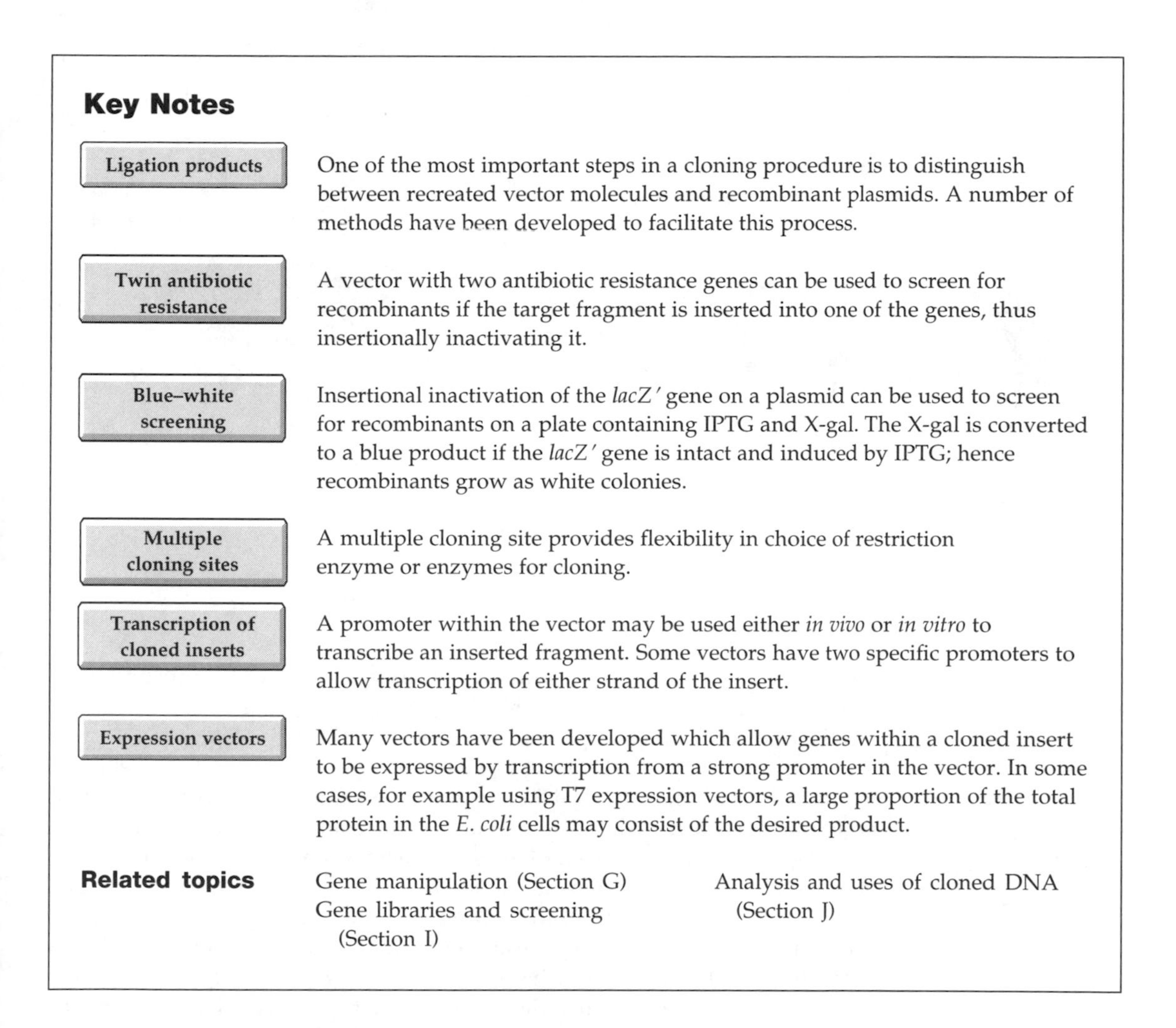

Key Notes

Ligation products — One of the most important steps in a cloning procedure is to distinguish between recreated vector molecules and recombinant plasmids. A number of methods have been developed to facilitate this process.

Twin antibiotic resistance — A vector with two antibiotic resistance genes can be used to screen for recombinants if the target fragment is inserted into one of the genes, thus insertionally inactivating it.

Blue–white screening — Insertional inactivation of the *lacZ′* gene on a plasmid can be used to screen for recombinants on a plate containing IPTG and X-gal. The X-gal is converted to a blue product if the *lacZ′* gene is intact and induced by IPTG; hence recombinants grow as white colonies.

Multiple cloning sites — A multiple cloning site provides flexibility in choice of restriction enzyme or enzymes for cloning.

Transcription of cloned inserts — A promoter within the vector may be used either *in vivo* or *in vitro* to transcribe an inserted fragment. Some vectors have two specific promoters to allow transcription of either strand of the insert.

Expression vectors — Many vectors have been developed which allow genes within a cloned insert to be expressed by transcription from a strong promoter in the vector. In some cases, for example using T7 expression vectors, a large proportion of the total protein in the *E. coli* cells may consist of the desired product.

Related topics

Gene manipulation (Section G)
Gene libraries and screening (Section I)
Analysis and uses of cloned DNA (Section J)

Ligation products

When ligating a target fragment into a plasmid vector, the most frequent unwanted product is the recreated vector plasmid formed by circularization of the linear vector fragment (see Topic G4). Religated vectors may be distinguished from recombinant products by performing minipreparations from a number of transformed colonies, and screening by digestion and agarose gel electrophoresis (see Topic G4), but this is impossibly inconvenient on a large scale, and more efficient methods based on specially developed vectors have been devised.

Twin antibiotic resistance

One of the earliest plasmid vectors to be developed was named **pBR322**. It and its derivatives contain two antibiotic resistance genes, *amp*^{r} and *tetA* (see Topic G2). If a target DNA fragment is ligated into the coding region of one of the resistance genes, say *tetA*, the gene will become **insertionally inactivated**, and

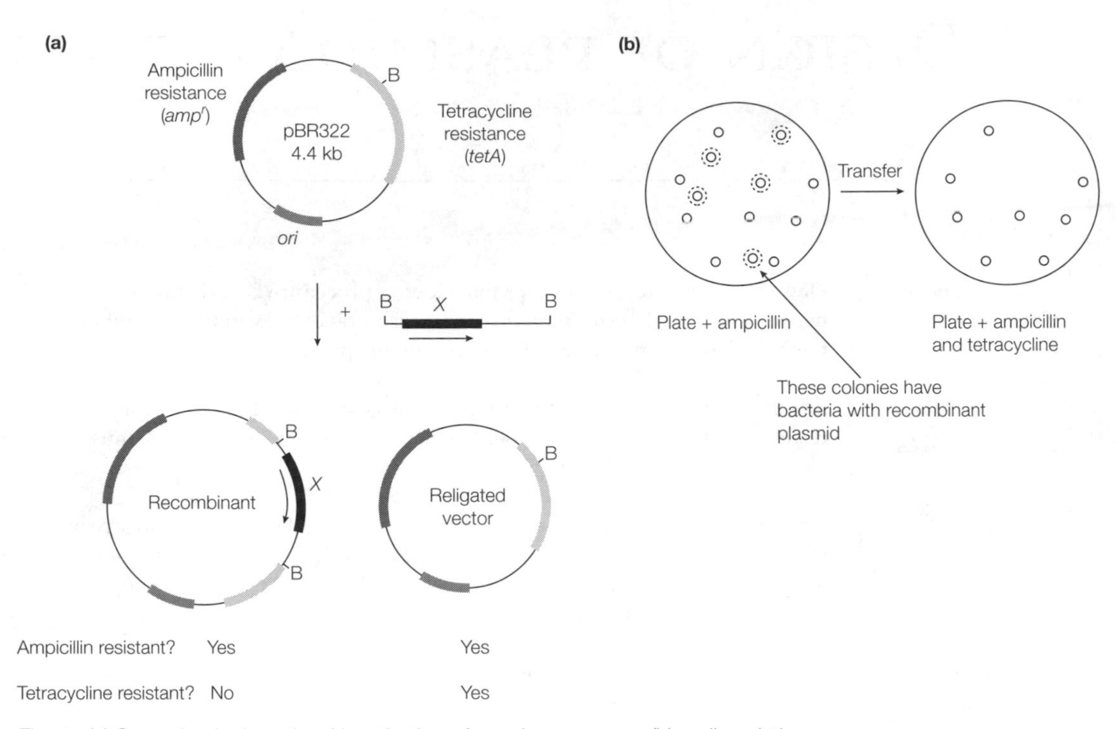

Fig. 1. (a) Screening by insertional inactivation of a resistance gene; (b) replica plating.

the presence or otherwise of a recombinant product can be determined by the antibiotic resistance exhibited by the transformants (*Fig. 1a*).

In the transformation and plating step (see Topic G4), all colonies which grow on an ampicillin plate must contain a plasmid with an intact *amp*^{r} gene. Religated vectors will also confer tetracycline resistance, whereas those with an inserted target fragment in the *tetA* gene will not produce an active TetA protein, and would be sensitive to tetracycline. Inclusion of tetracycline in the original selection plate will kill off the recombinant colonies in which we are interested, so a more roundabout method called **replica plating** must be used (*Fig. 1b*). The colonies grown on a normal ampicillin plate are transferred, using an absorbent pad, to a second plate containing tetracycline. Colonies which grow on this plate probably contain vector plasmids whereas those which do not are likely to be recombinants. The latter may be picked from the ampicillin plate for further analysis.

Blue–white screening

A more sophisticated procedure for screening for the presence of recombinant plasmids, which can be carried out on a single transformation plate, is called **blue–white** screening. This method also involves the insertional inactivation of a gene and, as the name implies, uses the production of a blue compound as an indicator. The gene in this case is ***lacZ***, which encodes the enzyme **β-galactosidase**, and is under the control of the ***lac*** promoter (see Topic L1). If the host *E. coli* strain is expressing the **lac repressor**, then expression of a *lacZ* gene on the vector may be induced using **isopropyl-β-D-thiogalactopyranoside (IPTG)** (Topic L1), and the expressed enzyme can utilize the synthetic substrate **5-bromo-4-chloro-3-indolyl-β-D-galactopyranoside (X-gal)** to yield a blue product. Insertional inactivation of *lacZ* in the production of a recombinant plasmid would prevent the development of the blue color. In this method

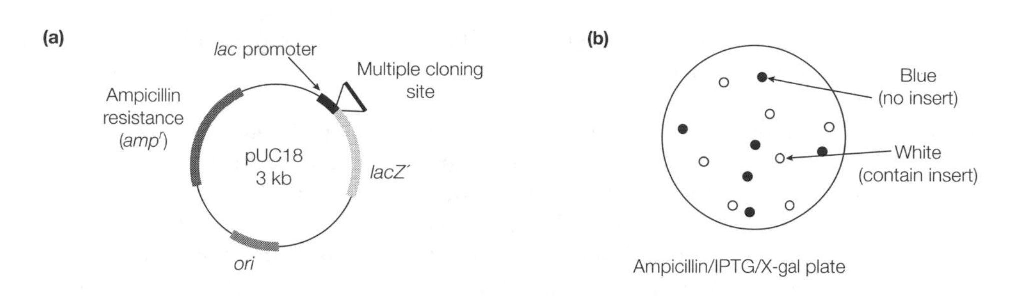

Fig. 2. (a) A plasmid vector designed for blue–white screening; (b) the colonies produced by blue–white screening.

(*Fig. 2*), the transformed cells are spread on to a plate containing ampicillin (to select for transformants in the usual way), IPTG and X-gal, to yield a mixture of blue and white colonies. The white colonies have no expressed β-galactosidase and are hence likely to contain the inserted target fragment. The blue colonies probably contain religated vector.

In practice, the vectors used in this method have a shortened derivative of *lacZ*, ***lacZ′***, which produces the N-terminal **α-peptide** of β-galactosidase. These vectors must be transformed into a special host strain which contains a mutant gene expressing only the C-terminal portion of β-galactosidase which can then complement the α-peptide to produce active enzyme. This reduces the size of the plasmid-borne gene, but does not alter the basis of the method.

Multiple cloning sites

The first vectors which utilized blue–white selection also pioneered the idea of the **multiple cloning site (MCS)**. These plasmids, the pUC series, contain an engineered version of the *lacZ′* gene, which has multiple restriction enzyme sites within the first part of the coding region of the gene (*Figs 2a* and *3*). This region is known as the MCS; insertion of target DNA in any of these sites, or between any pair, inactivates the *lacZ′* gene, to give a white colony on an appropriate plate. The use of an MCS allows flexibility in the choice of a restriction enzyme or enzymes for cloning.

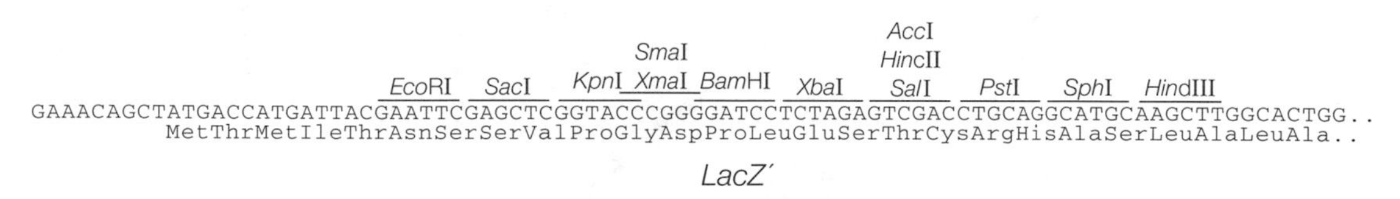

Fig. 3. A multiple cloning site at the 5′-end of lacZ′.

Transcription of cloned inserts

Since the pUC vectors above have a promoter (*lac*) adjacent to the site of insertion of a cloned fragment, it is a simple step to imagine that such a promoter could be used to transcribe the inserted DNA, either to produce an RNA transcript *in vitro*, which could be used as a hybridization probe (see Topic I3), or to express the protein product of a gene within the insert (see below).

Several varieties of transcriptional vector have been constructed, which allow the *in vitro* transcription of a cloned fragment. For example, the pGEM series has promoters from **bacteriophages T7** and **SP6** flanking an MCS (which itself is set up for *lacZ′* blue–white screening). The phage promoters are each recognized only by their corresponding bacteriophage RNA polymerases (see Topic

G1, *Table 1*), either of which may be used *in vitro* to transcribe the desired strand of the inserted fragment.

Expression vectors

The pUC vectors may be used to express cloned genes in *E. coli*. This requires the positioning of the cloned fragment within the MCS such that the coding region of the target gene is contiguous with and in the same **reading frame** (see Topic P1) as the *lacZ′* gene. This results, after induction of transcription from *lac*, in the production of a **fusion protein**, with a few amino acids derived from the N terminus of *lacZ′* followed directly by the protein sequence encoded by the insert.

Innumerable variants of this type of scheme have been developed, using strong (very active) promoters such as ***lacUV-5*** [a mutant *lac* promoter which is independent of cyclic AMP receptor protein (CRP); see Topic L1], the **phage λP_L** promoter (see Topic R2) or the **phage T7** promoter. Some vectors rely on the provision of a ribosome binding site (RBS) and translation initiation codon (see Topic Q1) within the cloned fragment, while some are designed to encode a fused sequence at the N terminus of the expressed protein which allows the protein to be purified easily by a specific binding step. An example of this is the **His-Tag**, a series of histidine residues, which will bind strongly to a chromatography column bearing Ni^{2+} ions (see Topic B3). Some vectors may even allow the fused N terminus to be removed from the protein by cleavage with a specific protease.

One example will serve as an illustration: the overexpression of many proteins in *E. coli* is now achieved using a T7 promoter system (*Fig. 4*). The vector contains a T7 promoter and an *E. coli* RBS, followed by a start codon, an MCS and, finally, a transcription terminator (see Topic K4). Expression is achieved in a special *E. coli* strain (a λ lysogen; see Topic H2) which produces the T7 RNA polymerase on induction with IPTG. The enzyme is very efficient, and only transcribes from the T7 promoter, resulting in favorable cases in the production of the target protein as up to 30% of the total *E. coli* protein.

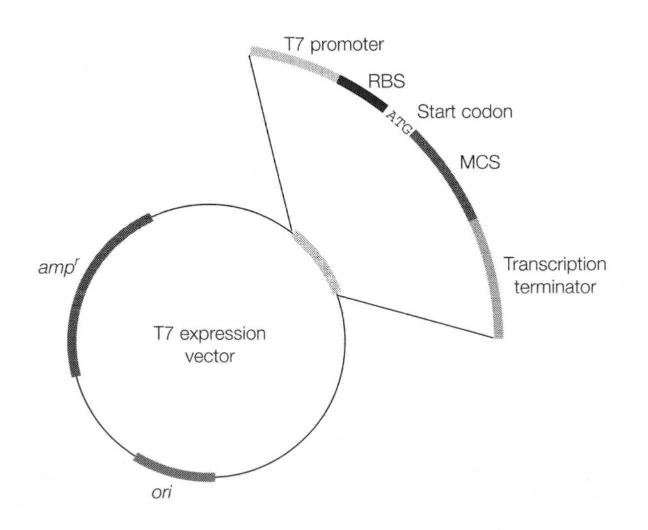

Fig. 4. A plasmid designed for expression of a gene using the T7 system.

H2 Bacteriophage vectors

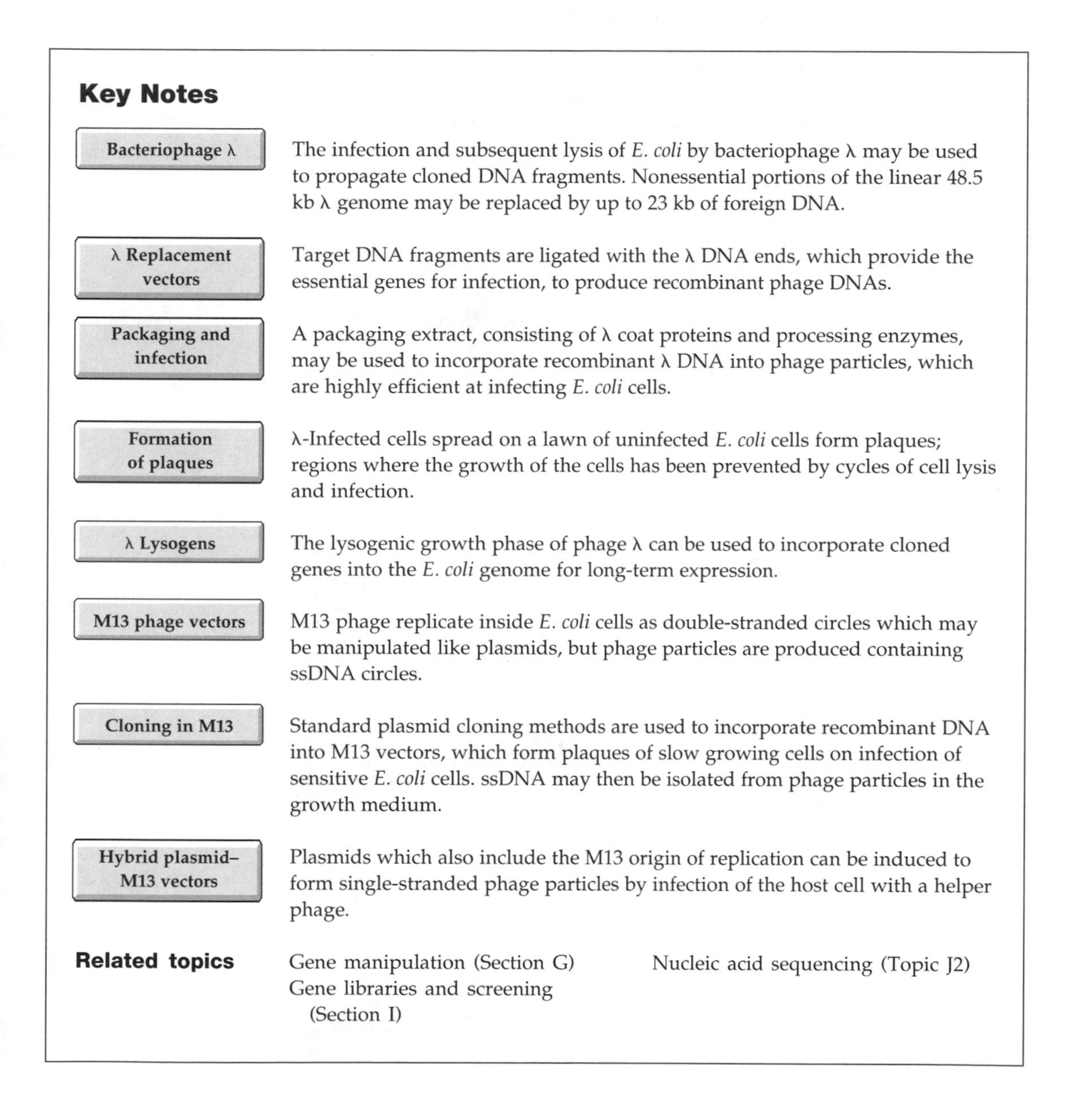

Key Notes

Bacteriophage λ

The infection and subsequent lysis of *E. coli* by bacteriophage λ may be used to propagate cloned DNA fragments. Nonessential portions of the linear 48.5 kb λ genome may be replaced by up to 23 kb of foreign DNA.

λ Replacement vectors

Target DNA fragments are ligated with the λ DNA ends, which provide the essential genes for infection, to produce recombinant phage DNAs.

Packaging and infection

A packaging extract, consisting of λ coat proteins and processing enzymes, may be used to incorporate recombinant λ DNA into phage particles, which are highly efficient at infecting *E. coli* cells.

Formation of plaques

λ-Infected cells spread on a lawn of uninfected *E. coli* cells form plaques; regions where the growth of the cells has been prevented by cycles of cell lysis and infection.

λ Lysogens

The lysogenic growth phase of phage λ can be used to incorporate cloned genes into the *E. coli* genome for long-term expression.

M13 phage vectors

M13 phage replicate inside *E. coli* cells as double-stranded circles which may be manipulated like plasmids, but phage particles are produced containing ssDNA circles.

Cloning in M13

Standard plasmid cloning methods are used to incorporate recombinant DNA into M13 vectors, which form plaques of slow growing cells on infection of sensitive *E. coli* cells. ssDNA may then be isolated from phage particles in the growth medium.

Hybrid plasmid–M13 vectors

Plasmids which also include the M13 origin of replication can be induced to form single-stranded phage particles by infection of the host cell with a helper phage.

Related topics

Gene manipulation (Section G)
Gene libraries and screening (Section I)
Nucleic acid sequencing (Topic J2)

Bacteriophage λ

Bacteriophage λ, which infects *E. coli* cells, can be used as a cloning vector. The process of infection by phage λ is described in Topic R2. In brief, the phage particle injects its **linear DNA** into the cell, where it is ligated into a circle. It may either replicate to form many phage particles, which are released from the cell by lysis and cell death (the **lytic** phase), or the DNA may integrate into the host genome by site-specific recombination (see Topic F4), where it may remain for a long period (the **lysogenic** phase).

The 48.5 kb λ genome is shown schematically in *Fig. 1*. At the ends are the ***cos*** (cohesive) sites, which consist of 12 bp cohesive ends. The *cos* sites are asymmetric, but in other respects are equivalent to very large (16 bp) restriction sites (*Fig. 1* or see Topic R2). The *cos* ends allow the DNA to be circularized in the cell. Much of the central region of the genome is dispensable for lytic infection, and may be replaced by unrelated DNA sequence. There are limits to the size of DNA which can be incorporated into a λ phage particle; the DNA must be between 75 and 105% of the natural length, that is 37–52 kb. Taking account of the essential regions, DNA fragments of around 20 kb (maximum 23 kb) can be cloned into λ, which is more than can be conveniently incorporated into a plasmid vector.

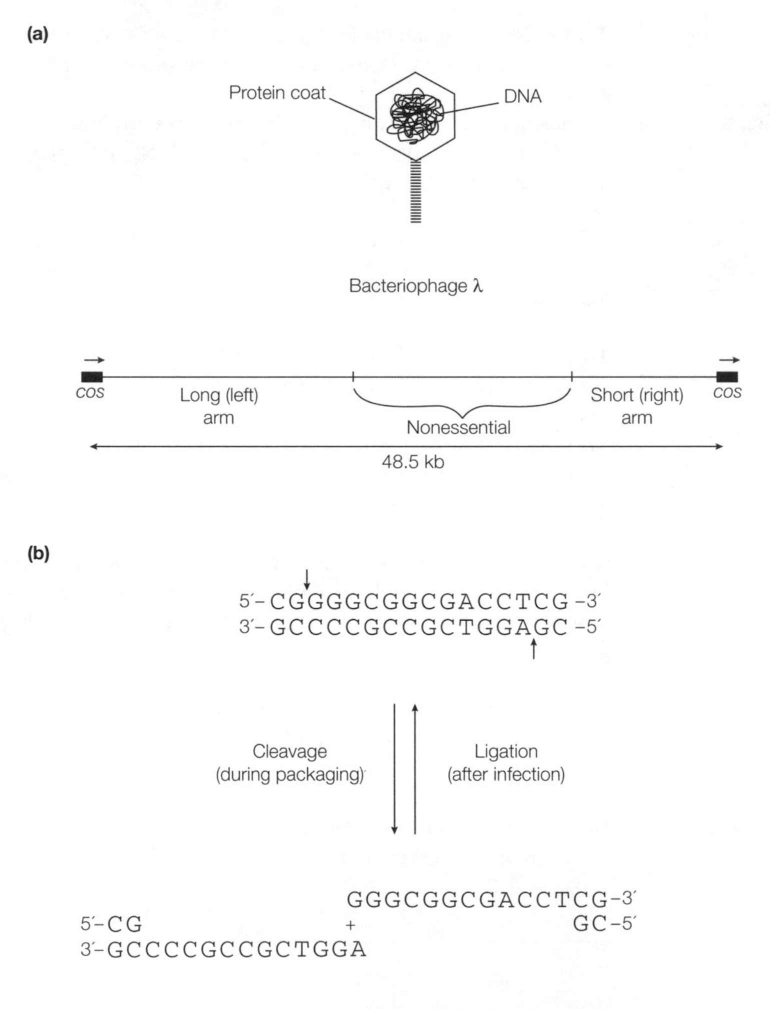

Fig. 1. (a) Phage λ and its genome; (b) the phage λ cos ends.

λ Replacement vectors

A number of so-called **replacement vectors** have been developed from phage λ; examples include **EMBL3** and **λDASH** (see Topic I1). A representative scheme for cloning using such a vector is shown in *Fig. 2*. The vector DNA is cleaved with *Bam*HI and the long (19 kb) and short (9 kb) ends (*Fig. 1*) are purified. The target fragment or fragments are prepared by digestion, also with *Bam*HI

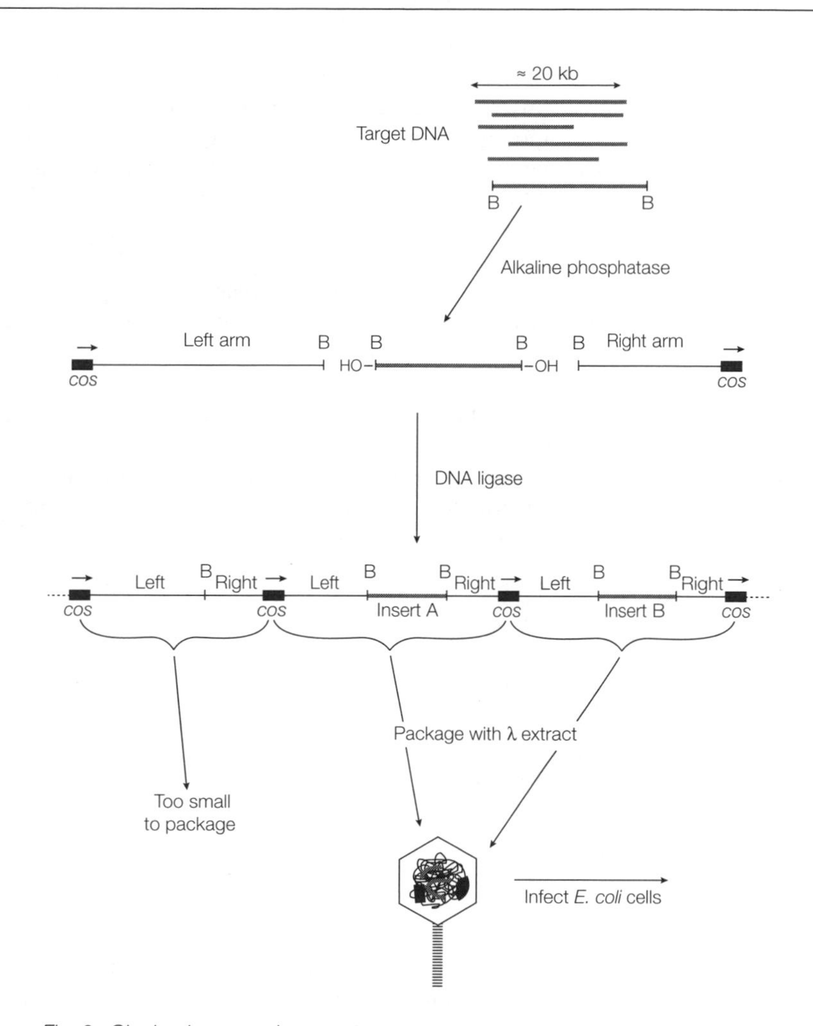

Fig. 2. Cloning in a λ replacement vector.

or a compatible enzyme (see Topics G3 and I1), and treated with alkaline phosphatase (see Topic G4) to prevent them ligating to each other. The λ arms and the target fragments are ligated together (see Topic G4) at relatively high concentration to form long linear products (*Fig. 2*).

Packaging and infection

Although pure circular λ DNA or derivatives of it can be transformed into competent cells, as described for plasmids (see Topic G4), the infective properties of the phage particle may be used to advantage, particularly in the formation of DNA libraries (see Section I). Replication of phage λ *in vivo* produces long linear molecules with multiple copies of the λ genome. These concatamers are then cleaved at the *cos* sites, to yield individual λ genomes, which are then packaged into the phage particles. A mixture of the phage coat proteins and the phage DNA-processing enzymes (a **packaging extract**) may be used *in vitro* to **package** the ligated linear molecules into phage particles (*Fig. 2*). The packaging extract is prepared from two bacterial strains each infected with a different **packaging-deficient** (mutant) λ phage, so that packaging proteins are abundant. In combination, the two extracts provide all the necessary proteins for packaging. The phage particles so produced may then be used to infect a culture of normal *E. coli* cells. Ligated λ ends which do not contain an insert (*Fig. 2*), or have one which is much smaller or larger than the 20 kb optimum, are too small or too large to be packaged, and recombinants with two left or right arms are likewise not viable. The infection process is very efficient and can produce up to 10^9 recombinants per microgram of vector DNA.

Formation of plaques

The infected cells from a packaging reaction are spread on an agar plate (see Topic G4), which has been pre-spread with a high concentration of uninfected cells, which will grow to form a continuous **lawn**. Single infected cells result in clear areas, or **plaques**, within the lawn after incubation, where cycles of lysis and re-infection have prevented the cells from growing (*Fig. 3*). These are the analogs of single bacterial colonies (see Topic G4). Recombinant λ DNA may be purified for further manipulation from phage particles isolated from plaques or from the supernatant broth of a culture infected with a specific recombinant plaque (see Topic J1).

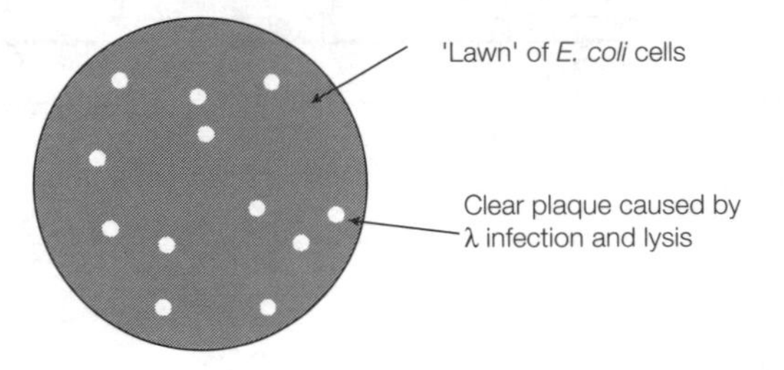

Fig. 3. The formation of plaques by λ infection.

λ Lysogens

The lysogenic phase of λ infection is also used in cloning technology. Genes or foreign sequences may be incorporated essentially permanently into the genome of *E. coli* by integration of a λ vector containing the sequence of interest. One example of the use of this method is a strain used for overexpression of proteins by the T7 method. The strain BL21(DE3) and derivatives include the gene for T7 RNA polymerase under control of the *lac* promoter as a λ lysogen, designated DE3. The gene can be induced by IPTG, and the polymerase will then transcribe the target gene in the expression vector (see Topic H1).

M13 phage vectors

The so-called filamentous phages (see Topic R2), specifically **M13**, are also used as *E. coli* vectors. The phage particles contain a 6.7 kb circular single strand of DNA. After infection of a sensitive *E. coli* host, the complementary strand is synthesized, and the DNA replicated as a double-stranded circle, the **replicative form (RF)**, with about 100 copies per cell. In contrast to the situation with phage λ, the cells are not lysed by M13, but continue to grow slowly, and single-stranded forms are continuously packaged and released from the cells as new phage particles (up to 1000 per cell generation). The same single strand of the complementary pair is always present in the phage particle.

The useful properties of M13 as a vector are that the RF can be purified and manipulated exactly like a plasmid, but the same DNA may be isolated in a single-stranded form from phage particles in the medium. ssDNA has a number of applications, including DNA sequencing (see Topic J2) and site-directed mutagenesis (see Topic J5).

Cloning in M13

M13 is not used as a primary vector to clone new DNA targets, but fragments are normally subcloned into M13 RF using standard plasmid methods (see Section G) when the single-stranded form of a fragment is required. Transfection of *E. coli* cells with the recombinant DNA, followed by plating on a lawn of cells, produces plaques as with λ infection, except the plaques are formed by the slow growth, rather than the lysis, of infected cells. Blue–white selection using MCSs and *lacZ′* (see Topic H1) has been engineered into M13 vectors. Examples include M13mp18 and 19, which are a related pair of vectors in which the MCSs are in opposite orientations relative to the M13 origin of replication. The origin defines the strand which is packaged into the phage, so either vector may be used depending on which strand of the insert is required to be produced in the phage particle.

Hybrid plasmid–M13 vectors

A number of small plasmid vectors, for example pBluescript, have been developed to incorporate M13 functionality. They contain both plasmid and M13 origins of replication, but do not possess the genes required for the full phage life cycle. Hence, they normally propagate as true plasmids, and have the advantages of rapid growth and easy manipulation of plasmid vectors, but they can be induced, when required, to produce single-stranded phage particles by co-infection with a fully functional **helper phage**, which provides the gene products required for single-strand production and packaging.

H3 Cosmids, YACs and BACs

Key Notes

Cloning large DNA fragments

Analysis of eukaryotic genes and the genome organization of eukaryotes requires vectors with a larger capacity for cloned DNA than plasmids or phage λ.

Cosmid vectors

Cosmids use the λ packaging system to package large DNA fragments bounded by λ *cos* sites, which circularize and replicate as plasmids after infection of *E. coli* cells. Some cosmid vectors have two *cos* sites, and are cleaved to produce two *cos* ends, which are ligated to the ends of target fragments and packaged into λ particles. Cosmids have a capacity for cloned DNA of 30–45 kb.

YAC vectors

Yeast artificial chromosomes can be constructed by ligating the components required for replication and segregation of natural yeast chromosomes to very large fragments of target DNA, which may be more than 1 Mb in length. Yeast artificial chromosome (YAC) vectors contain two telomeric sequences (TEL), one centromere (CEN), one autonomously replicating sequence (ARS) and genes which can act as selectable markers in yeast.

Selection in *S. cerevisiae*

Selection for the presence of YACs or other vectors in yeast is achieved by complementation of a mutant strain unable to produce an essential metabolite, with the correct copy of the mutant gene carried on the vector.

BAC vectors

Bacterial artificial chromosomes are based on the F factor of *E. coli* and can be used to clone up to 350 kb of genomic DNA in a conveniently-handled *E. coli* host. They are a more stable and easier to use alternative to YACs.

Related topics

Design of plasmid vectors (H1)
Bacteriophage vectors (H2)
Genomic libraries (I1)

Cloning large DNA fragments

The analysis of genome organization and the identification of genes, particularly in organisms with large genome sizes (human DNA is 3×10^9 bp, for example) is difficult to achieve using plasmid and bacteriophage λ vectors (see Section G and Topic H2), since the relatively small capacity of these vectors for cloned DNA means that an enormous number of clones would be required to represent the whole genome in a DNA library (see Topics G1 and I1). In addition, the very large size of some eukaryotic genes, due to their large intron sequences, means that an entire gene may not fit on a single cloned fragment. Vectors with much larger size capacity have been developed to circumvent these problems. **Pulsed field gel electrophoresis (PFGE)** has made it possible to separate, map and analyze very large DNA fragments. In Section G3 it was seen that the limitation of conventional agarose gel electrophoresis becomes apparent as large DNA fragments above a critical size do not separate, but instead, co-migrate. This is because nucleic acids alternate between folded (more globular) and extended (more linear) forms as they migrate through a porous matrix such as an agarose gel. However, when the DNA molecules become so large that their globular forms do not fit into the matrix pores, even in the lowest percentage agarose gels that can be easily handled (0.1–0.3%), then they co-migrate. This limitation can be overcome if the electric field is applied discontinuously (pulsed) and even greater separation can be achieved if the direction of the field is also made to vary. Each time the electric field changes the DNA molecules reorient their long axes and this process takes longer for larger molecules. A number of variations of PFGE have been developed which differ in the number of electrodes used and how the field is varied (e.g. field inversion gel electrophoresis, FIGE and contour clamped homogeneous electric field, CHEF). Separations of molecules up to 7 Mb have been achieved and it is now possible to resolve whole chromosome DNA fragments, including artificial chromosomes (see below).

Cosmid vectors

Cosmid vectors are so-called because they utilize the properties of the phage λ ***cos*** sites in a plasmid molecule. The *in vitro* packaging of DNA into λ particles (see Topic H2) requires only the presence of the λ *cos* sites spaced by the correct distance (37–52 kb) on linear DNA. The intervening DNA can have any sequence at all; it need not contain any λ genes, for example. The simplest cosmid vector is a normal small plasmid, containing a plasmid origin of replication (*ori*), and a selectable marker, which also contains a *cos* site and a suitable restriction site for cloning (*Fig. 1*). After cleavage with a restriction enzyme and ligation with target DNA fragments, the DNA is packaged into λ phage particles. The DNA is re-circularized by annealing of the *cos* sites after infection, and propagates as a normal plasmid, under selection by ampicillin.

In a real cloning situation, as for phage λ (see Topic H2), more sophisticated methods are used to ensure that multiple copies of the vector or the target DNA are not included in the recombinant. An example of a cosmid vector is C2XB (*Fig. 2*), which contains two *cos* sites, with a site for a blunt-cutting restriction enzyme (*Sma*I) between them. Cleavage with *Bam*HI and *Sma*I yields two ***cos* ends**, which are ligated with target DNA fragments which have been treated with alkaline phosphatase (see Topics G1 and G4) to prevent self-ligation. The blunt *Sma*I ends are only ligated inefficiently under the conditions used, and ultimately the only products which can be packaged are those shown, which yield recombinant cosmids with inserts in the size range 30–45 kb. Libraries of clones prepared with cosmid vectors can be screened as described in Topic I3. For a number of years, the entire *E. coli* genome has been available as a set of around 100 cosmid clones.

YAC vectors

The realization that the components of a eukaryotic chromosome that are required for stable replication and segregation, at least in the yeast *Saccharomyces cerevisiae*, consist of rather small and well-defined sequences (see Topic D3) has led to the construction of recombinant chromosomes (**yeast artificial chromosomes; YACs**). These were used initially for investigation of the maintenance of chromosomes, but latterly as vectors capable of carrying very large cloned fragments. The centromere, telomere and replication origin sequences (see Topics D3, E1 and E3) have been isolated and combined on plasmids constructed in *E. coli*. The structure of a typical pYAC vector is shown in *Fig. 3*. The method of

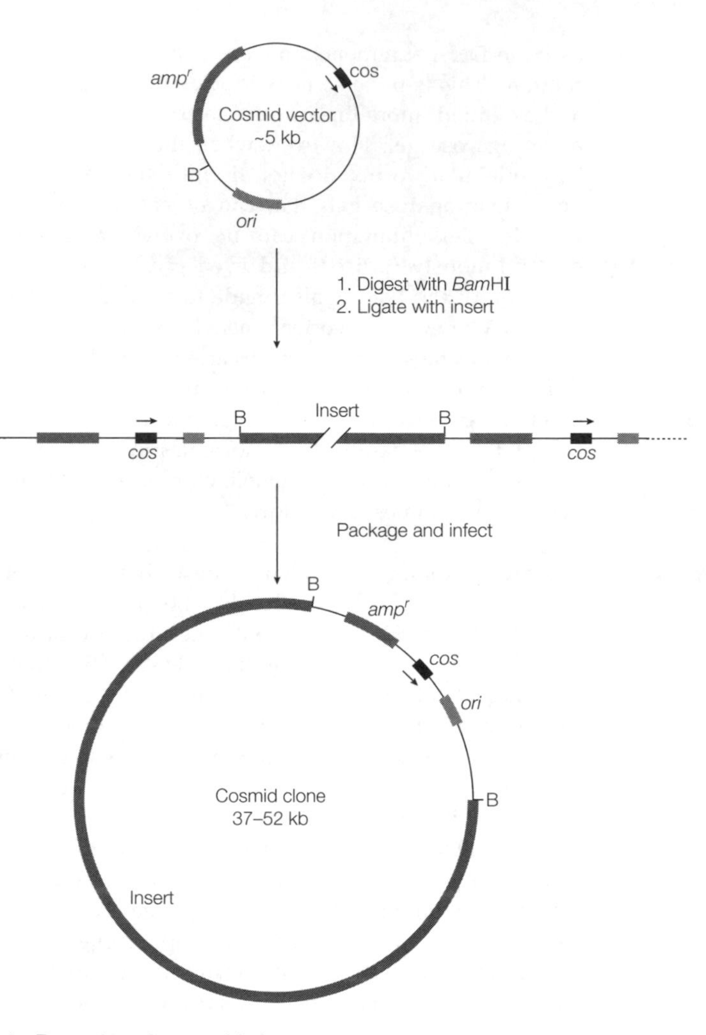

Fig. 1. Formation of a cosmid clone.

construction of the YAC clone is similar to that for cosmids, in that two end fragments are ligated with target DNA to yield the complete chromosome, which is then introduced (transfected) into yeast cells (see Topic H4). YAC vectors can accommodate genomic DNA fragments of more than 1 Mb, and hence can be used to clone entire human genes, such as the cystic fibrosis gene, which is 250 kb in length. YACs have been invaluable in mapping the large-scale structure of large genomes, for example in the Human Genome Project.

The yeast sequences that have been included on pYAC3 (*Fig. 3*) are as follows: **TEL** represents a segment of the telomeric DNA sequence, which is extended by the telomerase enzyme inside the yeast cell (see Topics D3 and E3). **CEN4** is the centromere sequence for chromosome 4 of *S. cerevisiae*. Despite its name, the centromere will function correctly to segregate the daughter chromosomes even if it is very close to one end of the artificial chromosome. The **ARS** (**autonomously replicating sequence**) functions as a yeast origin of replication (see Topic E3). ***TRP1*** and ***URA3*** are yeast selectable markers (see below), one for each end, to ensure that only properly reconstituted YACs survive in the yeast cells. ***SUP4***, which is insertionally inactivated in recombinants (*Fig. 3*), is

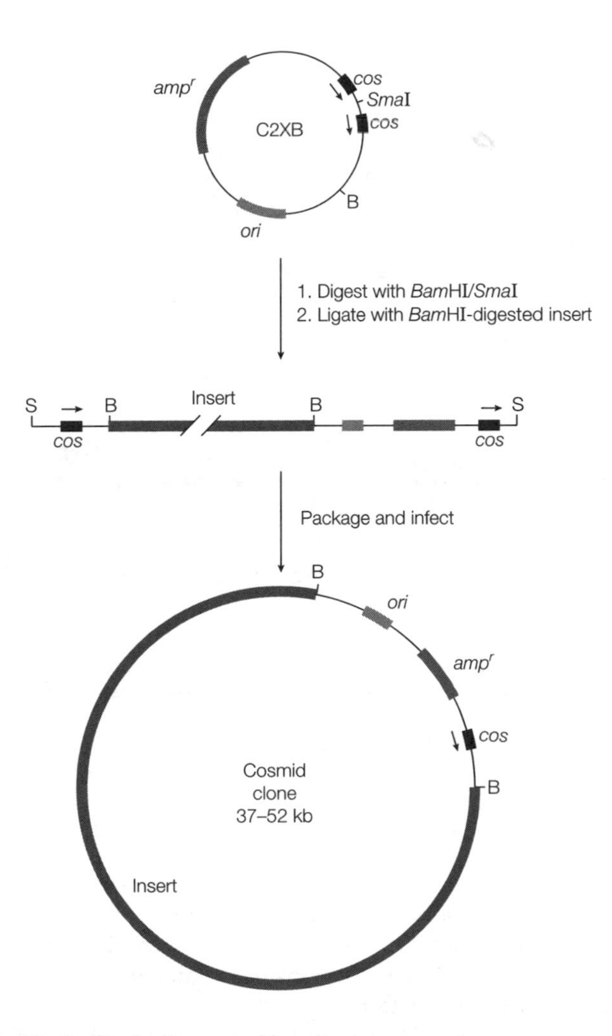

Fig. 2. Cloning in a cosmid vector.

a gene which is the basis of a red–white color test, which is analogous to blue–white screening in *E. coli* (see Topic H1).

Selection in *S. cerevisiae*

Saccharomyces cerevisiae selectable markers do not normally confer resistance to toxic substances, as in *E. coli* plasmids, but instead enable the growth of yeast on selective media lacking specific nutrients. **Auxotrophic** yeast mutants are unable to make a specific compound. *TRP1* mutants, for example, cannot make tryptophan, and can only grow on media supplemented with tryptophan. Transformation of the mutant yeast strain with a YAC (or other vector; see Topic H4) containing an intact *TRP1* gene **complements** this deficiency and hence only transfected cells can grow on media lacking tryptophan.

BAC vectors

BAC vectors, or **bacterial artificial chromosomes**, were developed to overcome one or two problems with the use of YACs to clone very large genomic DNA fragments. Although YACs can accommodate very large fragments, quite often these fragments turn out to comprise noncontiguous (nonadjacent) segments of the genome and they frequently lose parts of the DNA during propagation (i.e.

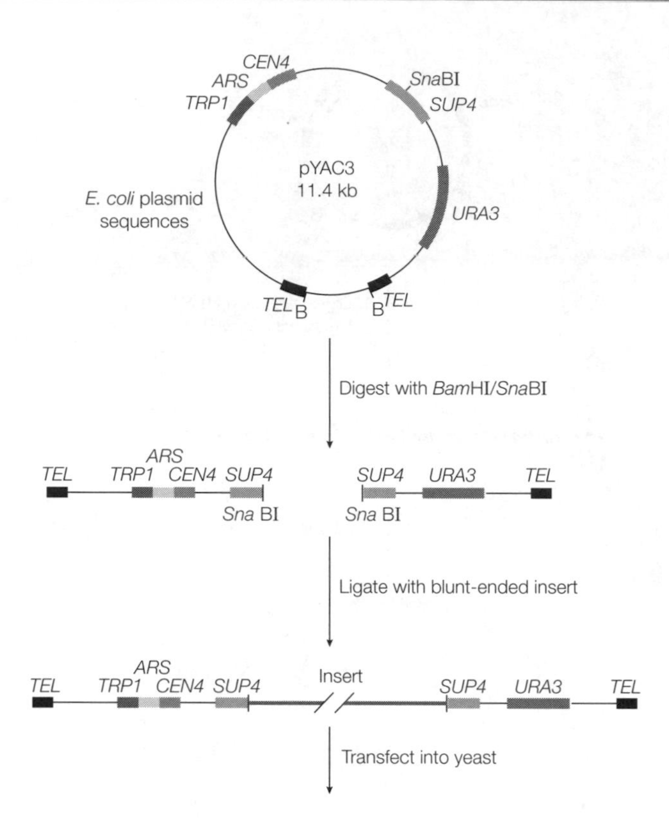

Fig. 3. Cloning in a YAC vector.

they are unstable). BACs are able to accommodate up to around 300–350 kb of insert sequence, less than YACs, but they have the advantages not only of stability, but also of the ease of transformation and speed of growth of their *E. coli* host, and are simpler to purify, using standard plasmid minipreparation techniques (see Section G2). The vectors are based on the natural extrachromosomal F factor of *E. coli*, which encodes its own DNA polymerase and is maintained in the cell at a level of one or two copies. A BAC vector incorporates the genes essential for replication and maintenance of the F factor, a selectable marker and a cloning site flanked by rare-cutting restriction enzyme sites and other specific cleavage sites, which serve to enable the clones to be linearized within the vector region, without the possibility of cutting within the very large insert region. BACs are a more user-friendly alternative to YACs and are now being used extensively in genomic mapping projects.

H4 EUKARYOTIC VECTORS

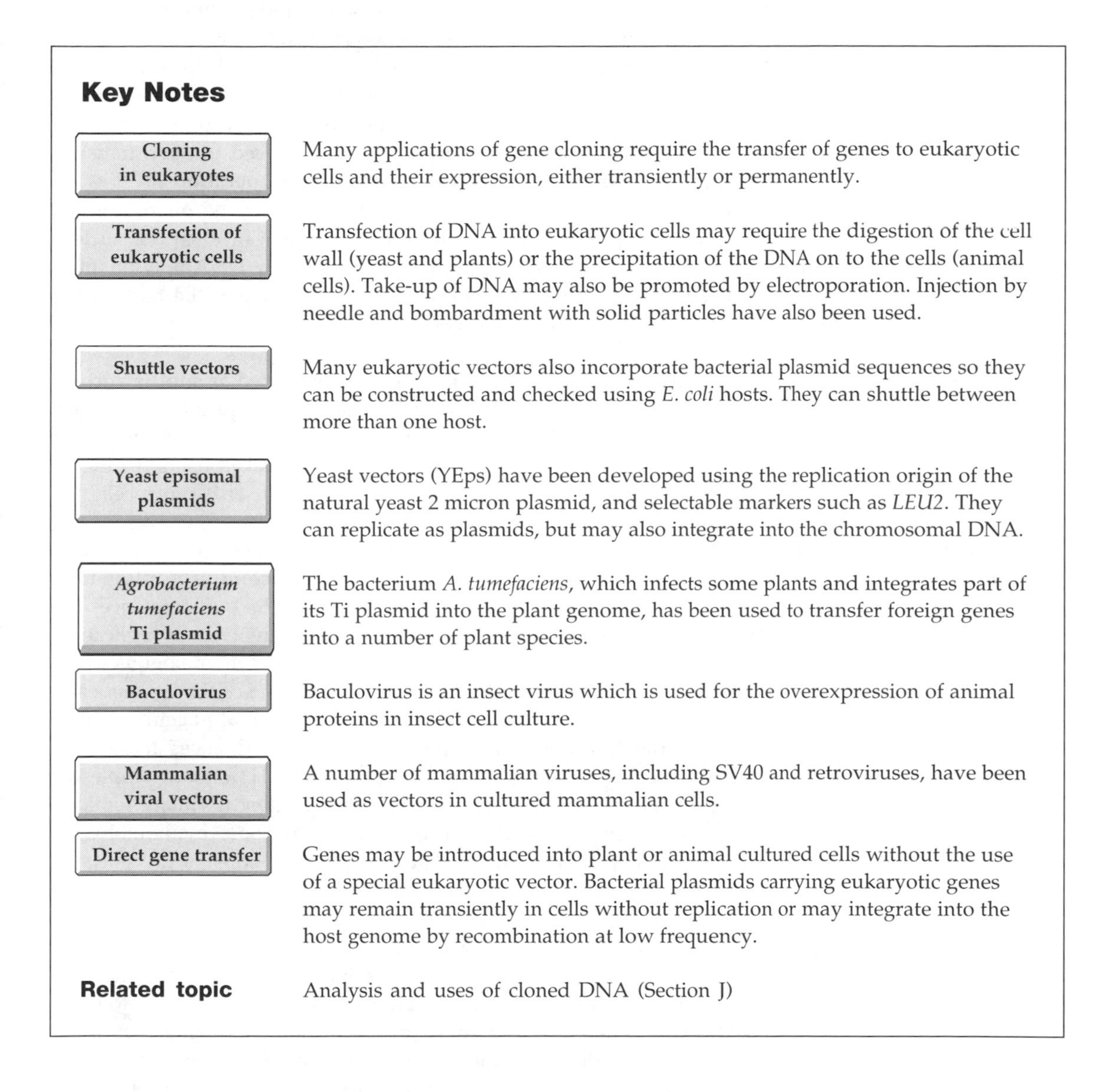

Key Notes

Cloning in eukaryotes	Many applications of gene cloning require the transfer of genes to eukaryotic cells and their expression, either transiently or permanently.
Transfection of eukaryotic cells	Transfection of DNA into eukaryotic cells may require the digestion of the cell wall (yeast and plants) or the precipitation of the DNA on to the cells (animal cells). Take-up of DNA may also be promoted by electroporation. Injection by needle and bombardment with solid particles have also been used.
Shuttle vectors	Many eukaryotic vectors also incorporate bacterial plasmid sequences so they can be constructed and checked using *E. coli* hosts. They can shuttle between more than one host.
Yeast episomal plasmids	Yeast vectors (YEps) have been developed using the replication origin of the natural yeast 2 micron plasmid, and selectable markers such as *LEU2*. They can replicate as plasmids, but may also integrate into the chromosomal DNA.
***Agrobacterium tumefaciens* Ti plasmid**	The bacterium *A. tumefaciens*, which infects some plants and integrates part of its Ti plasmid into the plant genome, has been used to transfer foreign genes into a number of plant species.
Baculovirus	Baculovirus is an insect virus which is used for the overexpression of animal proteins in insect cell culture.
Mammalian viral vectors	A number of mammalian viruses, including SV40 and retroviruses, have been used as vectors in cultured mammalian cells.
Direct gene transfer	Genes may be introduced into plant or animal cultured cells without the use of a special eukaryotic vector. Bacterial plasmids carrying eukaryotic genes may remain transiently in cells without replication or may integrate into the host genome by recombination at low frequency.

Related topic Analysis and uses of cloned DNA (Section J)

Cloning in eukaryotes

Many eukaryotic genes and their control sequences have been isolated and analyzed using gene cloning techniques based on *E. coli* as host. However, many applications of genetic engineering (see Section J), from the large-scale production of eukaryotic proteins to the engineering of new plants and gene therapy, require vectors for the expression of genes in diverse eukaryotic species. Examples of such vectors designed for a variety of hosts are discussed in this topic.

Transfection of eukaryotic cells

The take-up of DNA into eukaryotic cells (**transfection**) can be more problematic than bacterial transformation, and the efficiency of the process is much lower. In yeast and plant cells, for example, the cell wall must normally be digested with degradative enzymes to yield fragile **protoplasts**, which may then take up DNA fairly readily. The cell walls are re-synthesized once the degrading enzymes are removed. In contrast, animal cells in culture, which have no cell wall, will take up DNA at low efficiency if it is precipitated on to their surface with **calcium phosphate**. The efficiency of the process may be increased by treatment of the cells with a high voltage, which is believed to open transient pores in the cell membrane. This process is called **electroporation**.

Direct physical methods have also been used. DNA may be **microinjected**, using very fine glass pipettes, directly into the cytoplasm or even the nucleus of individual animal or plant cells in culture. Alternatively DNA may be introduced by firing metallic microprojectiles coated with DNA at the target cells, using what has become known as a '**gene gun**'.

Shuttle vectors

Most of the vectors for use in eukaryotic cells are constructed as **shuttle vectors**. This means that they incorporate the sequences required for replication and selection in *E. coli* (*ori*, *amp*^{r}) as well as in the desired host cells. This enables the vector with its target insert to be constructed and its integrity checked using the highly developed *E. coli* methods, before transfer to the appropriate eukaryotic cells.

Yeast episomal plasmids

Vectors for the cloning and expression of genes in *Saccharomyces cerevisiae* have been designed based on the natural **2 micron (2μ) plasmid**. The plasmid is named for the length of its DNA, which corresponds to 6 kb of sequence. The 2μ plasmid has an origin of replication and two genes involved in replication, and also encodes a site-specific recombination protein FLP, homologous to the phage λ integrase, Int (see Topic F4), which can invert part of the 2μ sequence. Vectors based on the 2μ plasmid (*Fig. 1*), called **yeast episomal plasmids (YEps)** normally contain the 2μ replication origin, *E. coli* shuttle sequences and a yeast gene which can act as a selectable marker (see Topic H3), for example the *LEU2* gene, involved in leucine biosynthesis. Although they normally replicate as plasmids, YEps may integrate into a yeast chromosome by homologous recombination (see Topic F4) with the defective genomic copy of the selection gene (*Fig. 1*).

Agrobacterium tumefaciens Ti plasmid

Plant cells do not contain natural plasmids that can be utilized as cloning vectors. However, the bacterium *A. tumefaciens* which primarily infects dicotyledenous plants (tomato, tobacco, peas, etc.), but has been shown recently to infect the monocot, rice, contains the 200 kb **Ti plasmid** (tumor inducing). On infection, part of the Ti plasmid, the **T-DNA**, is integrated into the plant chromosomal DNA (*Fig. 2*), resulting in uncontrolled growth of the plant cells directed by genes in the T-DNA, and the development of a **crown gall**, or tumor.

Recombinant Ti plasmids with a target gene inserted in the T-DNA region can integrate that gene into the plant DNA, where it may be expressed. In practice, however, several refinements are made to this simple scheme. The size of the Ti plasmid makes it difficult to manipulate, but it has been discovered that if the T-DNA and the remainder of the Ti plasmid are on separate molecules within the same bacterial cell, integration will still take place. The recombinant T-DNA can be constructed in a standard *E. coli* plasmid, then transformed into

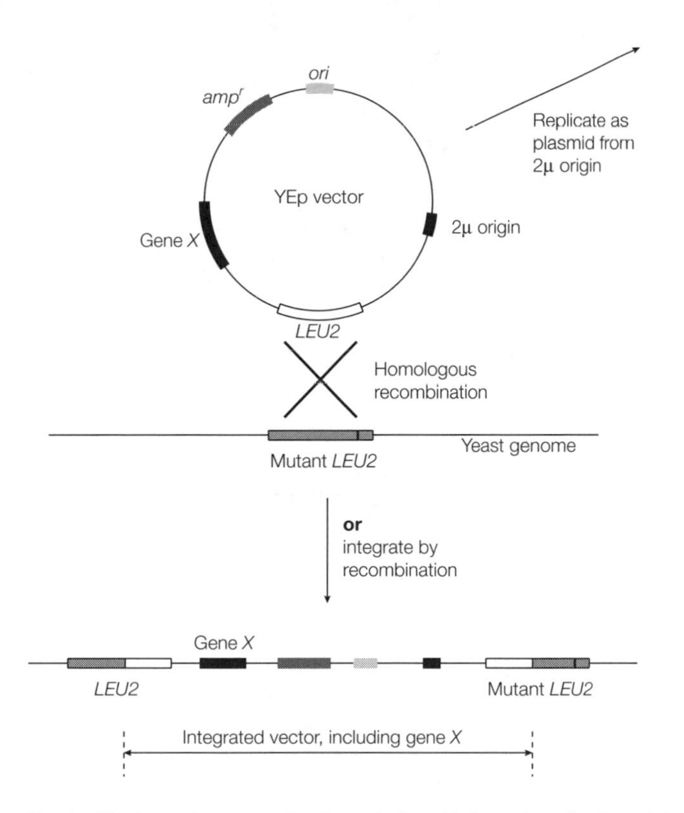

Fig. 1. Cloning using a yeast episomal plasmid, based on the 2μ origin.

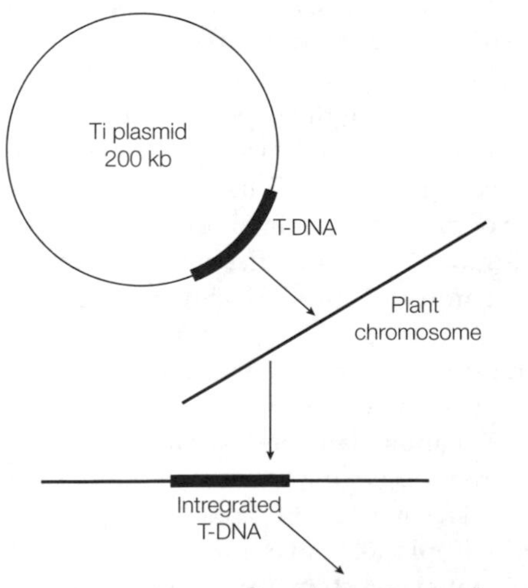

Fig. 2. Action of the A. tumefaciens *Ti plasmid.*

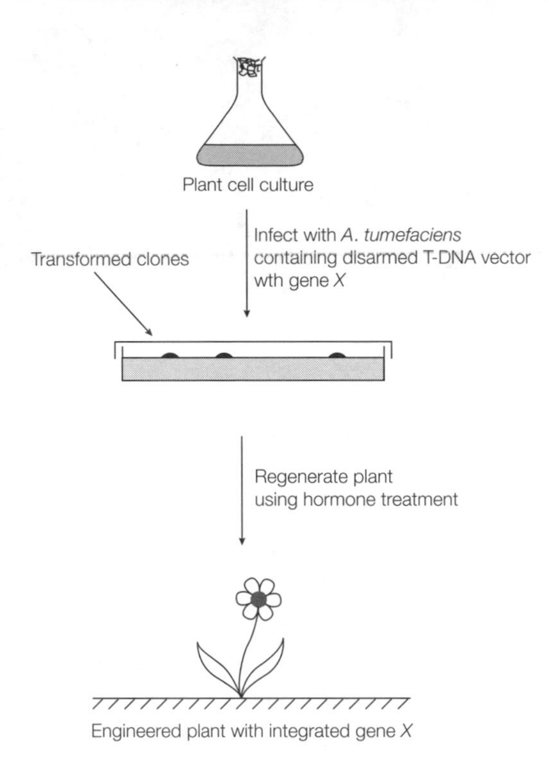

Fig. 3. Engineering new plants with A. tumefaciens.

the *A. tumefaciens* cell carrying a modified Ti plasmid without T-DNA. A further improvement is made by deleting the genes for crown gall formation from the T-DNA. So-called **disarmed** T-DNA shuttle vectors can integrate cloned genes benignly, and, if the host cells are growing in culture, complete recombinant plants can be reconstituted from the transformed cells (*Fig. 3*).

Baculovirus

Baculovirus infects insect cells. One of the major proteins encoded by the virus genome is polyhedrin, which accumulates in very large quantities in the nuclei of infected cells, since the gene has an extremely active promoter. The same promoter can be used to drive the overexpression of a foreign gene engineered into the baculovirus genome, and large quantities of protein can be produced in infected insect cells in culture. This method is being used increasingly for large-scale culture of proteins of animal origin, since the insect cells can produce many of the post-translational modifications of animal proteins which a bacterial expression system (see Topics H1 and Q4) cannot.

Mammalian viral vectors

Vectors for the transfer of genes into mammalian cells have also been based on viruses. One of the first to be used in this way was **SV40** (see Topic R4), which infects a number of mammalian species. The genome of SV40 is only 5.2 kb in size and it suffers from packaging constraints similar to phage λ (see Topic H2), so its utility for transferring large fragments is limited.

Retroviruses (see Topic R4) have a single-stranded RNA genome, which is copied into dsDNA after infection. The DNA is then stably integrated into the host genome by a transposition-like mechanism (see Topic F4). Retroviruses have some naturally strong promoters, and they have been considered as vectors for **gene therapy** (see Topic J6), since the foreign DNA will be incorporated into the host genome in a stable manner.

Direct gene transfer

Genes may be transiently or permanently introduced into cultured eukaryotic cells without the use of a vector in the strict sense. A eukaryotic gene on a bacterial plasmid, for example, may transiently express its product when transfected into a cell line, even if the plasmid does not replicate in that type of cell. Alternatively, DNA introduced by transfection or microinjection may become stably integrated into the cell's chromosomal DNA. This process normally requires significant sequence similarity between the incoming DNA and the genome in animal cells (analogous to *Fig. 1*) but, in plant cells, any supercoiled plasmid can randomly integrate into the genome in a process which is not understood in detail. Such stably transfected cells can be selected by the presence of a drug resistance gene in much the same way as bacterial transformants (see Topic G4), and can continue to express protein from foreign genes through many cell divisions.

I1 GENOMIC LIBRARIES

Key Notes

Representative gene libraries

Gene libraries made from genomic DNA are called genomic libraries and those made from complementary DNA are known as cDNA libraries. The latter lack nontranscribed genomic sequences (repetitive sequences, etc.). Good gene libraries are representative of the starting material and have not lost certain sequences due to cloning artifacts.

Size of library

A gene library must contain a certain number of recombinants for there to be a high probability of it containing any particular sequence. This value can be calculated if the genome size and the average size of the insert in the vector are known.

Genomic DNA

For making libraries, genomic DNA, usually prepared by protease digestion and phase extraction, is fragmented randomly by physical shearing or restriction enzyme digestion to give a size range appropriate for the chosen vector. Often combinations of restriction enzymes are used to partially digest the DNA.

Vectors

Plasmids, λ phage, cosmid, BAC or yeast artificial chromosome vectors can be used to construct genomic libraries, the choice depending on the genome size. The upper size limit of these vectors is about 10, 23, 45, 350 and 1000 kb respectively. The genomic DNA fragments are ligated to the prepared vector molecules using T4 DNA ligase.

Related topics

Design of plasmid vectors (H1)
Bacteriophage vectors (H2)
Cosmids, YACs and BACs (H3)
mRNA processing, hnRNPs and snRNPs (O3)

Representative gene libraries

A gene library is a collection of different DNA sequences from an organism each of which has been cloned into a vector for ease of purification, storage and analysis. There are essentially two types of gene library that can be made depending on the source of the DNA used. If the DNA is genomic DNA, the library is called a **genomic library**. If the DNA is a copy of an mRNA population, that is cDNA, then the library is called a **cDNA library**. When producing a gene library, an important consideration is how well it represents the starting material, that is does it contain all the original sequences (a **representative library**)? If certain sequences have not been cloned, for example repetitive sequences lacking restriction sites (see Topic D4), the library is not representative. Likewise, if the library does not contain a sufficient number of clones, then it is probable that some genes will be missing. cDNA libraries that are enriched for certain sequences (see Topic I2) will obviously lack others, but if correctly prepared and propagated they can be representative of the enriched mRNA starting material.

Size of library

It is possible to calculate the number (*N*) of recombinants (plaques or colonies) that must be in a gene library to give a particular probability of obtaining a given sequence. The formula is:

$$N = \frac{\ln(1-P)}{\ln(1-f)}$$

where *P* is the desired probability and *f* is the fraction of the genome in one insert. For example, for a probability of 0.99 with insert sizes of 20 kb, these values for the *E. coli* (4.6×10^6 bp) and human (3×10^9 bp) genomes are:

$$N_{E.\ coli} = \frac{\ln(1-0.99)}{\ln[1-(2 \times 10^4/4.6 \times 10^6)]} = 1.1 \times 10^3$$

$$N_{\text{human}} = \frac{\ln(1-0.99)}{\ln[1-(2 \times 10^4/3 \times 10^9)]} = 6.9 \times 10^5$$

These values explain why it is possible to make good genomic libraries from prokaryotes in plasmids where the insert size is 5–10 kb, as only a few thousand recombinants will be needed. For larger genomes, the larger the insert size, the fewer recombinants are needed, which is why cosmid and YAC vectors (see Topic I3) have been developed. However, the methods of cloning in λ and the efficiency of λ packaging still make it a good choice for constructing genomic libraries.

Genomic DNA

To make a representative genomic library, genomic DNA must be purified and then broken randomly into fragments that are the correct size for cloning into the chosen vector. Cell fractionation will reduce contamination from organelle DNA (mitochondria, chloroplasts). Hence, purification of genomic DNA from eukaryotes is usually carried out by first preparing cell nuclei and then removing proteins, lipids and other unwanted macromolecules by **protease digestion** and **phase extraction** (phenol–chloroform). Prokaryotic cells can be extracted directly. Genomic DNA prepared in this way is composed of long fragments of several hundred kilobases derived from chromosomes. There are two basic ways of fragmenting this DNA in an approximately random manner – **physical shearing** and **restriction enzyme digestion**. Physical shearing such as pipeting, mixing or **sonication** (see Topic C2) will break the DNA progressively into smaller fragments quite randomly. The choice of method and time of exposure depend on the size requirement of the chosen vector. The ends produced are likely to be blunt ends due to breakage across both DNA strands, but it would be advisable to repair the ends with Klenow polymerase (see Topic G1, *Table 1*) in case some ends are not blunt. This DNA polymerase will fill in any recessed 3′-ends on DNA molecules if dNTPs are provided.

The use of restriction enzymes (see Topic G3) to digest the genomic DNA is more prone to nonrandom results, due to the nonrandom distribution of restriction sites. To generate genomic DNA fragments of 15–25 kb or greater (a convenient size for λ and cosmid vectors), it is necessary to perform a **partial digest**, where, by using limiting amounts of restriction enzyme, the DNA is not digested at every recognition sequence that is present, thus producing molecules of lengths greater than in a complete digest. A common enzyme used is *Sau*3A (recognition sequence 5′-/GATC-3′, where / denotes the cleavage site) as it cleaves to produce a sticky end that is compatible with a vector that has been cut with *Bam*HI (5′-G/GATCC-3′). The choice of restriction enzyme must take into account the type of ends produced (sticky or blunt), whether they can be ligated directly to the cleaved vector and whether the enzyme is inhibited by DNA base modifications (such as **CpG methylation** in mammals; see Topic D3). The time of digestion and ratio of restriction enzyme to DNA are varied to produce fragments spanning the desired size range (i.e. the sizes that efficiently clone into the chosen vector). The correct sizes are then purified from an agarose gel (see Topic G3) or a sucrose gradient.

Vectors

In the case of organisms with small genome sizes, such as *E. coli*, a genomic library could be constructed in a plasmid vector (see Topic H1) as only 5000 clones (of average insert size 5 kb) would give a greater than 99% chance of cloning the entire genome (4.6×10^6 bp). Most libraries from organisms with larger genomes are constructed using **phage λ**, **cosmid**, **bacterial artificial chromosome (BAC)** or **yeast artificial chromosome** (**YAC**) vectors. These accept inserts of approximately 23, 45, 350 and 1000 kb respectively, and thus fewer recombinants are needed for complete genome coverage than if plasmids were used. The most commonly chosen genomic cloning vectors are **λ replacement vectors** (EMBL3, λDASH), and a typical cloning scheme is shown in Topic H2, *Fig.* 2. The λ vector DNA must be digested with restriction enzymes to produce the two λ end fragments, or λ arms, between which the genomic DNA will be ligated. With the original vectors, after digestion the arms would be purified from the central (stuffer) fragment on gradients, but newer λ vectors allow digestion with multiple enzymes which make the stuffer fragment unclonable. Following λ arm preparation, the genomic DNA prepared as described above is ligated to them using **T4 DNA ligase** (see Topics G1 and G4). Once ligated, the recombinant molecules are ready for **packaging** and **propagation** to create the library (see Topic H2).

I2 cDNA LIBRARIES

Key Notes

mRNA isolation, purification and fractionation

mRNA can be readily isolated from lysed eukaryotic cells by adding magnetic beads which have oligo(dT) covalently attached. The mRNA binds to the oligo(dT) via its poly(A) tail and can thus be isolated from the solution. The integrity of an mRNA preparation can be checked by translation in a wheat germ extract or reticulocyte lysate and then visualizing the translation products by polyacrylamide gel electrophoresis. Integrity can also be studied using gel electrophoresis, which allows the mRNA to be size fractionated by recovering chosen regions of the gel lane. Specific sequences can be removed from the mRNA by hybridization.

Synthesis of cDNA

In first strand synthesis, reverse transcriptase is used to make a cDNA copy of the mRNA by extending a primer, usually oligo(dT), by the addition of deoxyribonucleotides to the 3′-end. Synthesis can be detected by trace labeling. 3′-Tailing of the first strand cDNA using terminal transferase makes full-length second strand synthesis easier. Reverse transcriptase or Klenow enzyme can extend a primer [e.g. oligo(dG)] annealed to a homopolymeric tail [e.g. poly(dC)] to synthesize second strand cDNA.

Treatment of cDNA ends

To avoid blunt end ligation of cDNA to vector, linkers are usually added to the cDNA after the ends have been repaired (blunted) using a single strand-specific nuclease followed by Klenow enzyme. The cDNA may also be methylated to keep it from being digested when the added linkers are cleaved by a restriction enzyme. Adaptor molecules can be used as an alternative to linkers.

Ligation to vector

The vector is usually dephosphorylated using alkaline phosphatase to prevent self-ligation, and so promote the formation of recombinant molecules. Plasmid or phage vectors can be used to make cDNA libraries, but the phage λgt11 is preferred for the construction of expression libraries.

Related topics

DNA cloning: an overview (G1)
Design of plasmid vectors (H1)
Bacteriophage vectors (H2)
Genomic libraries (I1)
Screening procedures (I3)
mRNA processing, hnRNPs and snRNPs (O3)

mRNA isolation, purification and fractionation

Generally, cDNA libraries are not made using prokaryotic mRNA, since it is very unstable; genomic libraries are easier to make and contain all the genome sequences. Making cDNA libraries from eukaryotic mRNA is very useful because the cDNAs have no intron sequences and can thus be used to express the encoded protein in *E. coli*. Since they are derived from the mRNA, cDNAs represent the transcribed parts of the genome (i.e. the genes rather than the nontranscribed DNA). Furthermore, each cell type, or tissue, expresses a characteristic set of genes (which may alter after stimulation or during development), and mRNA preparations from particular tissues usually contain some specific sequences at higher abundance, for example globin mRNA in erythrocytes. Hence the choice of tissue as starting material can greatly facilitate cDNA cloning. In eukaryotes, most mRNAs are polyadenylated (see Topic O3) and this 3′-tail of about 200 adenine residues provides a useful method for isolating eukaryotic mRNA. **Oligo(dT)** can be bound to the poly(A) tail and used to recover the mRNA. Traditionally, this was done by passing a preparation of total RNA (made by extracting lysed cells with phenol–chloroform) down a column of **oligo(dT)-cellulose**. However, to keep damage by nucleases to a minimum, the more rapid procedure of adding oligo(dT) linked to **magnetic beads** directly to a cell lysate and 'pulling out' the mRNA using a strong magnet, is now the method of choice. In some circumstances, lysing cells and then preparing mRNA–ribosome complexes (polysomes, see Topic Q1) on **sucrose gradients** (see Topic A2) may provide an alternative route for isolating mRNA.

Before using an mRNA preparation for cDNA cloning, it is advisable to check that it is not degraded. This can be done by either **translating** the mRNA or by analyzing it by **gel electrophoresis**. **Cell-free translation systems,** such as **wheat germ extract** or **rabbit reticulocyte lysate**, are commonly used to check the integrity of an mRNA preparation. It is also possible to microinject mRNA preparations into cells to check whether they are translated. The use of agarose or polyacrylamide gels to analyze mRNA is also common. The mRNA preparation usually produces a smear of molecules from about 0.5 kb up to about 10 kb or more. Often the two largest rRNA species contaminate the mRNA preparations because of their abundance and appear as bands within the smear of mRNA.

Sometimes it can be useful to fractionate or enrich the mRNA prior to cDNA cloning, especially if one is trying to clone a particular gene rather than to make a complete cDNA library. Fractionation is usually performed on the basis of size, and mRNAs of different sizes are recovered from agarose gels. Enrichment is usually carried out by **hybridization**. For example, if one wanted to make a cDNA library of all the mRNA sequences that are induced by treating a cell with a hormone, cytokine or drug, one would prepare mRNA from both induced and noninduced cells. First strand cDNA (see below) made from the noninduced cell mRNA could be used to hybridize to the induced cell mRNA. Sequences common to both mRNA preparations would hybridize to form duplexes but the newly induced mRNAs would not. The nonduplex mRNA could then be isolated (e.g. using magnetic beads) for cDNA library construction. Such a library would be called a **subtracted cDNA library**.

Synthesis of cDNA

The scheme for making cDNA is shown in *Fig. 1*. During first strand synthesis, the enzyme **reverse transcriptase** (see Topic G1, *Table 1*) requires a **primer** to extend when making a copy of the mRNA template. This primer is usually oligo(dT) for synthesis of complete cDNA, but is sometimes a random mixture of all possible hexanucleotides if one wants to make **random-primed cDNA**. All four dNTPs must be added. If a small amount of one radioactive dNTP is added, the efficiency of the cDNA synthesis step can be determined by gel analysis. This is called **trace labeling**. The enzyme copies the template, by

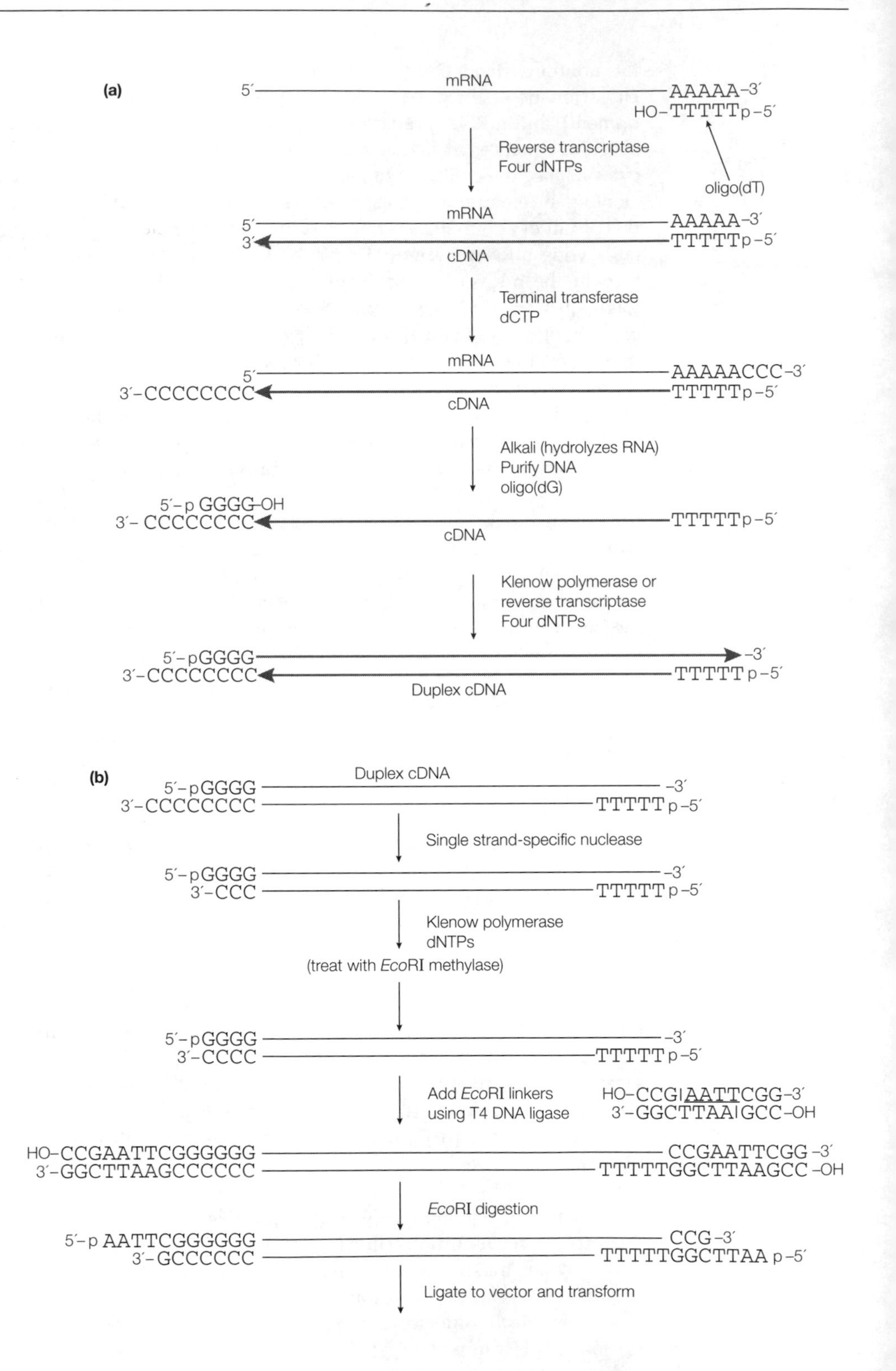

Fig. 1. cDNA cloning. (a) First and second strand synthesis; (b) end preparation and linker addition to duplex cDNA.

adding the complementary nucleotides to the 3′-end of the extending primer. Unfortunately, the enzyme has a tendency to dissociate from the template, especially at regions of extensive secondary structure, and thus it is sometimes hard to make complete (full-length) cDNA molecules in one step.

Second strand synthesis also requires a primer. Although there are a number of variations, perhaps the best way of making full-length cDNA is to 'tail' the 3′-end of the first strand and then use a complementary primer to make the second strand. **Terminal transferase** (see Topic G1, *Table 1*) adds nucleotides to the 3′-end of single- or double-stranded nucleic acids without the requirement for a template. If only provided with one dNTP, for example, dCTP (*Fig. 1*), it will add a **homopolymeric tail** of C residues to the 3′-ends of the mRNA–cDNA duplex. After destroying the mRNA strand with alkali (see Topic C2), oligo(dG) can be used to prime second strand synthesis, using either reverse transcriptase or the **Klenow fragment** of *E. coli* DNA polymerase I (see Topic G1, *Table 1*). The product of this reaction is duplex cDNA, but the ends of the molecules may not be suitable for cloning. The 3′-end of the first strand may protrude beyond the 5′-end of the second strand depending on the relative lengths of the poly(dC) tail and the oligo(dG) primer. It is therefore necessary to prepare the cDNA ends for cloning.

Treatment of cDNA ends

Blunt end ligation of large fragments is not efficient, and it is worthwhile manipulating the ends of the duplex cDNA to avoid blunt end cloning into the vector. Usually, special nucleic acid **linkers** are added to create sticky ends for cloning. Because they can be added in great excess, the blunt end reaction of joining linkers and cDNA proceeds reasonably well. The first step is to prepare the cDNA for the addition of linkers, which will involve using a single strand-specific nuclease (**S1** or **mung bean nuclease**) to remove protruding 3′-ends. This is followed by treatment with the Klenow fragment of DNA polymerase I and dNTPs to fill in any missing 3′ nucleotides. If the linkers contain a restriction enzyme site that is likely to be present in the cDNA (such as *Eco*RI, *Fig. 1b*), then the cDNA should be methylated using ***Eco*RI methylase** (see Topic G3) before the linkers are added. This will ensure that the restriction enzyme cannot cleave the cDNA internally. After this, the linkers can be ligated to the blunt-ended, duplex cDNA using **T4 DNA ligase** (see Topic G1, *Table 1*). This step will add one linker to each end of 5′-phosphorylated cDNA if the linkers are not phosphorylated. Multiple linkers would be added otherwise. Finally, restriction enzyme digestion with *Eco*RI generates sticky ends ready for ligation to the vector. Many alternative ways exist for preparing the cDNA ends for cloning. These include tailing with terminal transferase for cloning into a complementary tailed vector, or the use of **adaptor molecules** that have preformed 'sticky' ends to avoid the requirement for methylation of the cDNA.

Ligation to vector

Any vector with an *Eco*RI site would be suitable for cloning the cDNA in *Fig. 1*. As cDNAs are relatively short (0.5–10 kb), plasmid vectors are often used; however, for greater numbers of clones and especially for **expression cDNA libraries**, λ phage vectors (see Topic H2) are preferred. It is usual to **dephosphorylate** the vector with the enzyme **alkaline phosphatase** as this will prevent the vector fragments (ends) rejoining during ligation, helping to ensure that only recombinant molecules are produced by vector and cDNA joining. The vector **λgt11** has an *Eco*RI site placed near the C terminus of its *lacZ* gene, enabling expression of the cDNA as part of a large **β-galactosidase fusion**

protein. This aids screening of the library (see Topic I3). Ligation of vector to cDNA is carried out using T4 DNA ligase, and the recombinant molecules are either packaged (see Topic H2) or transformed (see Topic G2) to create the cDNA library.

I3 SCREENING PROCEDURES

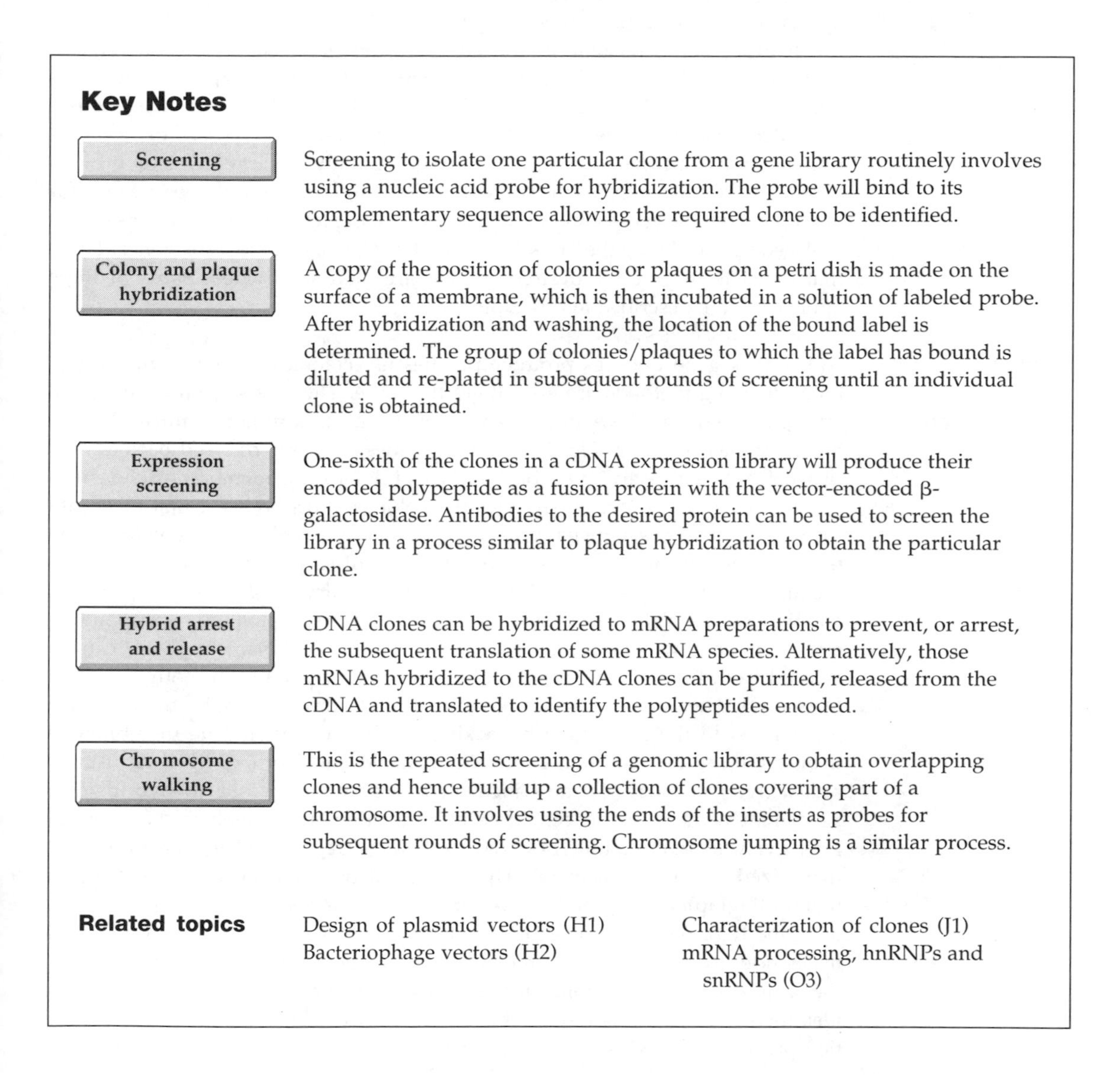

Key Notes

Screening

Screening to isolate one particular clone from a gene library routinely involves using a nucleic acid probe for hybridization. The probe will bind to its complementary sequence allowing the required clone to be identified.

Colony and plaque hybridization

A copy of the position of colonies or plaques on a petri dish is made on the surface of a membrane, which is then incubated in a solution of labeled probe. After hybridization and washing, the location of the bound label is determined. The group of colonies/plaques to which the label has bound is diluted and re-plated in subsequent rounds of screening until an individual clone is obtained.

Expression screening

One-sixth of the clones in a cDNA expression library will produce their encoded polypeptide as a fusion protein with the vector-encoded β-galactosidase. Antibodies to the desired protein can be used to screen the library in a process similar to plaque hybridization to obtain the particular clone.

Hybrid arrest and release

cDNA clones can be hybridized to mRNA preparations to prevent, or arrest, the subsequent translation of some mRNA species. Alternatively, those mRNAs hybridized to the cDNA clones can be purified, released from the cDNA and translated to identify the polypeptides encoded.

Chromosome walking

This is the repeated screening of a genomic library to obtain overlapping clones and hence build up a collection of clones covering part of a chromosome. It involves using the ends of the inserts as probes for subsequent rounds of screening. Chromosome jumping is a similar process.

Related topics

Design of plasmid vectors (H1)
Bacteriophage vectors (H2)
Characterization of clones (J1)
mRNA processing, hnRNPs and snRNPs (O3)

Screening

The process of identifying one particular clone containing the gene of interest from among the very large number of others in the gene library is called **screening**. Some knowledge of the gene, or its product, is required such as a cDNA fragment or a related sequence to use as a **nucleic acid probe**. If sufficient of the protein product is available to permit determination of some amino acid sequence (see Topic B2), this information could be used to derive a mixture of possible DNA sequences that would encode that amino acid sequence. This DNA sequence information could be used make a nucleic acid probe to screen the library by **hybridization**. One of the most common ways to make DNA probes for library screening uses the polymerase chain reaction (PCR, see Topic J3) to make what are called **PCR probes**. Short nucleic acid probes (**oligonucleotides**) are readily produced by automated chemical synthesis and these can be used directly for hybridization or to make longer PCR probes for hybridization. If a pair of primers can be designed, PCR can also be used as a technique to screen a library since a PCR product will only be detected if the sequence (i.e. clone) is present, and thus pools of clones can be successively subdivided till a single positive clone is isolated. If antibodies had been raised to the protein these could be used to detect the presence of a clone that expressed the protein, usually as part of a fusion protein. cDNA libraries that express the protein in a functional form could be screened for biological activity. A cDNA library that does not express encoded proteins can be screened by translating the mRNAs that bind (hybridize) to pools of clones. Those that translate to give the desired protein are then subdivided. Some of these approaches are described in this topic.

Colony and plaque hybridization

Although λ gene libraries produce plaques not colonies, after the initial step, both screening methods are essentially the same. The first step involves transferring some of the DNA in the plaque or colony to a **nylon** or **nitrocellulose membrane** as shown in *Fig. 1*. Because plaques are areas of lysed bacteria, the phage DNA is directly available and will bind to the membrane when it is placed on top of the petri dish. Bacterial colonies must be lysed first to release their DNA, and this is usually done by growing a replica of the colonies on the dish directly on the membrane surface (**replica plating**). The bacteria on the membrane are lysed by soaking in sodium dodecyl sulfate and a protease, and the original plate is kept to allow growth and isolation of the corresponding clone (i.e. colony with recombinant plasmid). Recombinant λ phage can be isolated from the remaining material on the dish of plaques. In both cases, the DNA on the membrane is denatured with alkali to produce single strands which are bonded to the membrane by baking or UV irradiation. The membrane is then immersed in a solution containing a nucleic acid probe, which is usually radioactive (see Topic J1) and incubated to allow the probe to hybridize to its complementary sequence. After hybridization, the membrane is washed extensively to remove unhybridized probe, and regions where the probe has hybridized are then visualized. This is carried out by exposure to X-ray film (**autoradiography**) if the probe was radioactively labeled, or by using a solution of antibody or enzyme and substrate if the probe was labeled with a modified nucleotide. By comparing the membrane with the original dish and lining up the regions of hybridization, the original group of colonies or plaques can be identified. This group is re-plated at a much lower density, and the hybridization process repeated until a single individual clone is isolated.

Expression screening

By cloning cDNAs into the *Eco*RI site of λgt11, there is a one in six chance that the cDNA will be in both the correct orientation and reading frame (see Topic P1) to be translated into its gene product. The *Eco*RI site is near the C terminus of the *lacZ* gene, and the coding region of the insert must be linked to that of *lacZ*, in the correct orientation and frame, for a fusion protein containing the cDNA gene product to be made. The β-gal fusion protein may contain regions of polypeptide (**epitopes**) that will be recognized by **antibodies** raised to the native protein. These antibodies can therefore be used to screen the **expression library**. The procedure has similarities to the plaque hybridization protocol (above) in that a '**plaque lift**' is taken by placing a membrane on the dish of plaques, though in this

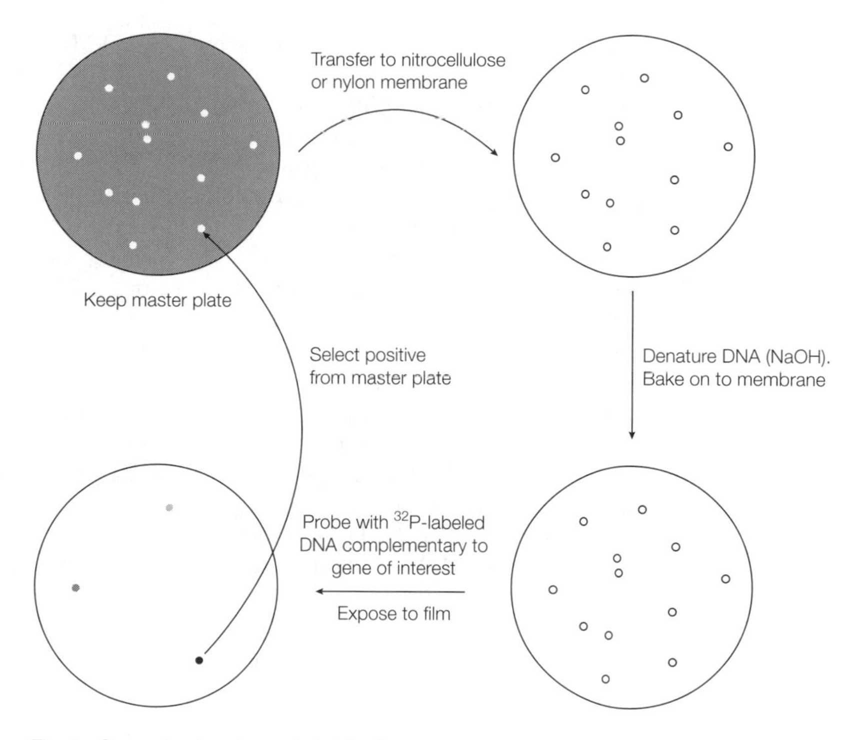

Fig. 1. Screening by plaque hybridization.

case it is the protein encoded by the cDNA rather than the DNA itself that is detected on the membrane. The membrane is treated to covalently attach the protein, and immersed in a solution of the antibody. When the antibody has bound to its epitope, it is detected by other antibodies and/or chemicals that recognize it. In this way, the location of the expressing plaque can be narrowed down. Repeat cycles of screening are again required to isolate pure plaques.

Hybrid arrest and release

Individual cDNA clones or pools of clones can be used to hybridize to mRNA preparations. After hybridization, the mRNA population can be translated directly, and the inhibition of translation of some products detected (**hybrid arrested translation**). Subdivision of the pools of cDNA into smaller numbers and repeating the experiment should ultimately allow the identification of a single cDNA that arrests the translation of one protein. Alternatively, after hybridization, the hybrids can be purified (e.g. if the cDNAs are attached to magnetic beads or are precipitable with an antibody to a modified nucleotide used to prime cDNA synthesis) and the hybridized mRNAs released from them (by heat and/or denaturant) and translated. This **hybrid release translation** identifies the protein encoded by the cDNA clone.

Chromosome walking

It is often necessary to find adjacent genomic clones from a library, perhaps because only part of a gene has been cloned, or the 5′- or 3′-ends, which contain control sequences, are missing. Sometimes, it is known from genetic mapping experiments that a particular gene is near a previously cloned gene, and it is possible to clone the desired gene (**positional cloning**) by repeatedly isolating adjacent genomic clones from the library. This process is called **chromosome walking** as one obtains overlapping genomic clones that represent progressively longer parts of a particular chromosome (*Fig. 2*). To isolate an overlapping genomic clone, one must prepare a probe from the end of the insert. This may involve **restriction mapping** (see Topic J1) the clone so that a particular fragment can be recovered by gel electrophoresis, but some vectors allow probes to be made by ***in vitro* transcription** of the vector–insert junctions. The probes are used to re-screen the library by colony or plaque hybridization and then the newly isolated clones are analyzed to allow them to be positioned relative to the starting clone. With luck, out of several new clones, some will overlap the starting clone by a small amount (e.g. 5 kb out of 25 kb), and then the whole process can be repeated using a probe from the distal end of the second clone (the furthest from the original insert end), and so on.

If the chromosome walk breaks down, perhaps because the library does not contain a suitable sequence, it is possible to use a technique called **chromosome jumping** to overcome this problem. Chromosome jumping can also be used when the 'walk' distance is quite large, for example when positional cloning from a relatively distant start point. Opposite ends of a long DNA fragment come together when the DNA is circularized or when cloned into a cosmid vector (see Topic H3). Without mapping the entire DNA, the end fragments can be (subcloned and) used as successive probes to screen a library and this can allow the isolation of clones that are around 50 kb from each other, so producing a more rapid 'walk' than the method shown in *Fig. 2*.

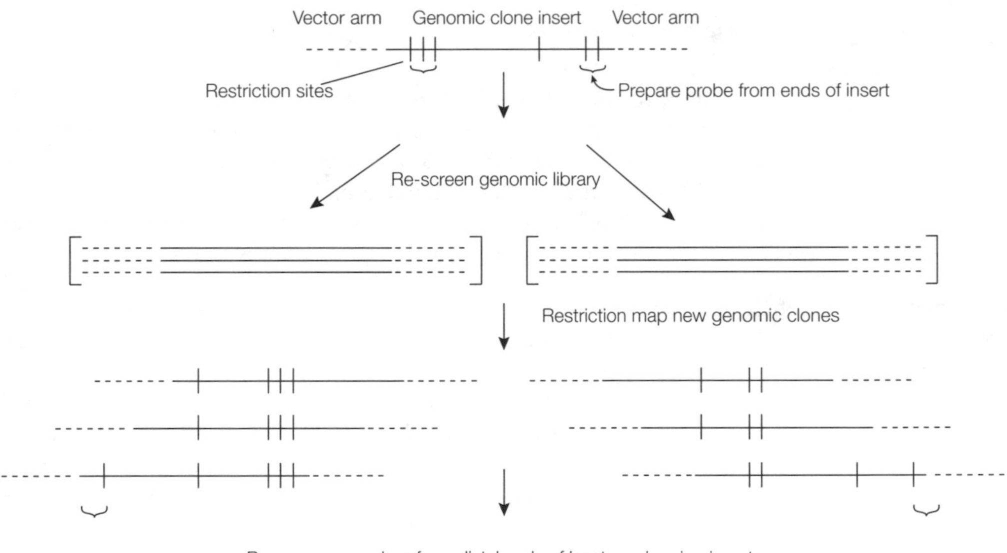

Fig. 2. Chromosome walking.

J1 Characterization of clones

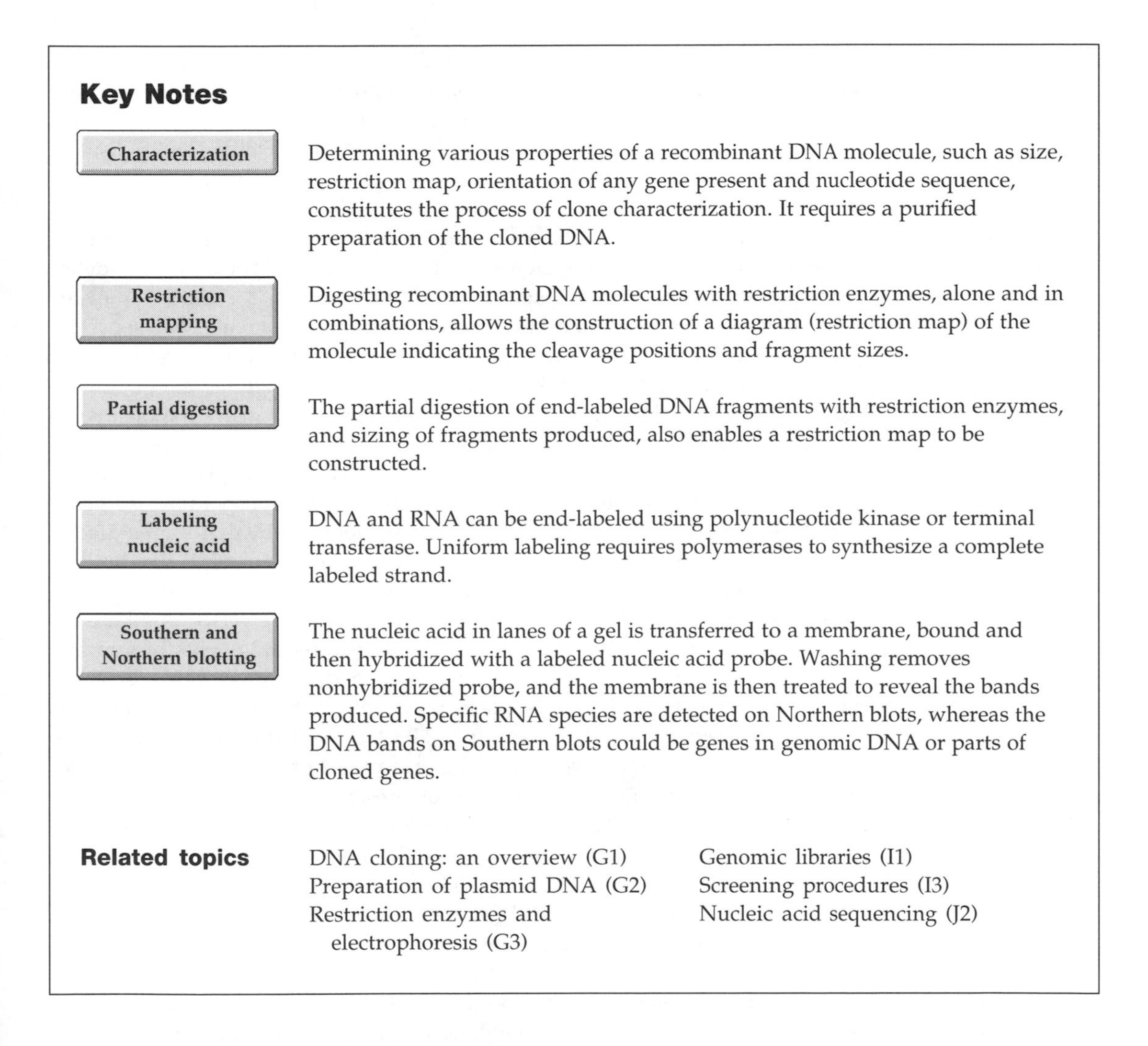

Key Notes

Characterization — Determining various properties of a recombinant DNA molecule, such as size, restriction map, orientation of any gene present and nucleotide sequence, constitutes the process of clone characterization. It requires a purified preparation of the cloned DNA.

Restriction mapping — Digesting recombinant DNA molecules with restriction enzymes, alone and in combinations, allows the construction of a diagram (restriction map) of the molecule indicating the cleavage positions and fragment sizes.

Partial digestion — The partial digestion of end-labeled DNA fragments with restriction enzymes, and sizing of fragments produced, also enables a restriction map to be constructed.

Labeling nucleic acid — DNA and RNA can be end-labeled using polynucleotide kinase or terminal transferase. Uniform labeling requires polymerases to synthesize a complete labeled strand.

Southern and Northern blotting — The nucleic acid in lanes of a gel is transferred to a membrane, bound and then hybridized with a labeled nucleic acid probe. Washing removes nonhybridized probe, and the membrane is then treated to reveal the bands produced. Specific RNA species are detected on Northern blots, whereas the DNA bands on Southern blots could be genes in genomic DNA or parts of cloned genes.

Related topics

DNA cloning: an overview (G1)
Preparation of plasmid DNA (G2)
Restriction enzymes and electrophoresis (G3)
Genomic libraries (I1)
Screening procedures (I3)
Nucleic acid sequencing (J2)

Characterization

The process of characterizing a genomic or cDNA clone begins by obtaining a pure preparation of the DNA. One may subsequently determine some, or all, of a range of properties including its size, the insert size, characteristics of the insert DNA such as its pattern of cleavage by restriction enzymes (**restriction map**), whether it contains a gene (transcribed sequence), the position and polarity of any gene and some or all of the sequence of the insert DNA. Before DNA sequencing (see Topic J2) became so rapid, it was usual to perform these steps of characterization roughly in the order listed, but nowadays sequencing all or part of a clone is often the first step. However, while the sequence will provide the size, restriction map and the orientation of the predicted gene(s), or **open reading frame(s) [ORF(s)]** (see Topic P1), experiments must be performed to verify that the predicted gene is actually transcribed into RNA in some cells of the organism and that, for protein-coding genes, a protein of the size predicted is made *in vivo*. The preparation of plasmid DNA from bacterial colonies has been described in Topic G2. DNA can readily be made from clones present in bacteriophage vectors by first isolating phage particles. This can be done by infecting a bacterial culture with the **plaque-purified phage** under conditions where cell lysis occurs. After removing lysed cell debris, the phage particles can be recovered from the liquid cell lysate by either direct centrifugation or by precipitation by polyethylene glycol in high salt, followed by centrifugation. The pellet of phage is then extracted with phenol–chloroform and the DNA precipitated with ethanol. Precipitated DNA is usually dissolved in TE (10 mM Tris.HCl, 1 mM EDTA, pH 8.0) (see Topic G2).

Restriction mapping

The sizes of linear DNA molecules can be determined by agarose gel electrophoresis using marker fragments of known sizes (see Topic G3). To determine the size of the cloned genomic or cDNA insert in a plasmid or phage clone, it is best to digest the DNA with a restriction enzyme that separates the insert and vector sequences. For a cDNA constructed using *Eco*RI linkers, digestion with *Eco*RI would achieve this. Running the digested sample on an agarose gel with size markers would give the size of the insert fragment(s) as well as the size of the vector (already known). This information is part of what is needed to draw a map of the recombinant DNA molecule, but the **orientation of the insert** relative to the vector (see Topic H2), and perhaps the order of multiple fragments, is not known. This can be obtained by performing digests with different restriction enzymes, particularly in combinations. *Figure 1a* illustrates the process for a λEMBL3 genomic clone. *Sal*I cuts adjacent to the *Eco*RI or *Bam*HI cloning sites and will release the insert from the vector. As there are no internal *Sal*I sites in the insert, there are only three fragments of approximately 9, 15 and 19 kb. *Hin*dIII cuts only once in the short arm of the vector, about 4 kb from the end. In this example, *Hin*dIII cuts twice in the insert as well, giving fragments of 4, 7, 11 and 21 kb. Although the 4 kb is part of the short arm and the 21 kb must be the long arm plus 2 kb (19 + 2 = 21), the order of the other two fragments is unknown at present. The *Sal*I + *Hin*dIII **double digest** gives fragments of 2, 4, 5, 6, 7 and 19 kb, of which the 19 kb and the 4 and 5 kb together are the two vector arms. Since *Sal*I cuts at the cloning sites, it cuts the 21 kb into 19 + 2 kb and the 11 kb into 6 + 5 kb, of which the 6 kb is part of the insert. The 7 kb *Hin*dIII fragment has not been cut by *Sal*I, and this confirms that it is in the central region of the insert, giving the map as shown.

Partial digestion

When a restriction enzyme cuts quite frequently and double digestions produce very complicated patterns, the technique of **partial restriction enzyme digestion** can provide the information needed for a complete map. This is best performed after adding a modified, or radioactive, nucleotide to one end of the DNA fragment producing an **end-labeled molecule** such as that in *Fig. 1b*. Total digestion of the unlabeled 10 kb fragment with *Eco*RI gives 1, 2, 3 and 4 kb pieces. Partial digestion, which can generate all possible contiguous fragments, shows additional bands of 6, 7 and 10 kb when the total DNA is viewed by staining with ethidium bromide. Note that fragments of 3, 4 and 6 kb can be generated in more than one way. However, if the partial digestion lane is autoradiographed to show up only those end-labeled molecules, the pattern is 3, 4, 6 and 10 kb. This allows the map shown to be generated.

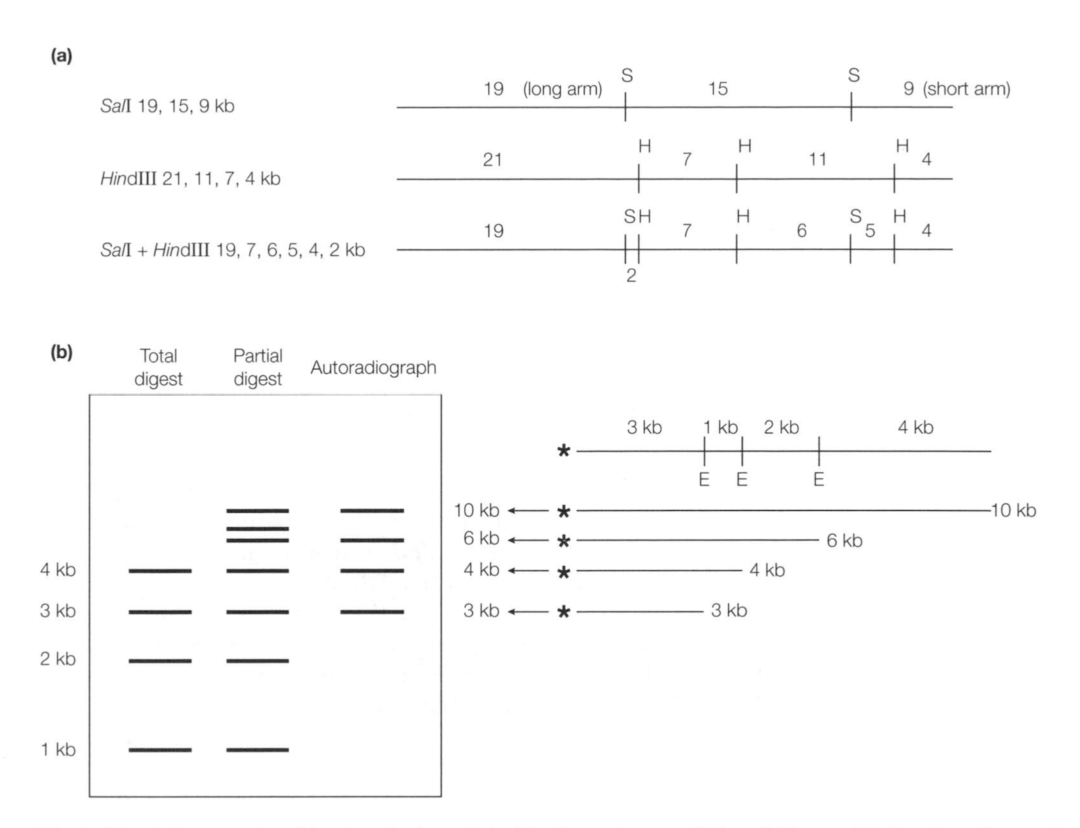

*Fig. 1. Restriction mapping (a) using single and double digests to completion; (b) by partial digestion of an end-labeled molecule. * is the labeled end.*

Labeling nucleic acid

DNA and RNA molecules can either be labeled at their ends (**end labeling**) or throughout their length (**uniform labeling**). It is also possible to begin labeling uniformly from one end, but for a limited time, thus creating a molecule that is not uniformly labeled throughout its entire length and which may behave more like an end-labeled molecule. Being double-stranded, DNA can also be labeled **strand-specifically**. Labeling is usually carried out using radioactive isotopes, but nonradioactive methods are also available, for example using biotinylated dUTP which is detected using biotin-specific antibodies.

5′-End labeling (*Fig. 2*) is performed using **polynucleotide kinase** to add a radioactive phosphate to nucleic acids with a free 5′-hydroxyl group, which can be created by **dephosphorylation** using **alkaline phosphatase**. The external γ-phosphate of ATP is the source of the label (see Topic G1, *Table 1*). **3′-End labeling** can be performed using the enzyme **terminal transferase** to add one or more nucleotides to the 3′-end of nucleic acids (see Topic I2).

For duplex DNA molecules with recessed 3′-ends such as those generated by some restriction enzymes (*Eco*RI, *Bam*HI, etc., see Topic G1), various **DNA polymerases** can be used to 'fill in' the 3′-end using labeled nucleotides (e.g. [α-^{32}P]dCTP). Because only 1–4 nt are added to each 3′-end, the molecules are essentially end-labeled, but at both ends on opposite strands. To use such molecules for restriction mapping they must be cut with another restriction enzyme and each end-labeled fragment isolated for independent restriction mapping. DNA polymerases can also be used to make uniformly labeled, **high specific activity** DNA for use as probes, etc. In **nick translation**, the duplex DNA is

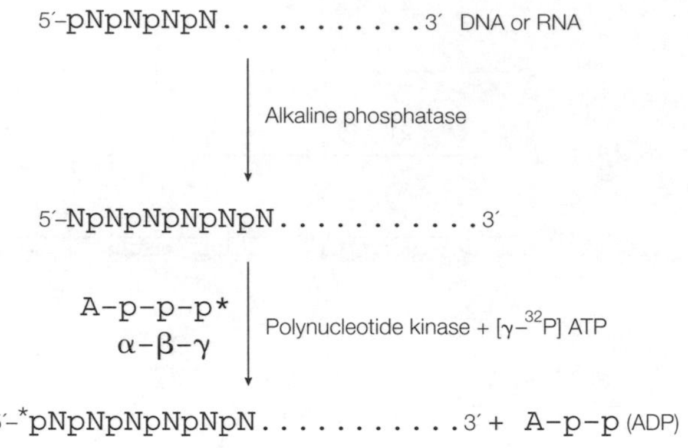

5′-pNpNpNpN . . . . 3′ DNA or RNA

↓ Alkaline phosphatase

5′-NpNpNpNpNpN . . . . 3′

A-p-p-p*
α-β-γ

↓ Polynucleotide kinase + [γ-^{32}P] ATP

5′-*pNpNpNpNpNpN . . . . 3′ + A-p-p (ADP)

Fig. 2. 5′-End labeling of a nucleic acid molecule.

treated with a tiny amount of **DNase I** which introduces random nicks along both strands. DNA polymerase I can find these nicks and remove dNTPs using its **5′→3′ exonuclease** activity (see Topic E2). As it removes a 5′-residue at the site of the nick, it adds a nucleotide to the 3′-end, incorporating radioactive nucleotides using its polymerizing activity. Hence the position of the nick moves along the molecule (i.e. it is vectorially translated). DNA polymerases can also be used to make probes by first denaturing the duplex DNA template in the presence of all possible hexanucleotides. These anneal at positions of complementarity along the single strands and act as **primers** for the polymerases to synthesize a complementary, radioactive strand. See also PCR (Topic J3).

Strand-specific DNA probes can be made by using single-stranded DNA obtained after cloning in an M13 phage vector (see Topic H2) or duplex DNA that has had part of one strand removed by a **3′→5′ exonuclease**. DNA polymerases can re-synthesize the missing strand incorporating radioactive nucleotides in the process as in second strand cDNA synthesis (see Topic I2, *Fig. 1a*). Strand-specific RNA probes are generated by ***in vitro* transcription** using RNA polymerases such as **SP6, T7 or T3 phage polymerases** (see Topic H1). If the desired sequences are cloned into an *in vitro* transcription vector (e.g. pGEM, pBluescript) which has one of these polymerase promoter sites at each end of the cloning site, then a **sense** RNA transcript (the same strand as the natural RNA transcript) can be made using one polymerase, and an **antisense** one (complementary to the natural transcript) using the other. Radioactive NTPs are used for labeling. Strand-specific probes are useful for Northern blots (see below) as the antisense strand will hybridize to cellular RNA while the sense probe will act as a control.

Southern and Northern blotting

To detect which of the many DNA (or RNA) molecules on an agarose gel hybridize to a particular probe, **Southern blots** (for DNA) or **Northern blots** (for RNA) are carried out (*Fig. 3*). The former is named after its inventor and the latter was extrapolated from the former. The nucleic acid molecules are separated by **agarose gel** electrophoresis and then transferred to a nylon or nitrocellulose membrane. This is often done by capillary action, but can be achieved by electrotransfer, vacuum transfer or centrifugation. In Southern blotting, before transfer, **DNA** is usually **denatured with alkali** (see Topic C2), so it is single stranded and ready for hybridization. For RNA in Northern blotting, this is not necessary and would in any case hydrolyze the molecules (see

Topic C2). Once transferred, the procedure for Southern and Northern blotting is identical. The nucleic acid must be bonded to the membrane, hybridized to labeled probe, washed extensively and then the hybridized probe must be detected (usually by autoradiography or antibody methods). The hybridization and washing conditions are critical. If the probe and target are 100% identical in sequence, then a high stringency hybridization can be carried out. The **stringency** is determined by the hybridization temperature and the salt concentration in the hybridization buffer (high temperature and low salt is more stringent as only perfectly matched hybrids will be stable). For probes that do not match the target completely, the stringency must be reduced to a level that allows imperfect hybrids to form. If the stringency of the hybridization (or washing) is too low, then the probe may bind to too many sequences to be useful. Formamide can be included in the hybridization buffer to reduce the actual hybridization temperature by about 25°C, from the usual 68°C to the more convenient 43°C. The washing step should be carried out at 12°C below the theoretical melting temperature (T_m) of the probe and target sequences, using the formula: $T_m = 69.3° + 0.41\ [\%(G + C)] - 650/l$, where l is the length of the probe molecule.

Northern blots give information about the size of the mRNA and any precursors, and can be useful to determine whether a cDNA clone used as a probe is full-length or whether it is one of a family of related (perhaps alternatively processed) transcripts. Northern blots can help to identify whether a genomic clone has regions that are transcribed and, if the RNA on the blot is made from different tissues, where these transcripts are made.

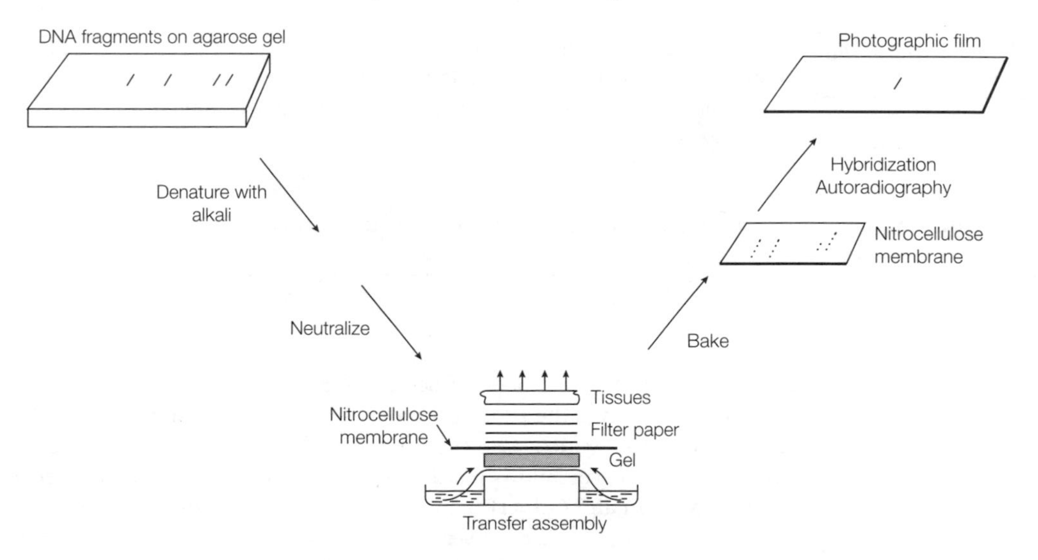

Fig. 3. Southern blotting.

Southern blots of cloned genomic DNA fragments can be probed with cDNA molecules to find which parts of the genomic clone correspond to the cDNA fragment. If the Southern blot contains genomic DNA fragments from the whole genome, the probe will give information about the size of the fragment the gene is on in the genome and how many copies of the gene are present in the genome. Blots with DNA or RNA samples from different organisms (**zoo blots**) can show how conserved a gene is between species.

J2 Nucleic acid sequencing

Key Notes

DNA sequencing

The two main methods of DNA sequencing are the Maxam and Gilbert chemical method in which end-labeled DNA is subjected to base-specific cleavage reactions prior to gel separation, and Sanger's enzymic method. The latter uses dideoxynucleotides as chain terminators to produce a ladder of molecules generated by polymerase extension of a primer.

RNA sequencing

A set of four RNases that cleave 3′ to specific nucleotides are used to produce a ladder of fragments from end-labeled RNA. Polyacrylamide gel electrophoresis analysis allows the sequence to be read.

Sequence databases

Newly determined DNA, RNA and protein sequences are entered into databases (EMBL and GenBank). These collections of all known sequences are available for analysis by computer.

Analysis of sequences

Special computer software is used to search nucleic acid and protein sequences for the presence of patterns (e.g. restriction enzyme sites) or similarities (e.g. to new sequences).

Genome sequencing projects

The entire genome sequences of several organisms have been determined (viruses, bacteria, yeast, worm and fly) and those of other organisms (plant, fish, mouse and human) are in progress. Often a genetic map is first produced to aid the project.

Related topics

DNA cloning: an overview (G1) Bacteriophage vectors (H2)

DNA sequencing

Each of the two main methods of sequencing long DNA molecules (**chemical** and **enzymic**) involves the production of a set of different sized molecules with one common end which are then separated by polyacrylamide gel electrophoresis (PAGE) to allow reading of the sequence. The earlier chemical method of **Maxam and Gilbert** requires that the DNA fragment to be sequenced is labeled at one end, usually by adding either a radioactive phosphate to the 5′- or 3′-end or a nucleotide to the 3′-end. The method works for both single- and double-stranded DNA and involves **base-specific cleavages** which occur in two steps. The base is first modified using specific chemicals (see below) after which **piperidine** can cleave the sugar–phosphate backbone of the DNA at that site. Limiting incubation times or concentrations of components in the base-modifying reactions ensures that a ladder of progressively longer molecules is created, rather than complete cleavage to short oligonucleotides. The specific base modification reactions use **dimethyl sulfate** (DMS) to methylate at G bases. **Formic acid** will attack the purines A and G. **Hydrazine** is used to

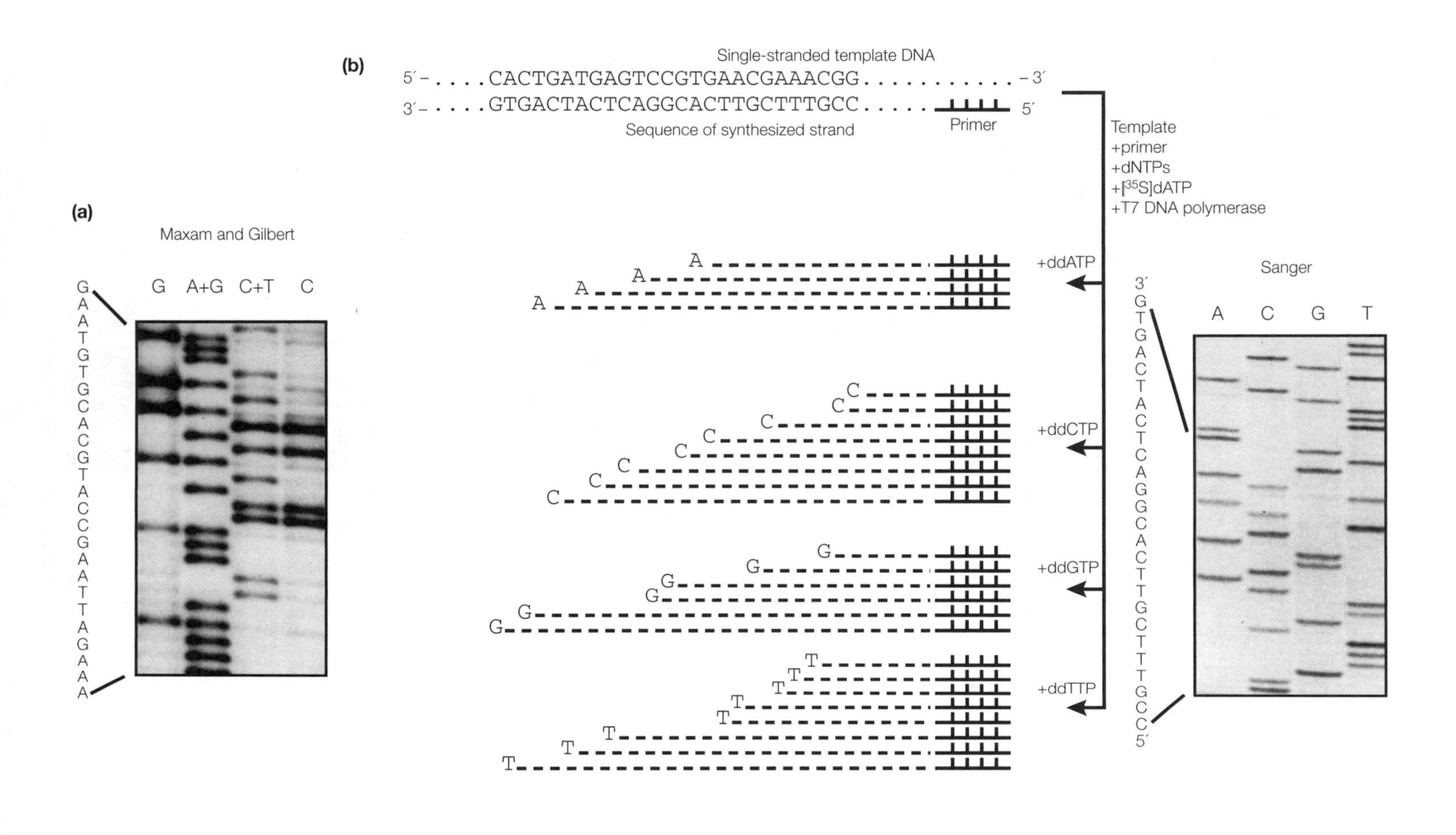

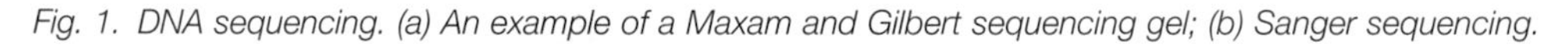
Fig. 1. DNA sequencing. (a) An example of a Maxam and Gilbert sequencing gel; (b) Sanger sequencing.

hydrolyze at pyrimidines (C + T) but high salt inhibits the T reaction. Thus four lanes on the sequencing gel (G, A+G, C+T and C) allow the sequence to be determined (*Fig. 1a*). This method has been adapted to sequence genomic DNA without cloning.

The chemical method of DNA sequencing has largely been superseded by the method of **Sanger**, which uses four specific **dideoxynucleotides (ddNTPs)** to terminate enzymically synthesized copies of a template (*Fig. 1b*). A sequencing primer is annealed to a ssDNA template molecule and a DNA polymerase extends the primer using dNTPs. The extension reaction is split into four and each quarter is terminated separately with one of the four specific ddNTPs, and the four samples (usually radioactive) are analyzed by PAGE. The dideoxynucleotides act as chain terminators since they have no 3′-OH group on the deoxyribose which is needed by the polymerase to extend the growing chain. The label can be incorporated during the synthesis step (e.g. [α-^{35}S]dATP) or the primer can first be end-labeled with either radioactivity or fluorescent dyes. The latter are used in some automatic DNA sequencers although it is more common to use fluorescent dideoxynucleotides.

The original method requires a ssDNA template on which to synthesize the complementary copies, which means that the DNA has to be cloned into the phage vector M13 (see Topic H2) before sequencing. The ssDNA recovered from the phage is **annealed to a primer** of 15–17 nt which is complementary to the region near the vector–insert junction. All sequences cloned into this vector can be sequenced using this universal primer. A DNA polymerase enzyme (usually **Klenow** or **T7 DNA polymerase**) is added to the annealed primer plus template along with a small amount of [α-^{35}S]thiodATP with one oxygen atom on the α phosphate replaced by sulfur (if the primer is not labeled). It is then divided into four tubes, each containing a different chain terminator mixed with normal dNTPs (i.e. tube C would contain ddCTP and dATP, dCTP, dGTP and dTTP) in specific ratios to ensure only a **limited** amount of **chain termination**. The four sets of reaction products, when analyzed by PAGE, usually result in fewer artifactual bands than with chemical sequencing (compare *Fig. 1a* and *b*). Many improvements have been made to this dideoxy method which can now be performed using double-stranded templates and polymerase chain reaction products (see Topic J3).

RNA sequencing

Although sequencing DNA is much easier than sequencing RNA due to its greater stability and the robust enzyme-based protocol, it is sometimes necessary to sequence RNA directly, especially to determine the positions of modified nucleotides present in, for example, tRNA and rRNA (see Topics O1 and O2). This is achieved by base-specific cleavage of 5′-end-labeled RNA using RNases that cleave 3′ to a particular nucleotide. Again, limiting amounts of enzyme and times of digestion are employed to generate a ladder of cleavage products which are analyzed by PAGE. The following RNases are used. **RNase T1** cleaves after G, **RNase U2** after A, **RNase Phy M** after A and U and ***Bacillus cereus*** **RNase** after U and C.

Sequence databases

Over the years, many nucleic acid sequences have been determined by scientists all over the world, and most scientific journals now require the prior submission of nucleic acid sequences to public databases before they will accept a paper for publication. The database managers share information and allow public access which makes these databases extremely valuable resources. New sequences are being added to the databases at an increasing rate, and special computer software is required to make good use of the data. The two largest DNA databases are **EMBL** in Europe and **GenBank** in the USA. There are other databases of protein and RNA sequences as well. Some companies have their own private sequence databases.

Analysis of sequences

When the sequence of a cDNA or genomic clone is determined, few features are immediately apparent without inspection or analysis of the sequence. In a cDNA clone, one end of the sequence should contain a run of A residues, if the cDNA was constructed by priming with oligo(dT). If present, this feature can indicate the orientation of the clone since the oligo(A) should be at the 3′-end. However, other features are hard to determine by eye, and genomic clones do not have this oligo(A) sequence to identify their orientation. Sequences are generally analyzed using computers and software packages, such as the **University of Wisconsin GCG package** or PC programs such as **DNAstar**. These programs can carry out two main operations. One is to identify important sequence features such as restriction sites, open reading frames, start and stop codons, as well as potential promoter sites, intron–exon junctions, etc. The second operation is to compare new sequence with all other known sequences in the databases, which can determine whether related sequences have been obtained before. **Bioinformatics** is the term used to describe the development and use of software such as this to analyze biological data.

Genome Sequencing Projects

Instead of standard ddNTPs, **automated DNA sequencing** makes use of four different, fluorescent dye-labeled chain terminators. It achieves much greater throughput than conventional sequencing because there is no time-consuming autoradiography as lasers read the sequence of the different colors directly off the bottom of the gel in real time. Furthermore, all four reactions can be performed in one tube and loaded in one gel lane and **robotic workstations** can prepare and process many samples at once. These developments have allowed the entire genome sequence of many organisms to be completely determined. Completely sequenced genomes include phage and virus sequences such as λ and the AIDS virus, HIV, at least 10 bacteria including *E. coli* (4.639×10^6 bp) and *Helicobacter pylori*, a causative agent of ulcers and stomach cancer (1.67×10^6 bp), and those of some eukaryotes including the yeast *Saccharomyces cerevisae* (12.1×10^6 bp), the nematode worm *Caenorhabditis elegans* (100×10^6 bp) and the fruit fly *Drosophila melanogaster* (160×10^6 bp). Many other genome sequencing projects are underway including plant, fish, mouse and human. In the latter case, chromosome 22 (34.5×10^6 bp) was 97% completed by the end of 1999, and a draft sequence of about 90% of the whole genome was expected by mid 2000. The problem known as **completion** refers to filling in gaps (such as the 11 small gaps in the human chromosome 22 sequence) which have been difficult to clone and/or sequence. Completion and full annotation for the human sequence may take a further year or so, but is helped by the prior construction of a detailed genetic map and the production of a panel of overlapping genomic clones covering each of the 23 pairs of chromosomes. Much of the human genome sequencing effort used BAC clones rather than YAC clones (see Topic H3) because the latter proved to be less stable during propagation and to contain non contiguous inserts. An alternative random (shotgun) sequencing strategy relies on enormous computing power to assemble the randomly generated sequences.

The availability of whole genome sequences has considerably advanced the area called **genomics** (the study of organism's genomes) because now the number, location, overall size and organization of all the genes needed to make up an organism can be known. Also the distribution of the various satellite sequences (see Topic D4) throughout the genome and the position of pseudogenes will aid our understanding of genome evolution. Perhaps most important is the information that can be gained by comparing the genomes of different organisms. For example in *S. cerevisae,* the genome sequence predicted at least 6200 genes, roughly half of which had no known function. The immediate challenge is to try to discover the function of the huge numbers of unknown genes predicted by genome sequencing projects. This will require large scale gene inactivation methods, i.e. **functional genomics,** coupled with proteomic approaches (see Topic B3).

Further advances in automated DNA sequencing may well use **DNA chips** which are high density arrays of different DNA sequences on a solid support such as glass or nylon. Like their lower density predecessors, **DNA microarrays,** they are used in parallel hybridization experiments to detect which sequences are present in a complex mixture. To be applied in DNA sequencing, DNA chips would need to contain every possible oligonucleotide sequence of a given length, because the maximum sequence 'read' possible is the square root of the number of oligonucleotide sequences on the chip. Hence all 65536 8-mers would allow a 256 bp 'read', but over 10^{12} 20-mers would be needed to "read" a 1 Mb sequence. Since about 10^6 oligos per cm^2 is currently possible, significant improvements in technology will be needed before 'reads' in excess of the current 1 kb limit are exceeded.

J3 POLYMERASE CHAIN REACTION

Key Notes

PCR

The polymerase chain reaction (PCR) is used to amplify a sequence of DNA using a pair of oligonucleotide primers each complementary to one end of the DNA target sequence. These are extended towards each other by a thermostable DNA polymerase in a reaction cycle of three steps: denaturation, primer annealing and polymerization.

The PCR cycle

The reaction cycle comprises a 95°C step to denature the duplex DNA, an annealing step of around 55°C to allow the primers to bind and a 72°C polymerization step. Mg^{2+} and dNTPs are required in addition to template, primers, buffer and enzyme.

Template

Almost any source that contains one or more intact target DNA molecule can, in theory, be amplified by PCR, providing appropriate primers can be designed.

Primers

A pair of oligonucleotides of about 18–30 nt with similar G+C content will serve as PCR primers as long as they direct DNA synthesis towards one another. Primers with some degeneracy can also be used if the target DNA sequence is not completely known.

Enzymes

Thermostable DNA polymerases (e.g. *Taq* polymerase) are used in PCR as they survive the hot denaturation step. Some are more error-prone than others.

PCR optimization

It may be necessary to vary the annealing temperature and/or the Mg^{2+} concentration to obtain faithful amplification. From complex mixtures, a second pair of nested primers can improve specificity.

PCR variations

Variations on basic PCR include quantitative PCR, degenerate oligonucleotide primer PCR (DOP-PCR), inverse PCR, multiplex PCR, rapid amplification of cDNA ends (RACE) and PCR mutagenesis.

Related topics

DNA cloning: an overview (G1)
Genomic libraries (I1)
Screening procedures (I3)
Characterization of clones (J1)
Applications of cloning (J6)

PCR

If a pair of **oligonucleotide primers** can be designed to be complementary to a target DNA molecule such that they can be extended by a DNA polymerase towards each other, then the region of the template bounded by the primers can be greatly amplified by carrying out cycles of **denaturation, primer annealing** and **polymerization**. This process is known as the **polymerase chain reaction (PCR)** and it has become an essential tool in molecular biology as an aid to cloning and gene analysis. The discovery of **thermostable DNA polymerases** has made the steps in the PCR cycle much more convenient. Its applications are finding their way into many areas of science (see Topic J6).

The PCR cycle

Figure 1 shows how PCR works. In the first cycle, the target DNA is separated into two strands by heating to **95°C typically for around 60 seconds**. The temperature is reduced to around **55°C (for about 30 sec)** to allow the primers to anneal to the template DNA. The actual temperature depends on the primer lengths and sequences. After annealing, the temperature is increased to **72°C (for 60–90 sec)** for optimal polymerization which uses up **dNTPs** in the reaction mix and requires $\mathbf{Mg^{2+}}$. In the first polymerization step, the target is copied from the primer sites for various distances on each target molecule until the beginning of cycle 2, when the reaction is heated to 95°C again which denatures the newly synthesized molecules. In the second annealing step, the other primer can bind to the newly synthesized strand and during polymerization can only copy till it reaches the end of the first primer. Thus at the end of cycle 2, some newly synthesized molecules of the correct length exist, though these are base paired to variable length molecules. In subsequent cycles, these soon outnumber the variable length molecules and increase two-fold with each cycle. If PCR was 100% efficient, one target molecule would become 2^n after n cycles. In practice, 20–40 cycles are commonly used.

Template

Because of the extreme amplification achievable, it has been demonstrated that PCR can sometimes amplify as little as one molecule of starting template. Therefore, any source of DNA that provides one or more target molecules can in principle be used as a template for PCR. This includes DNA prepared from blood, sperm or any other tissue, from older forensic specimens, from ancient biological samples or in the laboratory from bacterial colonies or phage plaques as well as purified DNA. Whatever the source of template DNA, PCR can only be applied if some sequence information is known so that primers can be designed.

Primers

Each one of a pair of PCR primers needs to be about 18–30 nt long and to have similar **G+C content** so that they anneal to their complementary sequences at similar temperatures. For short oligonucleotides (<25 nt), the annealing temperature (in °C) can be calculated using the formula: $Tm = 2(A+T) + 4(G+C)$, where Tm is the melting temperature and the annealing temperature is approximately 3–5°C lower. The primers are designed to anneal on opposite strands of the target sequence so that they will be extended towards each other by addition of nucleotides to their 3′-ends. Short target sequences amplify more easily, so often this distance is less than 500 bp, but, with optimization, PCR can amplify fragments over 10 kb in length. If the DNA sequence being amplified is known, then primer design is relatively easy. The region to be amplified should be inspected for two suitable sequences of about 20 nt with a similar G+C content, either side of the region to be amplified (e.g. the site of mutation in certain cancers, see Topic J6). If the PCR product is to be cloned, it is sensible to include the sequence of unique restriction enzyme sites within the 5′-ends of the primers.

If the DNA sequence of the target is not known, for example when trying to clone a cDNA for a protein for which there is only some limited amino acid sequence available, then primer design is more difficult. For this, **degenerate primers** are designed using the genetic code (see Topic P1) to work out what

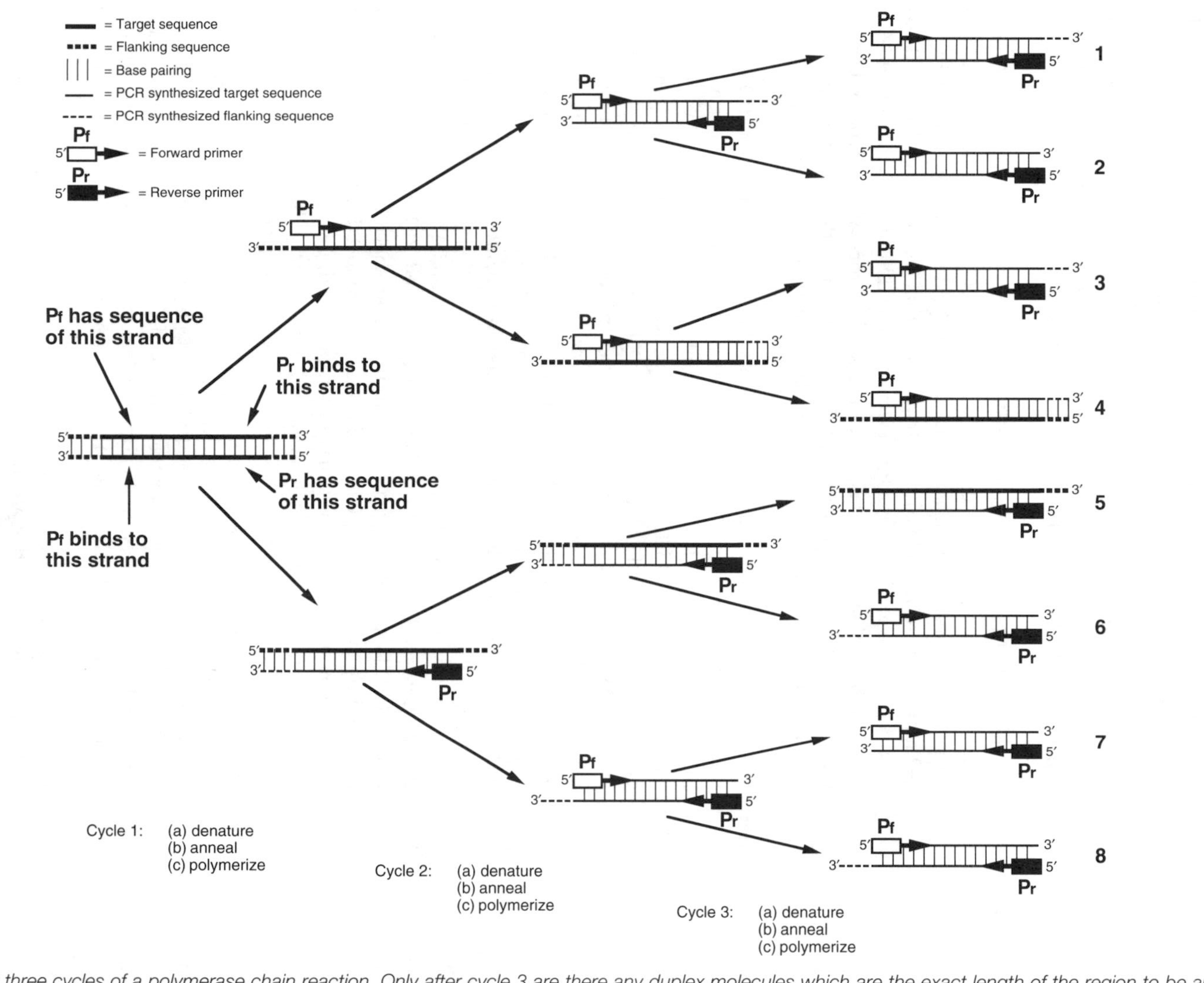

Fig. 1. The first three cycles of a polymerase chain reaction. Only after cycle 3 are there any duplex molecules which are the exact length of the region to be amplified (molecules 2 and 7). After a few more cycles these become the major product.

DNA sequences would encode the known amino acid sequence. For example HisPheProPheMetLys is encoded by the DNA sequence 5′-CAYTTYCCNTTY-ATGAAR-3′, where Y = pyrimidine, R = purine and N = any base. This sequence is 2×2×4×2×2=64-fold degenerate. Thus, if a mixture of all 64 sequences is made and used as a primer, then one of these sequences will be correct. A second primer must be made in a similar way. If one of the known peptide sequences is the N-terminal sequence, then the order of the sequences is known and thus the primer directions are defined. PCR using **degenerate oligonucleotide primers** is sometimes called **DOP-PCR.**

Enzymes

Thermostable DNA polymerases which have been isolated and cloned from a number of thermophilic bacteria are used for PCR. The most common is ***Taq* polymerase** from *Thermus aquaticus*. It survives the denaturation step of 95°C for 1–2 min, having a half-life of more than 2 h at this temperature. Because it has no associated 3′ to 5′ proofreading exonuclease activity (see Topics F1 and G1, *Table 1*), *Taq* polymerase is known to introduce errors when it copies DNA – roughly one per 250 nt polymerized. For this reason, other thermostable DNA polymerases with greater accuracy are used for certain applications.

PCR optimization

PCR reactions are not usually 100% efficient, even when using cloned DNA and primers of defined sequence. Usually the reaction conditions must be varied to improve the efficiency. This is very important when trying to amplify a particular target from a population of other sequences, for example one gene from genomic DNA, or one cDNA from either a cDNA library or the products of a first strand cDNA synthesis reaction. This latter method of reverse transcribing mRNA and then PCR amplifying the first strand cDNA is called **reverse transcriptase (RT)-PCR**. If the reaction is not optimal, PCR often generates a smear of products on a gel rather than a defined band. The usual parameters to vary include the **annealing temperature** and the **Mg^{2+} concentration**. Too low an annealing temperature favors mispairing. The optimal Mg^{2+} concentration varies with each new sequence, but is usually between 1 and 4 mM. The specificity of the reaction can be improved by carrying out **nested PCR**, where, in a second round of PCR, a new set of primers are used that anneal within the fragment amplified by the first pair, giving a shorter PCR product. If on the first round of PCR some nonspecific products have been produced, giving a smear or a number of bands, using nested PCR should ensure that only the desired product is amplified from this mixture as it should be the only sequence present containing both sets of primer-binding sites.

PCR variations

If multiple pairs of primers are added, PCR can be used to amplify more than one DNA fragment in the same reaction and these fragments can easily be distinguished on gels if they are of different lengths. This use of multiple sets of primers is called **multiplex PCR** and is often used as a quick test to detect the presence of microorganisms that may be contaminating food or water, or be infecting tissue. Modifications to the basic PCR make it possible to amplify (and hence clone) sequences that are upstream or downstream of the region amplified by the basic primer pair. For example, if genomic DNA is first digested by a restriction enzyme and then circularized by ligation, a pair of back-to-back primers can be used to amplify round the circle from the region of known sequence to obtain the 5′- and 3′-flanking regions up to the joined restriction sites. This is known as **inverse PCR.** When a fragment of cDNA has been produced by

RT-PCR it is possible to amplify the 5′-flanking sequence by first using terminal transferase to add a tail, e.g. oligo(dC), to the first strand cDNA (see *Fig. 1*, Topic I2). This allows a gene specific primer to be combined with oligo(dG) primer to amplify the 5′-region. This technique is called **rapid amplification of cDNA ends (RACE).** 3′-RACE to amplify the 3′-flanking sequence of eukaryotic mRNAs uses a gene specific primer and an oligo(dT) primer which will anneal to the poly(A) tail at the 3′-end of the mRNA. PCR can be used to make labeled probes to screen libraries (see Topic I3) or carry out blotting experiments (see Topic J1), by adding radioactive or modified nucleotides in the later stages of the PCR reaction, or labeling the PCR product generated as described in Topic J1. PCR can also be used to introduce specific mutations into a given DNA fragment and an example of this process of **PCR mutagenesis** is given in Topic J5.

Quantitative PCR can determine the amount (number of molecules) of DNA in a test sample. One of the best methods of quantitative PCR involves adding known amounts of a similar DNA fragment, such as one containing a short deletion, to the test sample before amplification. The ratio of the two products produced depends on the amount of the deleted fragment added and allows the quantity of the target molecule in the test sample to be calculated. In **asymmetric PCR** only one strand is amplified (in a linear fashion) and when applied to DNA sequencing (see Topic J2) it is known as **cycle sequencing.** PCR can also be used to increase the sensitivity of DNA fingerprinting (see Topics D4 and J6).

J4 ORGANIZATION OF CLONED GENES

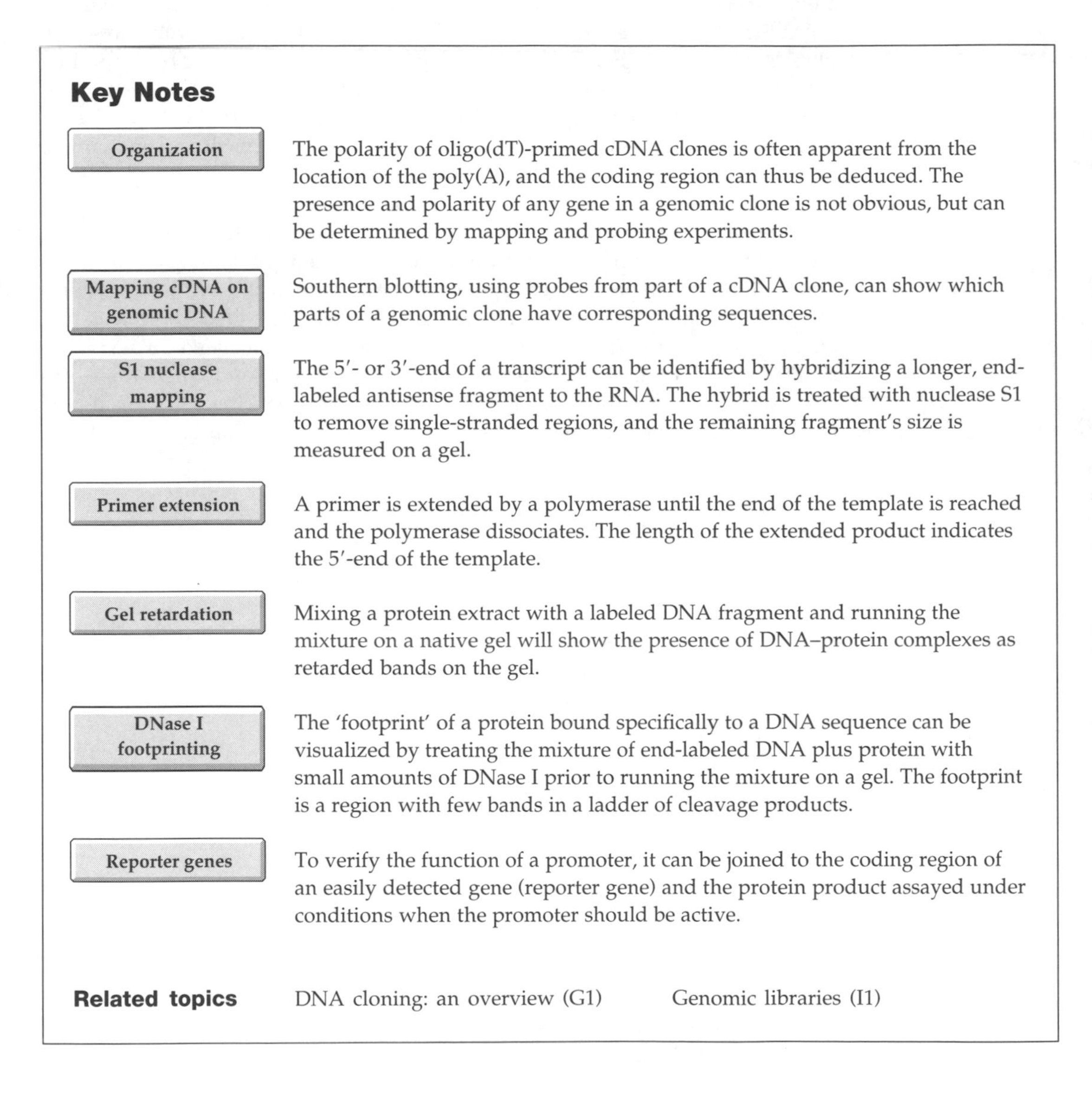

Key Notes

Organization	The polarity of oligo(dT)-primed cDNA clones is often apparent from the location of the poly(A), and the coding region can thus be deduced. The presence and polarity of any gene in a genomic clone is not obvious, but can be determined by mapping and probing experiments.
Mapping cDNA on genomic DNA	Southern blotting, using probes from part of a cDNA clone, can show which parts of a genomic clone have corresponding sequences.
S1 nuclease mapping	The 5′- or 3′-end of a transcript can be identified by hybridizing a longer, end-labeled antisense fragment to the RNA. The hybrid is treated with nuclease S1 to remove single-stranded regions, and the remaining fragment's size is measured on a gel.
Primer extension	A primer is extended by a polymerase until the end of the template is reached and the polymerase dissociates. The length of the extended product indicates the 5′-end of the template.
Gel retardation	Mixing a protein extract with a labeled DNA fragment and running the mixture on a native gel will show the presence of DNA–protein complexes as retarded bands on the gel.
DNase I footprinting	The 'footprint' of a protein bound specifically to a DNA sequence can be visualized by treating the mixture of end-labeled DNA plus protein with small amounts of DNase I prior to running the mixture on a gel. The footprint is a region with few bands in a ladder of cleavage products.
Reporter genes	To verify the function of a promoter, it can be joined to the coding region of an easily detected gene (reporter gene) and the protein product assayed under conditions when the promoter should be active.
Related topics	DNA cloning: an overview (G1) Genomic libraries (I1)

Organization

cDNA clones have a defined organization, especially those synthesized using oligo(dT) as primer. Usually a run of A residues is present at one end of the clone which defines its 3′-end, and at some variable distance upstream of this there will be an open reading frame (ORF) ending in a stop codon (see Topic P1). If the cDNA clone is complete, it will have an ATG start codon, preceded usually by only 20–100 nt. As genomic clones from eukaryotes are larger, and may contain intron sequences, as well as nontranscribed sequences, they provide a greater challenge to understand their organization. It is common, after isolating a cDNA clone, subsequently to obtain a genomic clone for the gene under study. The problem is then to find which parts of each clone correspond to one another. This means establishing which genomic sequences are present in the mature mRNA transcript. The genomic sequences absent in the cDNA clones are usually introns as well as sequences upstream of the transcription start site and downstream of the 3′-processing site (see Topics M4 and O3). Other important features to be identified are the start and stop sites for transcription and the sequences that regulate transcription (see Sections M and N).

Mapping cDNA on genomic DNA

If restriction maps are available for both the genomic and cDNA clones, an important experiment is to run a digest of the genomic clone on a gel and perform a Southern blot (see Topic J1) using all or part of the cDNA as a probe. Using all the cDNA as probe will show which genomic restriction fragments contain sequences also present in the cDNA. These may not be adjacent fragments in the restriction map if large introns are present. Use of a probe from one end of a cDNA will indicate the **polarity of the gene** in the genomic clone. Some of the restriction sites will be common to both clones but may be different distances apart. These can often help to determine the organization of the genomic clone. Based on this information, selected regions of the genomic clone can be sequenced to verify the conclusions.

S1 nuclease mapping

This technique determines the precise **5′- and 3′-ends of RNA transcripts**, although different probes are required in each case. As shown in *Fig. 1* for 5′-end mapping, an end-labeled **antisense DNA** molecule is hybridized to the **RNA** preparation. If duplex DNA is used (still with only the antisense strand labeled), 80% formamide is used in the hybridization buffer to favor RNA–DNA hybrids rather than DNA duplex formation. The hybrids are then treated with the single strand-specific **S1 nuclease**, which will remove the single strand protrusions at each end. The remaining material is analyzed by polyacrylamide gel electrophoresis (PAGE) next to size markers or a sequencing ladder. The size of the nuclease-resistant band, usually revealed by autoradiography, allows the end of the RNA molecule to be deduced.

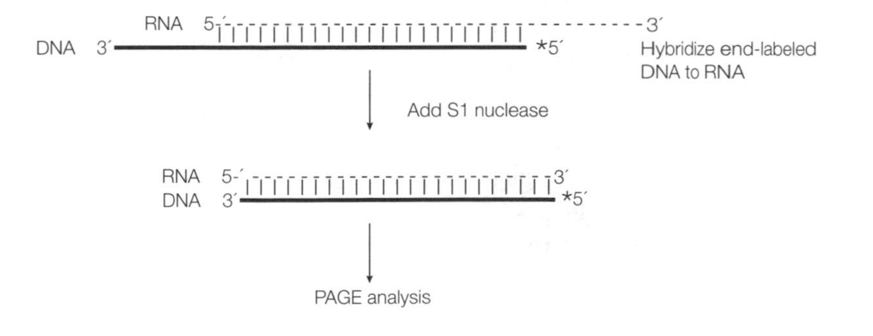

*Fig. 1. S1 nuclease mapping the 5′-end of an RNA. * = position of end label.*

Primer extension

The **5′-ends** of RNA molecules can be determined using **reverse transcriptase** to extend an antisense DNA **primer** in the 5′ to 3′ direction, from the site where it base-pairs on the target to where the polymerase dissociates at the end of the template (*Fig. 2*). The primer extension product is run on a gel next to size markers and/or a sequence ladder from which its length can be established.

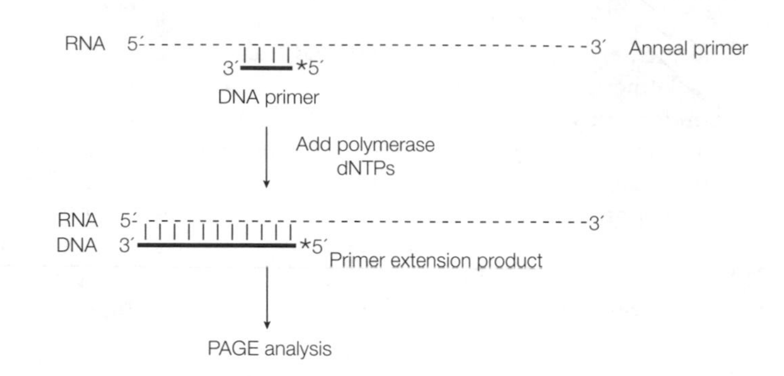

*Fig. 2. Primer extension. * = position of end label.*

Gel retardation

When the 5′-end of a gene transcript has been determined (e.g. by S1 mapping), the corresponding position in the genomic clone is the transcription start site. The DNA sequence upstream contains the regulatory sequences controlling when and where the gene is transcribed. Transcription factors (see Topic N1) bind to specific regulatory sequences and help transcription to occur. The technique of **gel retardation** (**gel shift analysis**) shows the effect of **protein binding** to a labeled nucleic acid and can be used to detect transcription factors binding to regulatory sequences. A short labeled nucleic acid, such as the region of a genomic clone upstream of the transcription start site, is mixed with a cell or nuclear extract expected to contain the binding protein. Then, samples of labeled nucleic acid, with and without extract, are run on a nondenaturing gel, either agarose or polyacrylamide. If a large excess of nonlabeled nucleic acid of different sequence is also present, which will bind proteins that interact nonspecifically, then the specific binding of a factor to the labeled molecule to form one or more DNA–protein complexes is shown by the presence of slowly migrating (**retarded**) **bands** on the gel by autoradiography.

DNase I footprinting

Although gel retardation shows that a protein is binding to a DNA molecule, it does not provide the sequence of the binding site which could be anywhere in the fragment used. DNase footprinting shows the actual region of sequence with which the protein interacts. Again, an end-labeled DNA fragment is required which is mixed with the protein preparation (e.g. a nuclear extract). After binding, the complex is very gently digested with **DNase I** to produce on average one cleavage per molecule. In the region of protein binding, the nuclease cannot easily gain access to the DNA backbone, and fewer cuts take place there. When the partially digested DNA is analyzed by PAGE, a ladder of bands is seen showing all the random nuclease cleavage positions in control DNA. In the lane where protein was added, the ladder will have a gap, or region of reduced cleavage, corresponding to the **protein-binding site** ('**footprint**') where the protein has protected the DNA from nuclease digestion. Other DNA cleaving reagents may also be used in footprinting experiments, for example hydroxyl radical (•OH) and dimethyl sulfate.

Reporter genes

When the region of a gene that controls its transcription (**promoter**; see Topic K1) has been identified by sequencing, S1 mapping and DNA–protein binding experiments, it is common to attach the promoter region to a **reporter gene** to study its action and verify that the promoter has the properties being ascribed to it. For example, the promoter of the heat-shock gene, *HSP70*, could be attached to the coding region of the β-galactosidase gene. When this gene construct is expressed, and if the chromogenic substrate (X-gal, see Topic H1) is present, a blue color is produced. If the *HSP70* promoter–reporter construct is introduced into a cell, and the cell, or cell line, is subjected to a heat shock, β-gal transcripts are made and the protein product can be detected by the blue color. This would show that the normally inactive promoter is activated after a heat shock. This is because a special transcription factor binds to a regulatory sequence in the promoter and activates the gene (see Topic N2).

J5 Mutagenesis of cloned genes

Key Notes

Deletion mutagenesis

Progressively deleting DNA from one end is very useful for defining the importance of particular sequences. Unidirectional deletions can be created using exonuclease III which removes one strand in a 3′ to 5′ direction from a recessed 3′-end. A single strand-specific nuclease then creates blunt end molecules for ligation, and transformation generates the deleted clones.

Site-directed mutagenesis

Changing one or a few nucleotides at a particular site usually involves annealing a mutagenic primer to a template followed by complementary strand synthesis by a DNA polymerase. Formerly, single-stranded templates prepared using M13 were used, but polymerase chain reaction (PCR) techniques are now preferred.

PCR mutagenesis

By making forward and reverse mutagenic primers and using other primers that anneal to common vector sequences, two PCR reactions are carried out to amplify 5′- and 3′-portions of the DNA to be mutated. The two PCR products are mixed and used for another PCR using the outer primers only. Part of this product is then subcloned to replace the region to be mutated in the starting molecule.

Related topics

DNA cloning: an overview (G1)
Ligation, transformation and analysis of recombinants (G4)
Bacteriophage vectors (H2)
Polymerase chain reaction (J3)

Deletion mutagenesis

In organisms with small genomes, and/or rapid generation times, it is possible to create and analyze mutants produced *in vivo*, but this is not easy or acceptable in organisms such as man. However, if cloned genes have been isolated, it is quite feasible to mutate them *in vitro* and then assay for the effects by expressing the mutant gene *in vitro* or *in vivo*. For both cDNA and genomic clones, creating deletion mutants is useful. In the case of cDNA clones, it is common to delete progressively from the ends of the coding region to produce either N-terminally or C-terminally **truncated proteins** (after expression) to discover which parts (domains) of the whole protein have particular properties. For example, the N-terminal domain of a given protein could be a DNA-binding domain, the central region an ATP-binding site and the C-terminal region could help the protein to interact to form dimers. In genomic clones, when the transcription start site has been identified, sequences upstream are removed progressively to discover the minimum length of upstream sequence that has promoter and regulatory function. Although it is possible to create deletion mutants using restriction enzymes if their sites fall in convenient positions, the general method of creating **unidirectional deletions** using an exonuclease is more versatile.

Figure 1 shows a cDNA cloned into a plasmid vector in the multiple cloning site (MCS). There are several restriction sites in the MCS at each side of the insert, and by choosing two that are unique and that give a 5′- and a 3′-recessed end on the vector and insert respectively, unidirectional deletions can be created. The enzyme **exonuclease III** can remove one strand of nucleotides in a 3′ to 5′ direction from a recessed 3′-end, but not from a 3′-protruding end. Therefore in *Fig. 1* it will remove the lower strand of the insert only progressively with time. At various times, aliquots are treated with **S1 or mung bean nuclease** (see Topic G1) which will remove the single strand protrusions at both ends, creating blunt ends. When these are ligated to re-circularize the plasmid, and transformed (see Topic G4), a population of subclones will be produced, each having a different amount of the insert removed. This technique is often used to obtain the complete sequence of clones as the one sequencing primer can be used to derive 200–300 nt of sequence from a series of clones that differ in size by about 200 bp. Approximately one in three of the deletions will be in-frame and could be used to express truncated protein. Deletion from the other end of the cDNA is performed in a similar way.

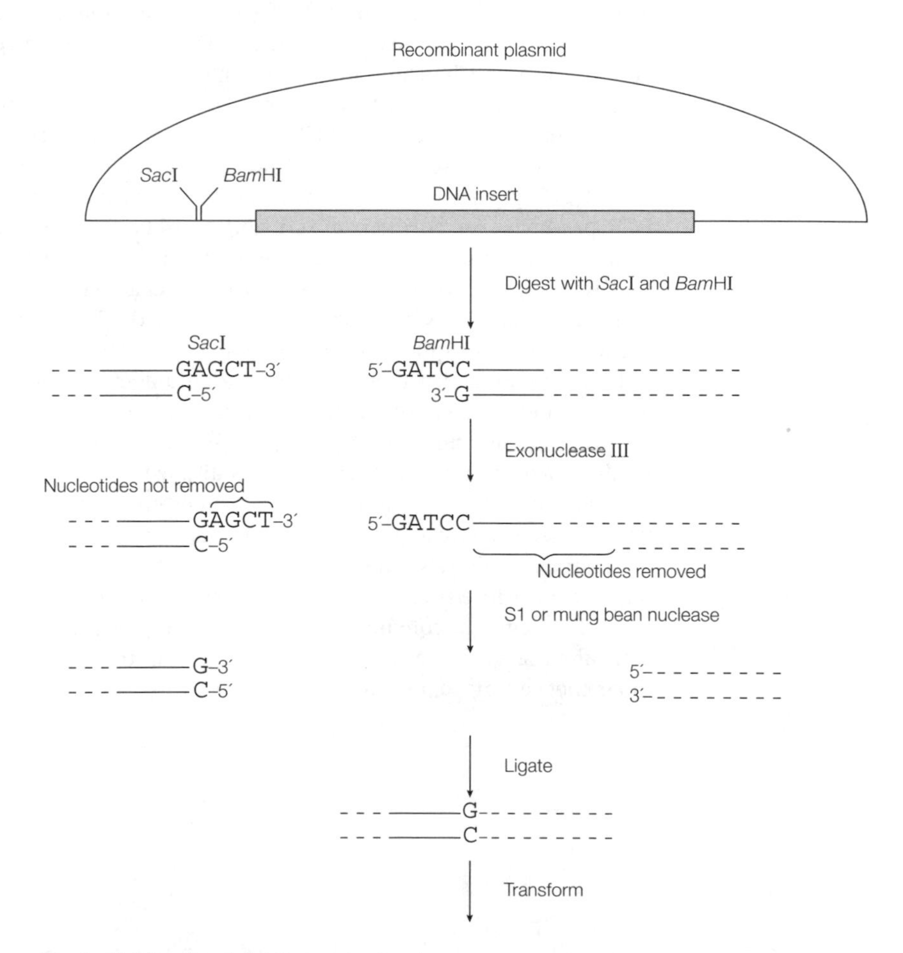

Fig. 1. Unidirectional deletion mutagenesis.

Site-directed mutagenesis

It is very useful to be able to change just one, or a few specific nucleotides in a sequence to test a hypothesis. The importance of each residue in a transcription factor-binding site could be examined by changing each one in turn. Suspected critical amino acids in a protein could be changed by altering the cDNA sequence so that by a one nucleotide change an amino acid substitution is made, the effect of which is examined by assaying for function using the mutant protein. Originally, site-directed mutagenesis used a **single-stranded template** (created by subcloning in M13) and a **primer oligonucleotide** with the desired mutation in it. The primer was annealed to the template and then extended using a **DNA polymerase**, ligated using DNA ligase to seal the nick and the mismatched duplex transformed into bacteria. Some bacteria would remove the mismatch to give the desired point mutation, but some clones with the original sequence would be produced. Because this method was not very efficient (not all clones were mutant) and because subcloning in M13 to prepare single-stranded DNA was time consuming, much site-directed mutagenesis is now carried out using the polymerase chain reaction (PCR).

PCR mutagenesis

There are several ways in which PCR (see Topic J3) can be used to create both deletion and point mutations. In *Fig. 2*, one method of creating point mutations is illustrated. A pair of primers is designed that have the altered sequence and which overlap by at least 20 nt. If the DNA to be mutated is in a standard vector, there will be standard primer sites in the vector such as the SP6 and T7 promoter sites of pGEM which can be used in combination with the **mutagenic primers**. Two separate PCR reactions are performed, one amplifying the 5′-portion of the insert using SP6 and the reverse primer, and the other amplifying the 3′-portion of the insert using the forward and T7 primers. If the two PCR products are purified, mixed and amplified using SP6 and T7 primers, then a full-length, mutated molecule is the only product that should be made. This will happen when a 5′-sense strand anneals via the 20 nt overlap to a 3′-antisense strand or vice versa, and is extended. This method is nearly 100% efficient, and very quick. An alternative even more convenient version, which is the basis of at least one commercial mutagenesis kit, involves using the forward and reverse mutagenic primers (*Fig. 2*) to extend all the way round the plasmid containing the target gene. At the end of the reaction, some of the product will consist of circular molecules containing the mutation in both strands and nicks at either end of the primer sequence. These molecules can be transformed directly to give clones containing mutant plasmid, without the need for restriction digests and ligation. As all mutants need to be checked by sequencing before use, instead of using the whole of the PCR-generated DNA, a smaller fragment containing the mutated region could be subcloned into the equivalent sites in the original clone. Only the transferred region would need to be checked by sequencing.

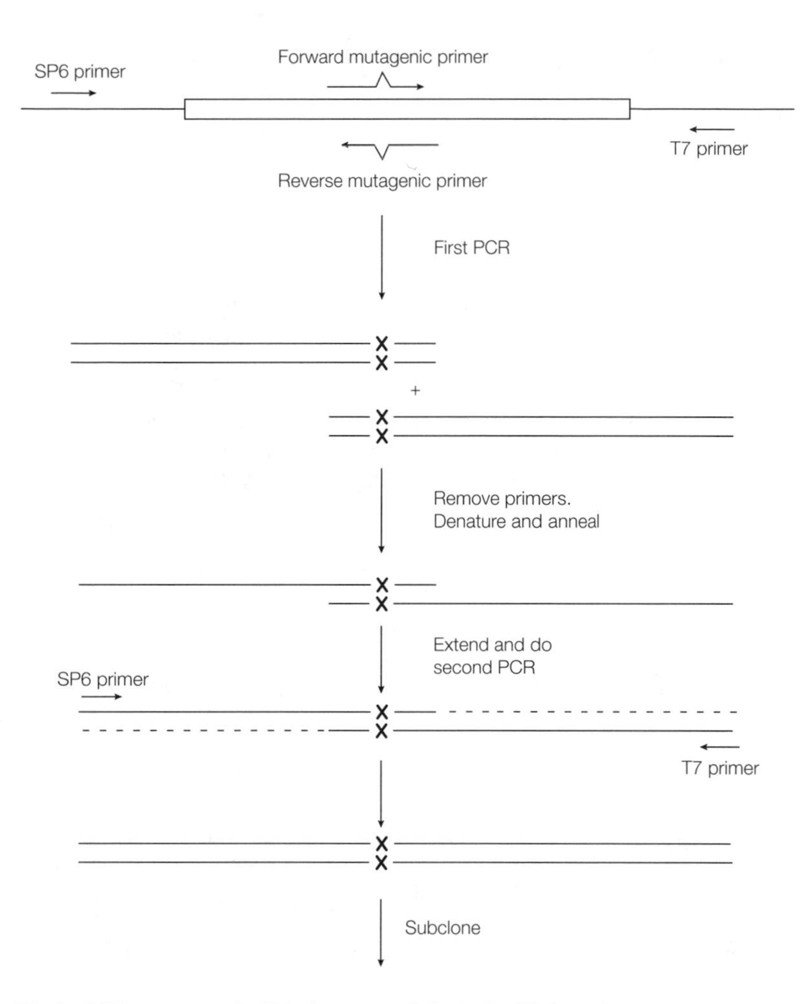

Fig. 2. PCR mutagenesis. **X** *is the mutated site in the PCR product.*

J6 APPLICATIONS OF CLONING

Key Notes

Applications
The various applications of gene cloning include recombinant protein production, genetically modified organisms, DNA fingerprinting, diagnostic kits and gene therapy.

Recombinant protein
By inserting the gene for a rare protein into a plasmid and expressing it in bacteria, large amounts of recombinant protein can be produced. If post-translational modifications are critical, the gene may have to be expressed in a eukaryotic cell.

Genetically modified organisms
Introducing a foreign gene into an organism which can propagate creates a genetically modified organism. Transgenic sheep have been created to produce foreign proteins in their milk.

DNA fingerprinting
Hybridizing Southern blots of genomic DNA with probes that recognize simple nucleotide repeats gives a pattern that is unique to an individual and can be used as a fingerprint. This has applications in forensic science, animal and plant breeding and evolutionary studies.

Medical diagnosis
The sequence information derived from cloning medically important genes has allowed the design of many diagnostic test kits which can help predict and confirm a wide range of disorders.

Gene therapy
Attempts to correct a genetic disorder by delivering a gene to a patient are described as gene therapy.

Related topics
Genome complexity (D4)
Design of plasmid vectors (H1)
Eukaryotic vectors (H4)
Categories of oncogenes (S2)

Applications

Gene cloning has made a phenomenal impact on the speed of biological research and it is increasing its presence in several areas of everyday life. These include the biotechnological production of proteins as therapeutics and for nontherapeutic use, the generation of modified organisms, especially for improved food production, the development of test kits for medical diagnosis, the application of the polymerase chain reaction (PCR) and cloning in forensic science and studies of evolution, and the attempts to correct genetic disorders by gene therapy. This topic describes some of these applications.

Recombinant protein

Many proteins that are normally produced in very small amounts are known to be missing or defective in various disorders. These include growth hormone, insulin in diabetes, interferon in some immune disorders and blood clotting Factor VIII in hemophilia. Prior to the advent of gene cloning and protein production via recombinant DNA techniques, it was necessary to purify these molecules from animal tissues or donated human blood. Both sources have drawbacks, including slight functional differences in the nonhuman proteins and possible viral contamination (e.g. HIV, CJD). Production of protein from a cloned gene in a defined, nonpathogenic organism would circumvent these problems, and so pharmaceutical and biotechnology companies have developed this technology. Initially, production in bacteria was the only route available and cDNA clones were used as they contained no introns. The cDNAs had to be linked to prokaryotic transcription and translation signals and inserted into multicopy plasmids (see Topic H1). However, often the overproduced proteins, which could represent up to 30% of total cell protein, were precipitated or insoluble and they lacked eukaryotic post-translational modifications (see Topic Q4). Sometimes these problems could be overcome by making fusion proteins which were later cleaved to give the desired protein, but the subsequent availability of eukaryotic cells for production (yeast or mammalian cell lines) has helped greatly. The human Factor VIII protein, which is administered to hemophiliacs, is produced in a hamster cell line which has been transfected with a 186 kb human genomic DNA fragment. Such a cell line has been **genetically modified** and it is now possible to genetically modify whole organisms. Recombinant proteins can also be modified by introducing amino acid substitutions by mutagenesis (see Topic J5). This can result in improvements such as more stable enzymes for inclusion in washing powders, etc.

Genetically modified organisms

Genetically modified organisms (**GMOs**) are created when cloned genes are introduced into cells that give rise to whole organisms. In eukaryotes, if the introduced genes are derived from another organism, the resulting **transgenic** plants or animals can be propagated by normal breeding. Several types of transgenic plant have been created and tested for safety in the production of foodstuffs. One example of a GMO is a tomato that has had a gene for a ripening enzyme inactivated. The strain of tomato takes longer to soften, and ultimately rot, due to the absence of the enzyme, and so has a longer shelf life and other improved qualities. Transgenic sheep have been produced with the intention of producing valuable proteins in their milk. The desired gene requires a sheep promoter (e.g. from caesin or lactalbumin) to be attached to ensure expression in the mammary gland. Purification of the protein from milk is easier than from cultured cells or blood. The definition of a transgenic organism is one containing a foreign gene, but the term is often now applied to organisms that have been genetically manipulated to contain multiple foreign genes, extra copies of an endogenous gene or that have had a gene disrupted (**gene knock out**). The term is not usually applied to the most extreme form of adding foreign genes seen in some forms of animal cloning (production of identical individuals) where all the genes are replaced by those from another nucleus. This procedure (**nuclear transfer**) of replacing the nucleus of an egg with the nucleus from an adult cell was used to create Dolly the sheep in 1997. This famous example of animal cloning created much controversy because it raised the possibility of human cloning from adult cells. Many countries have now introduced laws to ban most types of human cloning and this may well hinder the development of replacement cells/organs for therapeutic use. Identical twins are human clones that arise naturally.

DNA fingerprinting

Cloning and genomic sequencing projects have identified many repetitive sequences in the human genome (see Topic D4). Some of these are simple nucleotide repeats that vary in number between individuals but are inherited (VNTRs). If Southern blots of restriction enzyme-digested genomic DNA from members of a family are hybridized with a probe that detects one of these types of repeats, each sample will show a set of bands of varying lengths (the length of the repeats between the two flanking restriction enzyme sites). One hybridizing locus (pair of alleles) is shown in *Fig. 1*. Some of these bands will be in common with those of the mother and some with those of the father and the pattern of bands will be different for an unrelated individual. The different patterns in individuals at each of these kinds of simple repeat sites means that, by using a small number of probes, the likelihood of two individuals having the same pattern becomes vanishingly small. This is the technique of **DNA fingerprinting** which is used in forensic science to eliminate the innocent and convict criminals. It is also applied for maternity and paternity testing in humans. It can also be used to show pedigree in animals bred commercially and to discover mating habits in wild animals. *Fig. 1* also shows how DNA fingerprinting can be carried out on small DNA samples such as a blood spot or hair follicle left at a crime scene. Instead of digesting the DNA with restriction enzyme E at each end of the VNTR and Southern blotting, a pair of PCR primers can be designed based on the unique sequences flanking the repeats (shown as arrows in *Fig. 1*). The VNTRs can thus be amplified and directly visualized by staining after agarose gel electrophoresis.

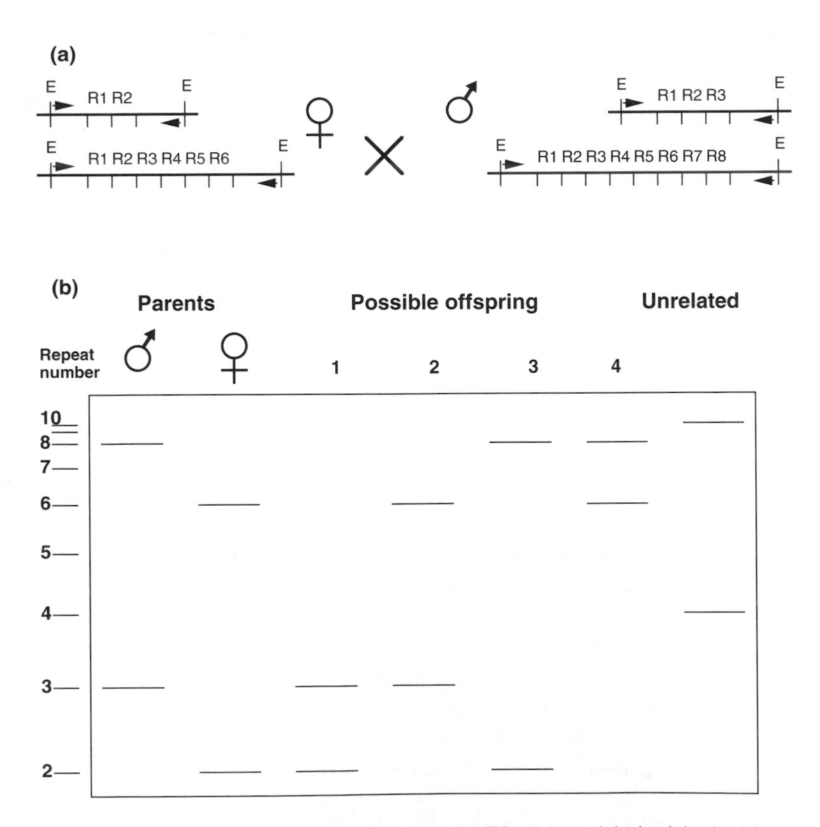

Fig. 1. DNA fingerprinting showing how two VNTR alleles might be inherited (see text). (a) Parental VNTR alleles. (b) Agarose gel analysis of VNTR alleles.

Medical diagnosis

A great variety of medical conditions arise from mutation. In genetic disorders such as muscular dystrophy or cystic fibrosis, individuals are born with faulty genes that cause the symptoms of the disorder. Many cancers arise due to spontaneous mutations in somatic cells in genes whose normal role is the regulation of cell growth (see Topic S2). Cloning of the genes involved in both genetic disorders and cancers has shown that certain mutations are more common and some correlate with more aggresive disorders. By using sequence information to design PCR primers and probes, many tests have been developed to screen patients for these clinically important mutations. Using these tests, parents who are both heterozygous for a mutation can now be advised whether an unborn child is going to suffer from a genetic disorder such as muscular dystrophy or cystic fibrosis (by inheriting one faulty gene from each parent) and can consider termination. Checking for the presence of mutations in a gene can confirm a diagnosis that is based on other clinical presentations. In cancer cases, knowing which oncogene is mutated, and in what way, can help decide the best course of treatment as well as providing information for the development of new therapies.

Gene therapy

Attempts have been made to treat some genetic disorders by delivering a normal copy of the defective gene to patients. This is known as gene therapy. In, ***in vivo* gene therapy**, the gene can be directly administered to the patient on its own or cloned into a defective virus used as a vector that can replicate but not cause infection (see Topic H4). For some disorders, the bone marrow is destroyed and replaced with treated cells that have had a normal gene, or a protective gene (e.g. for a ribozyme, see Topic O2), introduced (***ex vivo* gene therapy**). Gene therapy will produce transgenic somatic cells (see above) in the patient. Gene therapy is in its infancy, but it seems to have great potential.

K1 BASIC PRINCIPLES OF TRANSCRIPTION

Key Notes

Transcription: an overview

Transcription is the synthesis of a single-stranded RNA from a double-stranded DNA template. RNA synthesis occurs in the 5′→3′ direction and its sequence corresponds to that of the DNA strand which is known as the sense strand.

Initiation

RNA polymerase is the enzyme responsible for transcription. It binds to specific DNA sequences called promoters to initiate RNA synthesis. These sequences are upstream (to the 5′- end) of the region that codes for protein, and they contain short, conserved DNA sequences which are common to different promoters. The RNA polymerase binds to the dsDNA at a promoter sequence, resulting in local DNA unwinding. The position of the first synthesized base of the RNA is called the start site and is designated as position +1.

Elongation

RNA polymerase moves along the DNA and sequentially synthesizes the RNA chain. DNA is unwound ahead of the moving polymerase, and the helix is reformed behind it.

Termination

RNA polymerase recognizes the terminator which causes no further ribonucleotides to be incorporated. This sequence is commonly a hairpin structure. Some terminators require an accessory factor called rho for termination.

Related topics

Nucleic acid structure (C1)
Escherichia coli RNA polymerase (K2)
The *E. coli* σ^{70} promoter (K3)
Transcription initiation, elongation and termination (K4)
The *trp* operon (L2)

Transcription: an overview

Transcription is the enzymic synthesis of RNA on a DNA template. This is the first stage in the overall process of gene expression and ultimately leads to synthesis of the protein encoded by a gene. Transcription is catalyzed by an **RNA polymerase** which requires a dsDNA template as well as the precursor ribonucleotides ATP, GTP, CTP and UTP (*Fig. 1*). RNA synthesis always occurs in a fixed direction, from the 5′- to the 3′-end of the RNA molecule (see Topic C1). Usually, only one of the two strands of DNA becomes transcribed into RNA. One strand is known as the **sense strand**. The sequence of the RNA is a direct copy of the sequence of the deoxynucleotides in the sense strand (with U in place of T). The other strand is known as the **antisense strand**. This strand

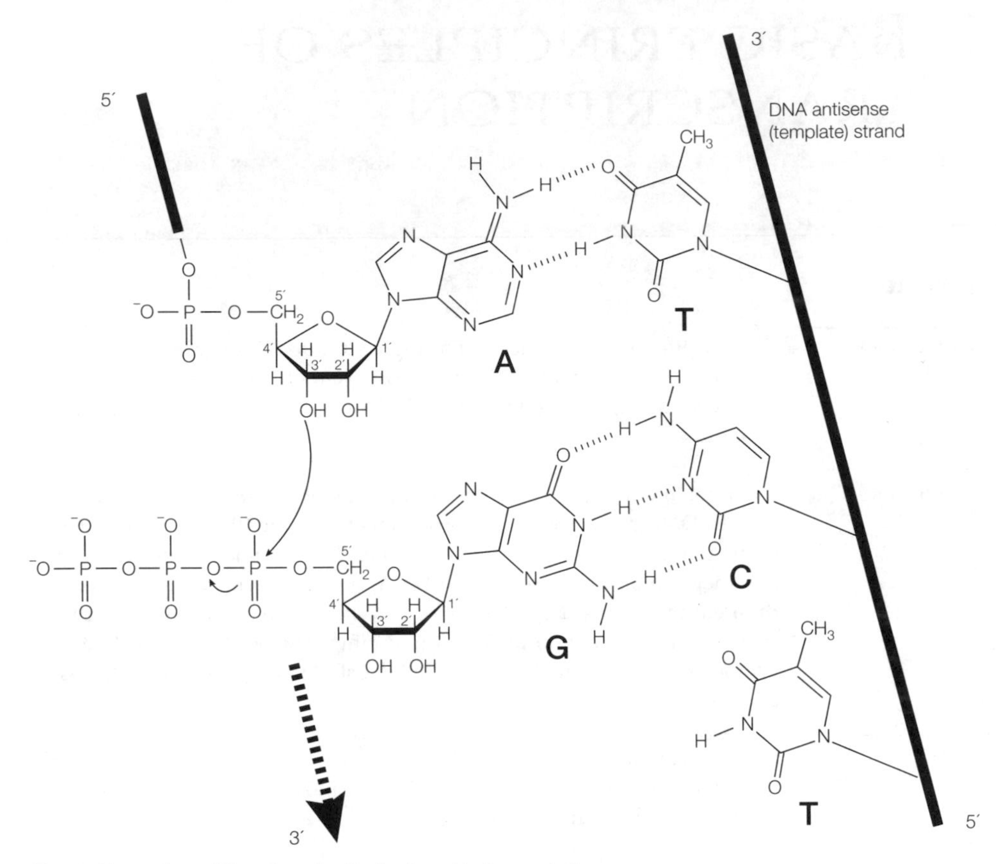

Fig. 1. Formation of the phosphodiester bond in transcription.

may also be called the **template strand** since it is used as the template to which ribonucleotides base-pair for the synthesis of the RNA.

Initiation

Initiation of transcription involves the binding of an RNA polymerase to the dsDNA. RNA polymerases are usually multisubunit enzymes. They bind to the dsDNA and initiate transcription at sites called **promoters** (*Fig. 2*). Promoters are sequences of DNA at the start of genes, that is to the 5′-side (**upstream**) of the coding region. Sequence elements of promoters are often conserved between different genes. Differences between the promoters of different genes give rise to differing efficiencies of transcription initiation and are involved in their regulation (see Section L). The short conserved sequences within promoters are the sites at which the polymerase or other DNA-binding proteins bind to initiate or regulate transcription.

In order to allow the template strand to be used for base pairing, the DNA helix must be locally unwound. Unwinding begins at the promoter site to which the RNA polymerase binds. The polymerase then initiates the synthesis of the RNA strand at a specific nucleotide called the **start site (initiation site)**. This is defined as position +1 of the gene sequence (*Fig. 2*). The RNA polymerase and its co-factors, when assembled on the DNA template, are often referred to as the **transcription complex**.

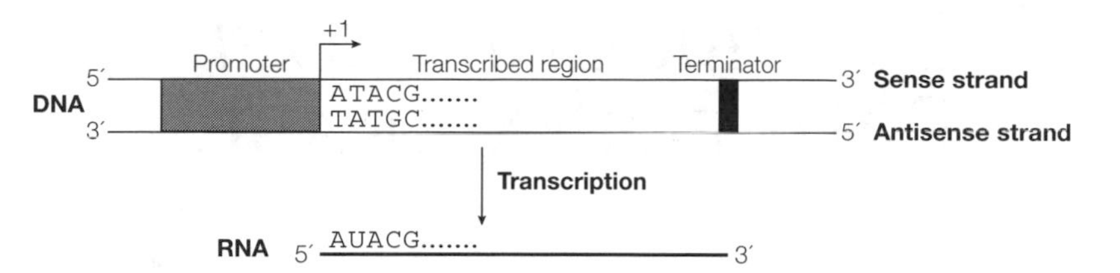

Fig. 2. Structure of a typical transcription unit showing promoter and terminator sequences, and the RNA product.

Elongation

The RNA polymerase covalently adds ribonucleotides to the 3′-end of the growing RNA chain (*Fig. 1*). The polymerase therefore extends the growing RNA chain in a 5′→3′ direction. This occurs while the enzyme itself moves in a 3′→5′ direction along the antisense DNA strand (template). As the enzyme moves, it locally unwinds the DNA, separating the DNA strands, to expose the template strand for ribonucleotide base pairing and covalent addition to the 3′-end of the growing RNA chain. The helix is reformed behind the polymerase. The *E. coli* RNA polymerase performs this reaction at a rate of around 40 bases per second at 37°C.

Termination

The termination of transcription, namely the dissociation of the transcription complex and the ending of RNA synthesis, occurs at a specific DNA sequence known as the **terminator** (see *Fig. 2* and Topics K2 and K3). These sequences often contain self-complementary regions which can form a **stem–loop** or **hairpin** secondary structure in the RNA product (*Fig. 3*). These cause the polymerase to pause and subsequently cease transcription.

Some terminator sequences can terminate transcription without the requirement for accessory factors, whereas other terminator sequences require the rho protein (ρ) as an accessory factor. In the termination reaction, the RNA–DNA hybrid is separated allowing the reformation of the dsDNA, and the RNA polymerase and synthesized RNA are released from the DNA.

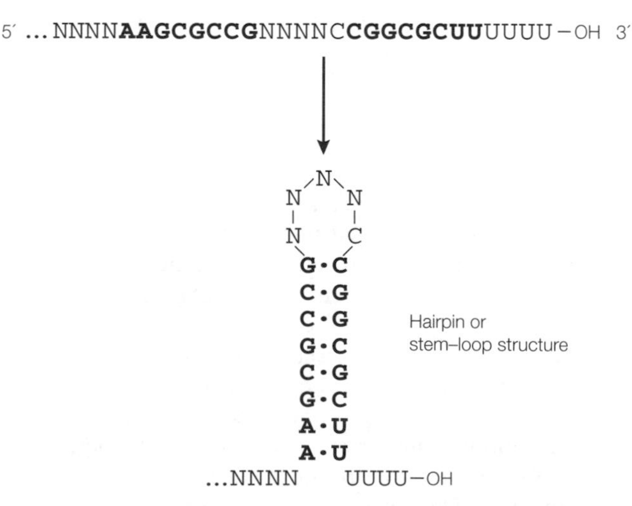

Fig. 3. RNA hairpin structure.

K2 *ESCHERICHIA COLI* RNA POLYMERASE

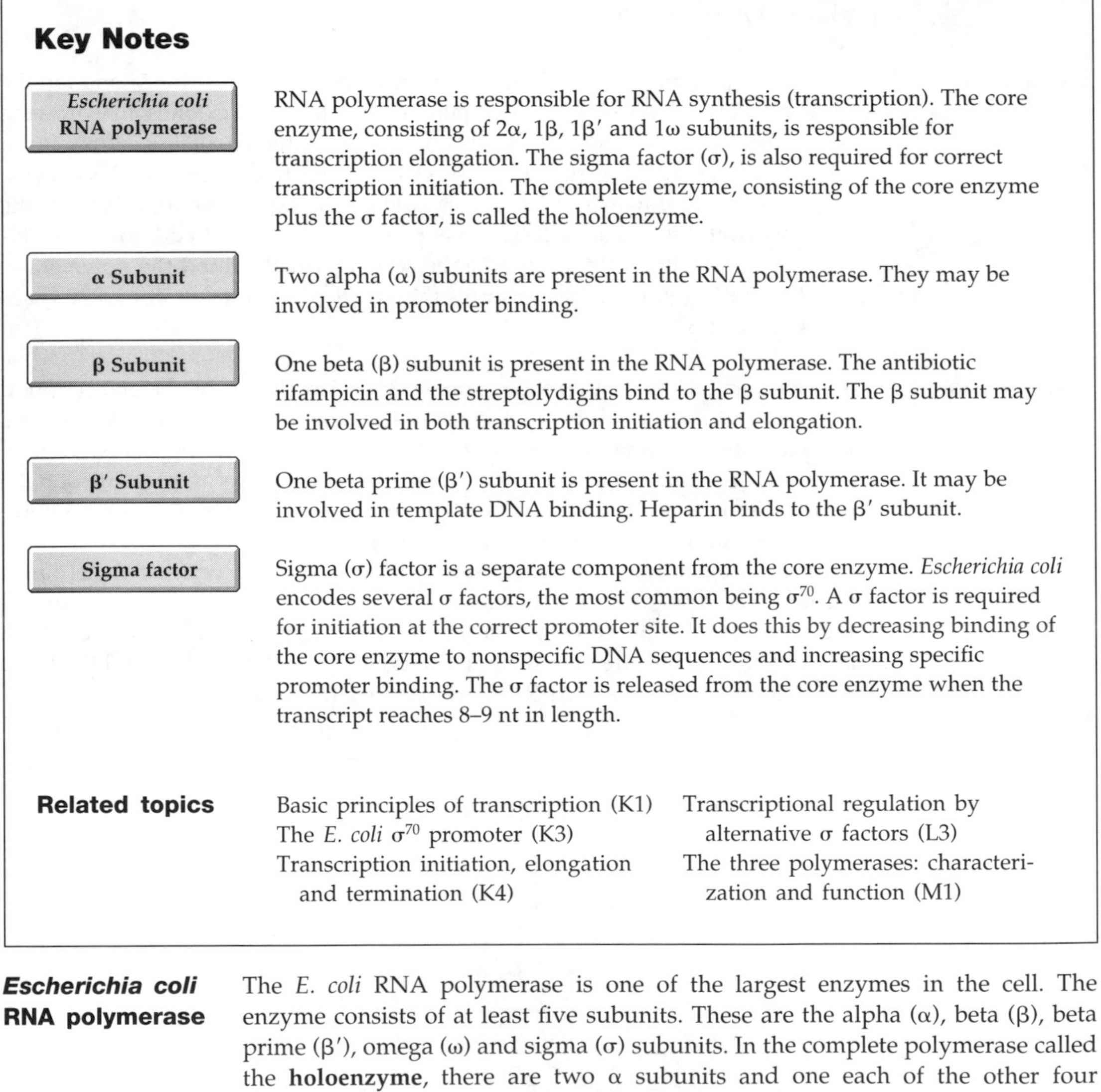

Key Notes

Escherichia coli RNA polymerase: RNA polymerase is responsible for RNA synthesis (transcription). The core enzyme, consisting of 2α, 1β, 1β′ and 1ω subunits, is responsible for transcription elongation. The sigma factor (σ), is also required for correct transcription initiation. The complete enzyme, consisting of the core enzyme plus the σ factor, is called the holoenzyme.

α Subunit: Two alpha (α) subunits are present in the RNA polymerase. They may be involved in promoter binding.

β Subunit: One beta (β) subunit is present in the RNA polymerase. The antibiotic rifampicin and the streptolydigins bind to the β subunit. The β subunit may be involved in both transcription initiation and elongation.

β′ Subunit: One beta prime (β′) subunit is present in the RNA polymerase. It may be involved in template DNA binding. Heparin binds to the β′ subunit.

Sigma factor: Sigma (σ) factor is a separate component from the core enzyme. *Escherichia coli* encodes several σ factors, the most common being σ^{70}. A σ factor is required for initiation at the correct promoter site. It does this by decreasing binding of the core enzyme to nonspecific DNA sequences and increasing specific promoter binding. The σ factor is released from the core enzyme when the transcript reaches 8–9 nt in length.

Related topics

Basic principles of transcription (K1)
The *E. coli* σ^{70} promoter (K3)
Transcription initiation, elongation and termination (K4)
Transcriptional regulation by alternative σ factors (L3)
The three polymerases: characterization and function (M1)

Escherichia coli RNA polymerase

The *E. coli* RNA polymerase is one of the largest enzymes in the cell. The enzyme consists of at least five subunits. These are the alpha (α), beta (β), beta prime (β′), omega (ω) and sigma (σ) subunits. In the complete polymerase called the **holoenzyme**, there are two α subunits and one each of the other four subunits (i.e. $\alpha_2\beta\beta'\omega\sigma$). The complete enzyme is required for transcription initiation. However, the σ factor is not required for transcription elongation and is released from the transcription complex after transcription initiation. The remaining enzyme, which translocates along the DNA, is known as the **core enzyme** and has the structure $\alpha_2\beta\beta'\omega$. The *E. coli* RNA polymerase can synthesize RNA at a rate of around 40 nt per sec at 37°C and requires Mg^{2+} for its activity. The enzyme has a nonspherical structure with a projection flanking a cylindrical channel. The size of the channel suggests that it can bind directly to 16 bp of DNA. The whole polymerase binds over a region of DNA covering around 60 bp.

Although most RNA polymerases like the *E. coli* polymerase have a multi-subunit structure, it is important to note that this is not an absolute requirement. The RNA polymerases encoded by bacteriophages T3 and T7 (see Topic H1) are single polypeptide chains which are much smaller than the bacterial multi-subunit enzymes. They synthesize RNA rapidly (200 nt per sec at 37°C) and recognize their own specific DNA-binding sequences.

α Subunit

Two identical α subunits are present in the core RNA polymerase enzyme. The subunit is encoded by the *rpoA* gene. The α subunit is required for core protein assembly, but has had no clear transcriptional role assigned to it. When phage T4 infects *E. coli* the α subunit is modified by adenosine diphosphate (ADP) ribosylation of an arginine. This is associated with a reduced affinity for binding to promoters, suggesting that the α subunit may play a role in promoter recognition.

β Subunit

One β subunit is present in the core enzyme. This subunit is thought to be the catalytic center of the RNA polymerase. Strong evidence for this has come from studies with antibiotics which inhibit transcription by RNA polymerase. The important antibiotic **rifampicin** is a potent inhibitor of RNA polymerase that blocks initiation but not elongation. This class of antibiotic does not inhibit eukaryotic polymerases and has, therefore, been used medically for treatment of Gram-positive bacteria infections and tuberculosis. Rifampicin has been shown to bind to the β subunit. Mutations that give rise to resistance to rifampicin map to *rpoB*, the gene that encodes the β subunit. A further class of antibiotic, the **streptolydigins**, inhibit transcription elongation, and mutations that confer resistance to these antibiotics also map to *rpoB*. These studies suggest that the β subunit may contain two domains responsible for transcription initiation and elongation.

β′ Subunit

One β′ subunit is present in the core enzyme. It is encoded by the *rpoC* gene. This subunit binds two Zn^{2+} ions which are thought to participate in the catalytic function of the polymerase. A polyanion, heparin, has been shown to bind to the β′ subunit. **Heparin** inhibits transcription *in vitro* and also competes with DNA for binding to the polymerase. This suggests that the β′ subunit may be responsible for binding to the template DNA.

Sigma factor

The most common sigma factor in *E. coli* is σ^{70} (since it has a molecular mass of 70 kDa). Binding of the σ factor converts the core RNA polymerase enzyme into the holoenzyme. The σ factor has a critical role in promoter recognition, but is not required for transcription elongation. The σ factor contributes to promoter recognition by decreasing the affinity of the core enzyme for nonspecific DNA sites by a factor of 10^4 and increasing affinity for the promoter. Many prokaryotes (including *E. coli*) have multiple σ factors. They are involved in the recognition of specific classes of promoter sequences (see Topic L3). The σ factor is released from the RNA polymerase when the RNA chain reaches 8–9 nt in length. The core enzyme then moves along the DNA synthesizing the growing RNA strand. The σ factor can then complex with a further core enzyme complex and re-initiate transcription. There is only 30% of the amount of σ factor present in the cell compared with core enzyme complexes. Therefore only one-third of the polymerase complexes can exist as holoenzyme at any one time.

K3 THE *E. COLI* σ^{70} PROMOTER

Key Notes

Promoter sequences

Promoters contain conserved sequences which are required for specific binding of RNA polymerase and transcription initiation.

Promoter size

The promoter region extends for around 40 bp. Within this sequence, there are short regions of extensive conservation which are critical for promoter function.

-10 sequence

The –10 sequence is a 6 bp region present in almost all promoters. This hexamer is generally 10 bp upstream from the start site. The consensus –10 sequence is TATAAT.

-35 sequence

The –35 sequence is a further 6 bp region recognizable in most promoters. This hexamer is typically 35 bp upstream from the start site. The consensus –35 sequence is TTGACA

Transcription start site

The base at the start site is almost always a purine. G is more common than A.

Promoter efficiency

There is considerable variation between different promoter sequences and in the rates at which different genes are transcribed. Regulated promoters (e.g. *lac* promoter) are activated by the binding of accessory activation factors such as cAMP receptor protein (CRP). Alternative classes of consensus promoter sequences (e.g. heat-shock promoters) are recognized only by an RNA polymerase enzyme containing an alternative σ factor.

Related topics

Organization of cloned genes (J4)
Basic principles of transcription (K1)
Escherichia coli RNA polymerase (K2)
Transcription initiation, elongation and termination (K4)
The *lac* operon (L1)
Transcriptional regulation by alternation σ factors (L3)

Promoter sequences

RNA polymerase binds to specific initiation sites upstream from transcribed sequences. These are called promoters. Although different promoters are recognized by different σ factors which interact with the RNA polymerase core enzyme, the most common σ factor in *E. coli* is σ^{70}. Promoters were first characterized through mutations that enhance or diminish the rate of transcription of genes such as those in the *lac* operon (see Topic L1). The promoter lies upstream of the start site of transcription, generally assigned as position +1 (see Topic K1). In accordance with this, promoter sequences are assigned a negative number reflecting the distance upstream from the start of transcription.

Mutagenesis of *E. coli* promoters has shown that only very short conserved sequences are critical for promoter function.

Promoter size

The σ^{70} promoter consists of a sequence of between 40 and 60 bp. The region from around –55 to +20 has been shown to be bound by the polymerase, and the region from –20 to +20 is strongly protected from nuclease digestion by DNase I (see Topic J4). This suggests that this region is tightly associated with the polymerase which blocks access of the nuclease to the DNA. Mutagenesis of promoter sequences showed that sequences up to around position –40 are critical for promoter function. Two 6 bp sequences at around positions –10 and –35 have been shown to be particularly important for promoter function in *E. coli*.

–10 sequence

The most conserved sequence in σ^{70} promoters is a 6 bp sequence which is found in the promoters of many different *E. coli* genes. This sequence is centered at around the –10 position with respect to the transcription start site (*Fig. 1*). This is sometimes referred to as the **Pribnow box**, having been first recognized by Pribnow in 1975. It has a consensus sequence of **TATAAT**, where the consensus sequence is made up of the most frequently occurring nucleotide at each position when many sequences are compared. The first two bases (TA) and the final T are most highly conserved. This hexamer is separated by between 5 and 8 bp from the transcription start site. This intervening sequence is not conserved, although the distance is critical. The –10 sequence appears to be the sequence at which DNA unwinding is initiated by the polymerase (see Topic K4).

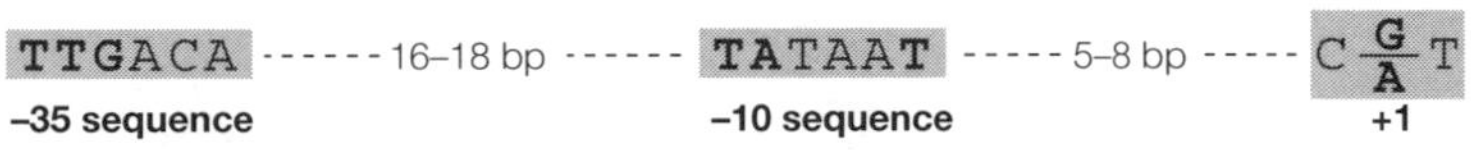

TTGACA ------ 16–18 bp ------ TATAAT ----- 5–8 bp ----- C $\frac{G}{A}$ T

–35 sequence –10 sequence +1

Fig. 1. Consensus sequences of E. coli *promoters (the most conserved sequences are shown in bold).*

–35 sequence

Upstream regions around position –35 also have a conserved hexamer sequence (see *Fig. 1*). This has a consensus sequence of TTGACA, which is most conserved in efficient promoters. The first three positions of this hexamer are the most conserved. This sequence is separated by 16–18 bp from the –10 box in 90% of all promoters. The intervening sequence between these conserved elements is not important.

Transcription start site

The transcription start site is a purine in 90% of all genes (*Fig. 1*). G is more common at the transcription start site than A. Often, there are C and T bases on either side of the start site nucleotide (i.e. CGT or CAT).

Promoter efficiency

The sequences described above are consensus sequences typical of strong promoters. However, there is considerable variation in sequence between different promoters, and they may vary in transcriptional efficiency by up to 1000-fold. Overall, the functions of different promoter regions can be defined as follows:

- the –35 sequence constitutes a recognition region which enhances recognition and interaction with the polymerase σ factor;
- the –10 region is important for DNA unwinding;
- the sequence around the start site influences initiation.

The sequence of the first 30 bases to be transcribed also influences transcription. This sequence controls the rate at which the RNA polymerase clears the promoter, allowing re-initiation of another polymerase complex, thus influencing the rate of transcription and hence the overall promoter strength. The importance of strand separation in the initiation reaction is shown by the effect of negative supercoiling of the DNA template which generally enhances transcription initiation, presumably because the supercoiled structure requires less energy to unwind the DNA. Some promoter sequences are not sufficiently similar to the consensus sequence to be strongly transcribed under normal conditions. An example is the ***lac* promoter P_{lac}**, which requires an accessory activating factor called **cAMP receptor protein (CRP)** to bind to a site on the DNA close to the promoter sequence in order to enhance polymerase binding and transcription initiation (see Topic L1). Other promoters, such as those of genes associated with heat shock, contain different consensus promoter sequences that can only be recognized by an RNA polymerase which is bound to a σ factor different from the general factor σ^{70} (see Topic L3).

K4 TRANSCRIPTION, INITIATION, ELONGATION AND TERMINATION

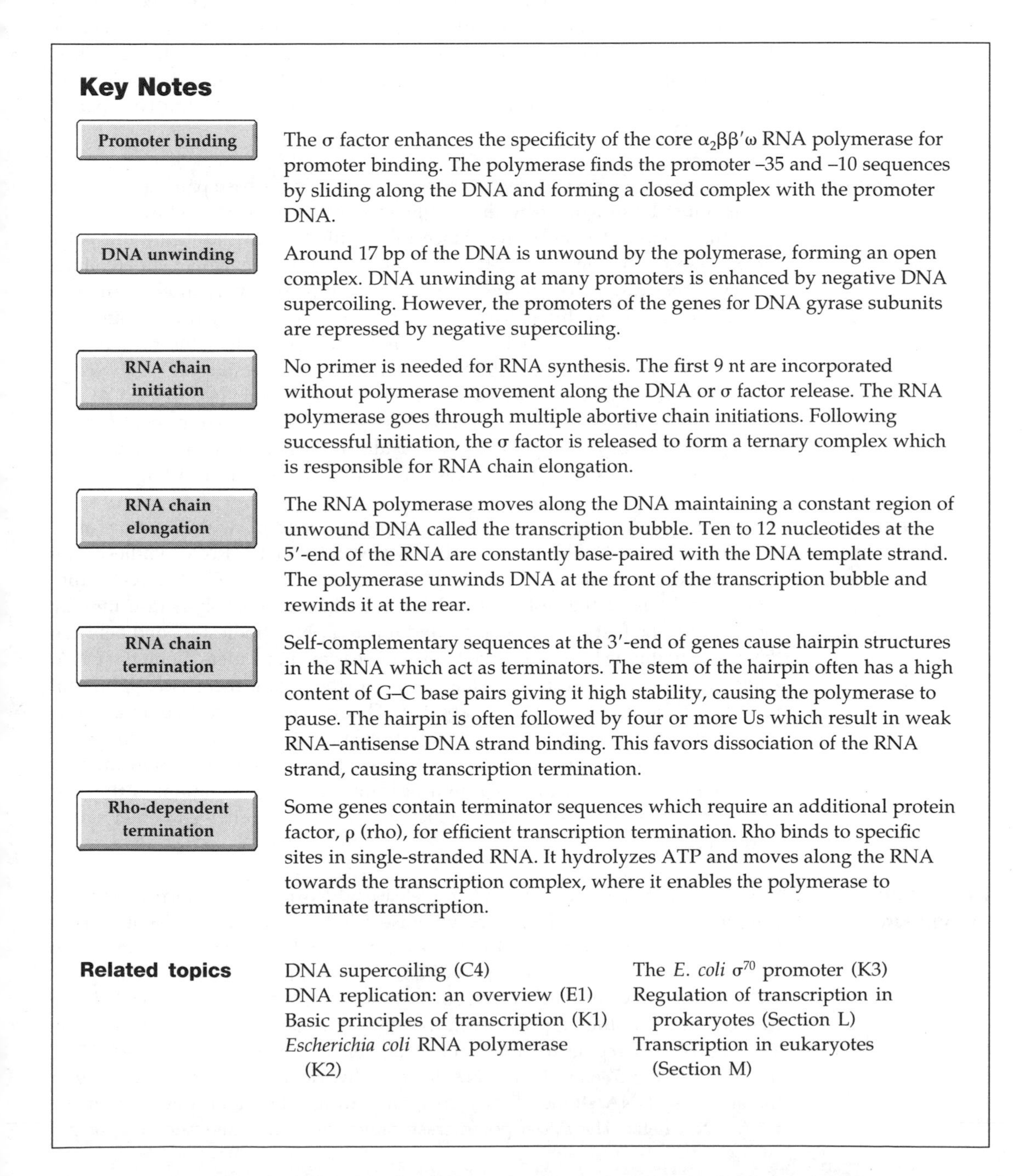

Key Notes

Promoter binding

The σ factor enhances the specificity of the core $\alpha_2\beta\beta'\omega$ RNA polymerase for promoter binding. The polymerase finds the promoter –35 and –10 sequences by sliding along the DNA and forming a closed complex with the promoter DNA.

DNA unwinding

Around 17 bp of the DNA is unwound by the polymerase, forming an open complex. DNA unwinding at many promoters is enhanced by negative DNA supercoiling. However, the promoters of the genes for DNA gyrase subunits are repressed by negative supercoiling.

RNA chain initiation

No primer is needed for RNA synthesis. The first 9 nt are incorporated without polymerase movement along the DNA or σ factor release. The RNA polymerase goes through multiple abortive chain initiations. Following successful initiation, the σ factor is released to form a ternary complex which is responsible for RNA chain elongation.

RNA chain elongation

The RNA polymerase moves along the DNA maintaining a constant region of unwound DNA called the transcription bubble. Ten to 12 nucleotides at the 5′-end of the RNA are constantly base-paired with the DNA template strand. The polymerase unwinds DNA at the front of the transcription bubble and rewinds it at the rear.

RNA chain termination

Self-complementary sequences at the 3′-end of genes cause hairpin structures in the RNA which act as terminators. The stem of the hairpin often has a high content of G–C base pairs giving it high stability, causing the polymerase to pause. The hairpin is often followed by four or more Us which result in weak RNA–antisense DNA strand binding. This favors dissociation of the RNA strand, causing transcription termination.

Rho-dependent termination

Some genes contain terminator sequences which require an additional protein factor, ρ (rho), for efficient transcription termination. Rho binds to specific sites in single-stranded RNA. It hydrolyzes ATP and moves along the RNA towards the transcription complex, where it enables the polymerase to terminate transcription.

Related topics

DNA supercoiling (C4)
DNA replication: an overview (E1)
Basic principles of transcription (K1)
Escherichia coli RNA polymerase (K2)
The *E. coli* σ^{70} promoter (K3)
Regulation of transcription in prokaryotes (Section L)
Transcription in eukaryotes (Section M)

Promoter binding

The RNA polymerase core enzyme, $\alpha_2\beta\beta'\omega$, has a general nonspecific affinity for DNA. This is referred to as **loose binding** and it is fairly stable. When σ factor is added to the core enzyme to form the holoenzyme, it markedly reduces the affinity for nonspecific sites on DNA by 20 000-fold. In addition, σ factor enhances holoenzyme binding to correct promoter-binding sites 100 times. Overall, this dramatically increases the specificity of the holoenzyme for correct promoter-binding sites. The holoenzyme searches out and binds to promoters in the *E. coli* genome extremely rapidly. This process is too fast to be achieved by repeated binding and dissociation from DNA, and is believed to occur by the polymerase sliding along the DNA until it reaches the promoter sequence. At the promoter, the polymerase recognizes the double-stranded –35 and –10 DNA sequences. The initial complex of the polymerase with the base-paired promoter DNA is referred to as a **closed complex.**

DNA unwinding

In order for the antisense strand to become accessible for base pairing, the DNA duplex must be unwound by the polymerase. Negative supercoiling enhances the transcription of many genes, since this facilitates unwinding by the polymerase. However, some promoters are not activated by negative supercoiling, implying that differences in the natural DNA topology may affect transcription, perhaps due to differences in the steric relationship of the –35 and –10 sequences in the double helix. For example, the promoters for the enzyme subunits of DNA gyrase are inhibited by negative supercoiling. DNA gyrase is responsible for negative supercoiling of the *E. coli* genome (see Topic C4) and so this may serve as an elegant feedback loop for DNA gyrase protein expression. The initial unwinding of the DNA results in formation of an **open complex** with the polymerase; and this process is referred to as **tight binding**.

RNA chain initiation

Almost all RNA start sites consist of a purine residue, with G being more common than A. Unlike DNA synthesis (see Section E), RNA synthesis can occur without a primer (*Fig. 1*). The chain is started with a GTP or ATP, from which synthesis of the rest of the chain is initiated. The polymerase initially incorporates the first two nucleotides and forms a phosphodiester bond between them. The first nine bases are added without enzyme movement along the DNA. After each one of these first 9 nt is added to the chain, there is a significant probability that the chain will be aborted. This process of abortive initiation is important for the overall rate of transcription since it has a major role in determining how long the polymerase takes to leave the promoter and allow another polymerase to initiate a further round of transcription. The minimum time for promoter clearance is 1–2 seconds, which is a long event relative to other stages of transcription.

RNA chain elongation

When initiation succeeds, the enzyme releases the σ factor and forms a **ternary complex** (three components) of polymerase–DNA–nascent (newly synthesized) RNA, causing the polymerase to progress along the DNA (**promoter clearance**) allowing re-initiation of transcription from the promoter by a further RNA polymerase holoenzyme. The region of unwound DNA, which is called the **transcription bubble**, appears to move along the DNA with the polymerase. The size of this region of unwound DNA remains constant at around 17 bp (*Fig. 2*), and the 5′-end of the RNA forms a hybrid helix of about 12 bp with the antisense DNA strand. This corresponds to just less than one turn of the RNA–DNA helix. The *E. coli* polymerase moves at an average rate of 40 nt per

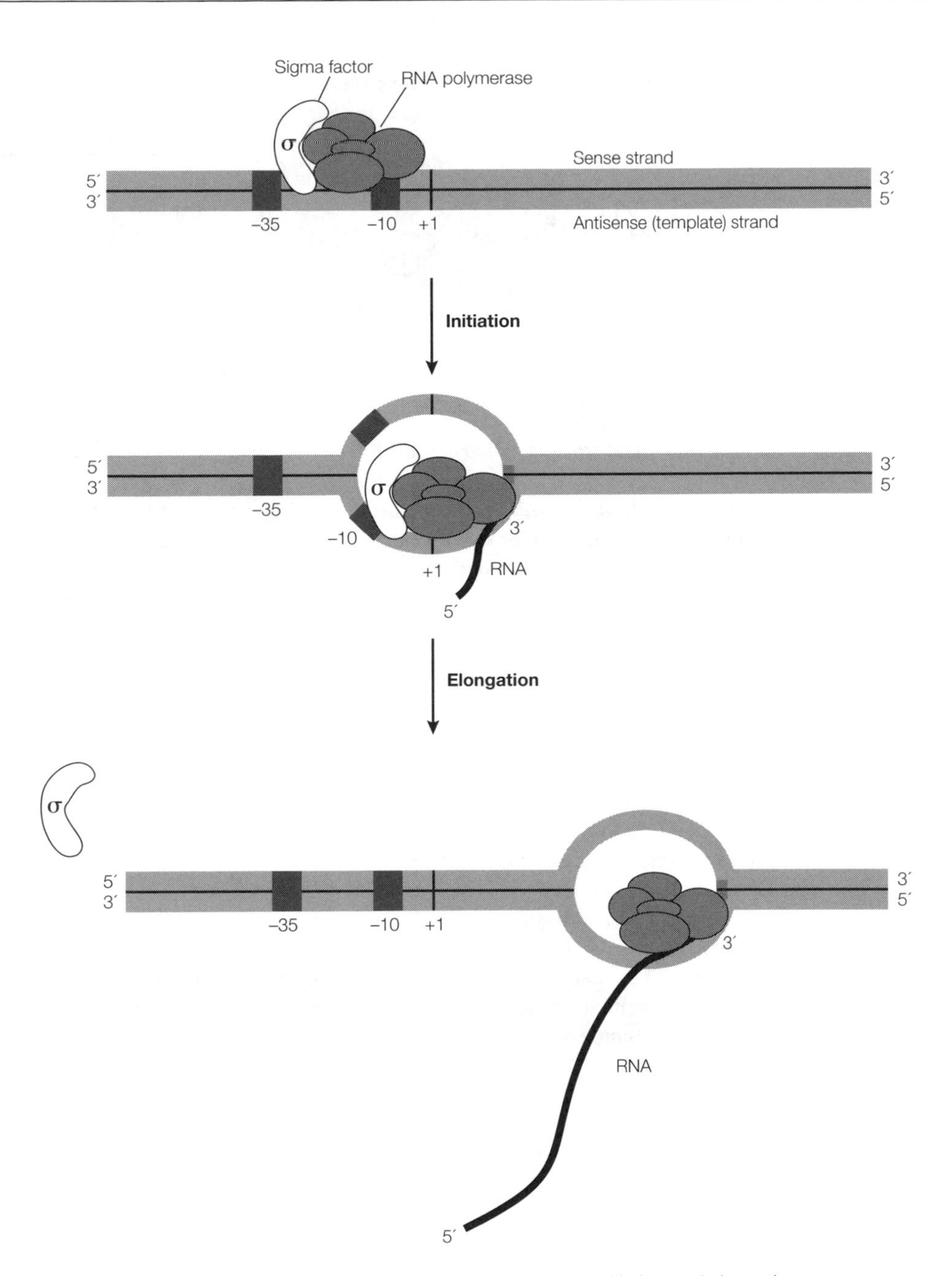

Fig. 1. Formation of the transcription complex: initiation and elongation.

sec, but the rate can vary depending on local DNA sequence. Maintenance of the short region of unwound DNA indicates that the polymerase unwinds DNA in front of the transcription bubble and rewinds DNA at its rear. The RNA–DNA helix must rotate each time a nucleotide is added to the RNA.

RNA chain termination

The RNA polymerase remains bound to the DNA and continues transcription until it reaches a **terminator sequence (stop signal)** at the end of the tran-

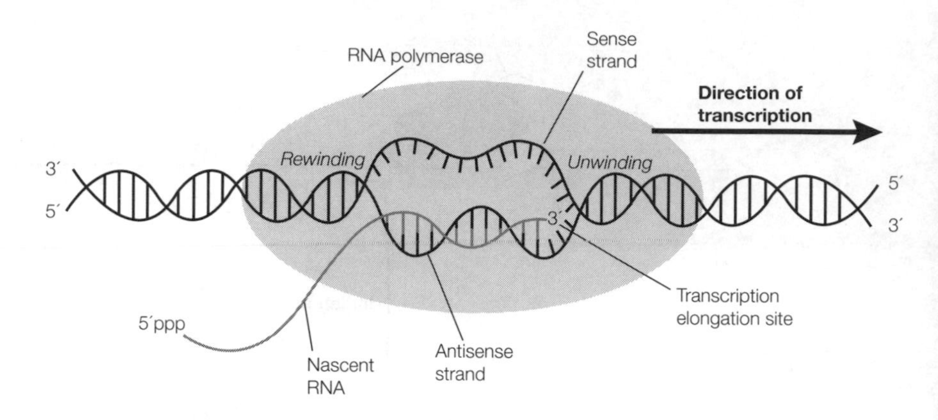

Fig. 2. Schematic structure of the transcription bubble during elongation.

scription unit (*Fig. 3*). The most common stop signal is an **RNA hairpin** in which the RNA transcript is self-complementary. As a result, the RNA can form a stable hairpin structure with a stem and a loop. Commonly the stem structure is very GC-rich, favoring its base pairing stability due to the additional stability of G–C base pairs over A–U base pairs. The RNA hairpin is often followed by a sequence of four or more U residues. It seems that the polymerase pauses immediately after it has synthesized the hairpin RNA. The subsequent stretch of U residues in the RNA base-pairs only weakly with the corresponding A residues in the antisense DNA strand. This favors dissociation of the RNA from the complex with the template strand of the DNA. The RNA is therefore released from the transcription complex. The non-base-paired antisense strand of the DNA then re-anneals with the sense DNA strand and the core enzyme disassociates from the DNA.

Rho-dependent termination

While the RNA polymerase can self-terminate at a hairpin structure followed by a stretch of U residues, other known terminator sites may not form strong hairpins. They use an accessory factor, the **rho protein** (**ρ**) to mediate transcription termination. Rho is a hexameric protein that hydrolyzes ATP in the presence of single-stranded RNA. The protein appears to bind to a stretch of 72 nucleotides in RNA, probably through recognition of a specific structural feature rather than a consensus sequence. Rho moves along the nascent RNA towards the transcription complex. There, it enables the RNA polymerase to terminate at rho-dependent transcriptional terminators. Like rho-independent terminators, these signals are recognized in the newly synthesized RNA rather than in the template DNA. Sometimes, the rho-dependent terminators are hairpin structures which lack the subsequent stretch of U residues which are required for rho-independent termination.

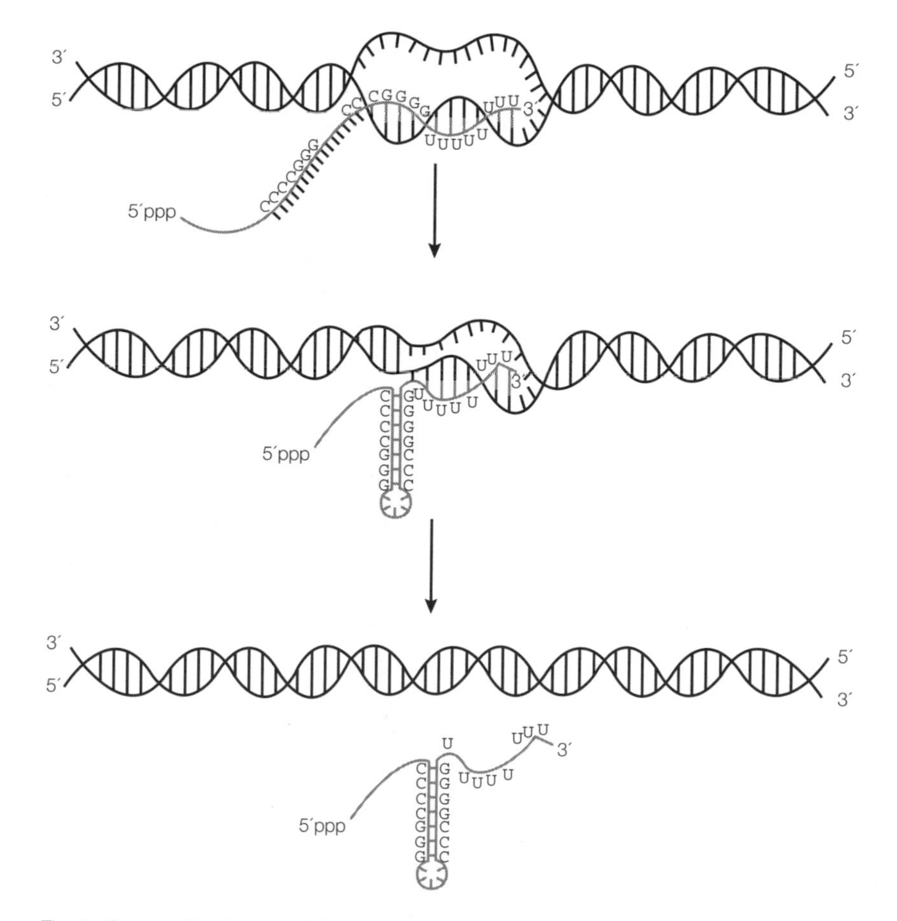

Fig. 3. Schematic diagram of rho-independent transcription termination.

L1 THE *LAC* OPERON

Key Notes

The operon

The concept of the operon was first proposed in 1961 by Jacob and Monod. An operon is a unit of prokaryotic gene expression which includes co-ordinately regulated (structural) genes and control elements which are recognized by regulatory gene products.

The lactose operon

The *lacZ*, *lacY* and *lacA* genes are transcribed from a *lacZYA* transcription unit under the control of a single promoter P_{lac}. They encode enzymes required for the use of lactose as a carbon source. The *lacI* gene product, the lac repressor, is expressed from a separate transcription unit upstream from P_{lac}.

The lac repressor

The lac repressor is made up of four identical protein subunits. It therefore has a symmetrical structure and binds to a palindromic (symmetrical) 28 bp operator DNA sequence O_{lac} that overlaps the *lacZYA* RNA start site. Bound repressor blocks transcription from P_{lac}.

Induction

When lac repressor binds to the inducer (whose presence is dependent on lactose), it changes conformation and cannot bind to the O_{lac} operator sequence. This allows rapid induction of *lacZYA* transcription.

cAMP receptor protein

The cAMP receptor protein (CRP) is a transcriptional activator which is activated by binding to cAMP. cAMP levels rise when glucose is lacking. This complex binds to a site upstream from P_{lac} and induces a 90° bend in the DNA. This induces RNA polymerase binding to the promoter and transcription initiation. The CRP activator mediates the global regulation of gene expression from catabolic operons in response to glucose levels.

Related topics

Basic principles of transcription (K1)
Escherichia coli RNA polymerase (K2)
The *E. coli* σ^{70} promoter (K3)
Transcription initiation, elongation and termination (K4)
The *trp* operon (L2)

The operon

Jacob and Monod proposed the operon model in 1961 for the co-ordinate regulation of transcription of genes involved in specific metabolic pathways. The **operon** is a unit of gene expression and regulation which typically includes:

- The **structural genes** (any gene other than a regulator) for enzymes involved in a specific biosynthetic pathway whose expression is co-ordinately controlled.
- Control elements such as an **operator sequence**, which is a DNA sequence that regulates transcription of the structural genes.
- **Regulator gene(s)** whose products recognize the control elements, for example a repressor which binds to and regulates an operator sequence.

The lactose operon

Escherichia coli can use lactose as a source of carbon. The enzymes required for the use of lactose as a carbon source are only synthesized when lactose is available as the sole carbon source. The **lactose operon** (or *lac* operon, *Fig. 1*) consists of three structural genes: ***lacZ***, which codes for β-galactosidase, an enzyme responsible for hydrolysis of lactose to galactose and glucose; ***lacY*** which encodes a galactoside permease which is responsible for lactose transport across the bacterial cell wall; and ***lacA***, which encodes a thiogalactoside transacetylase. The three structural genes are encoded in a single transcription unit, *lacZYA*, which has a single promoter, P_{lac}. This organization means that the three lactose operon structural proteins are expressed together as a **polycistronic mRNA** containing more than one coding region under the same regulatory control. The *lacZYA* transcription unit contains an **operator site** O_{lac} which is positioned between bases –5 and +21 at the 5′-end of the P_{lac} promoter region. This site binds a protein called the **lac repressor** which is a potent inhibitor of transcription when it is bound to the operator. The lac repressor is encoded by a separate regulatory gene ***lacI*** which is also a part of the lactose operon; *lacI* is situated just upstream from P_{lac}.

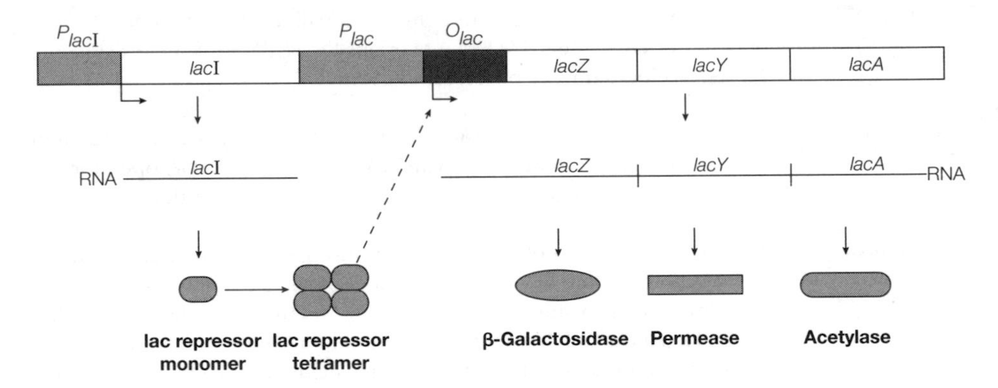

Fig. 1. Structure of the lactose operon.

The lac repressor

The *lacI* gene encodes the lac repressor, which is active as a tetramer of identical subunits. It has a very strong affinity for the *lac* operator-binding site, O_{lac}, and also has a generally high affinity for DNA. The *lac* operator site consists of 28 bp which is **palindromic**. (A **palindrome** has the same DNA sequence when one strand is read left to right in a 5′ to 3′ direction and the complementary strand is read right to left in a 5′ to 3′ direction, see Topic G3). This inverted repeat symmetry of the operator matches the inherent symmetry of the lac repressor which is made up of four identical subunits. In the absence of lactose, the repressor occupies the operator-binding site. It seems that both the lac repressor and the RNA polymerase can bind simultaneously to the *lac* promoter and operator sites. The lac repressor actually increases the binding of the polymerase to the *lac* promoter by two orders of magnitude. This means that when lac repressor is bound to the O_{lac} operator DNA sequence, polymerase is also likely to be bound to the adjacent P_{lac} promoter sequence.

Induction

In the absence of an inducer, the lac repressor blocks all but a very low level of transcription of *lacZYA*. When lactose is added to cells, the low basal level of the permease allows its uptake, and β-galactosidase catalyzes the conversion of some lactose to **allolactose** (*Fig. 2*).

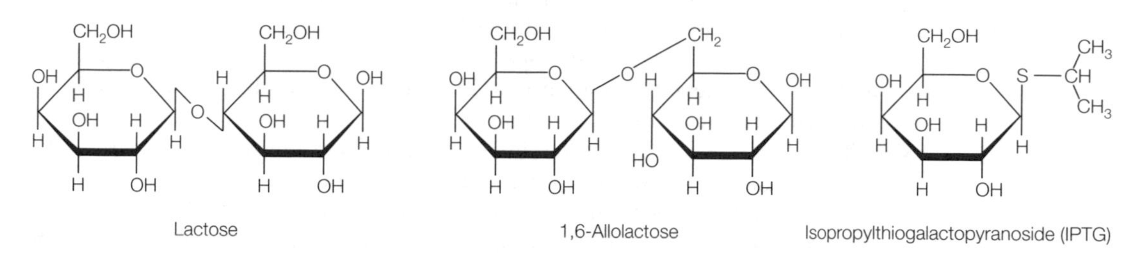

Fig. 2. Structures of lactose, allolactose and IPTG.

Allolactose acts as an inducer and binds to the lac repressor. This causes a change in the conformation of the repressor tetramer, reducing its affinity for the *lac* operator (*Fig. 3*). The removal of the lac repressor from the operator site allows the polymerase (which is already sited at the adjacent promoter) to rapidly begin transcription of the *lacZYA* genes. Thus, the addition of lactose, or a synthetic inducer such as **isopropyl-β-D-thiogalactopyranoside (IPTG)** (*Fig. 2*), very rapidly stimulates transcription of the lactose operon structural genes. The subsequent removal of the inducer leads to an almost immediate inhibition of this induced transcription, since the free lac repressor rapidly re-occupies the operator site and the *lacZYA* RNA transcript is extremely unstable.

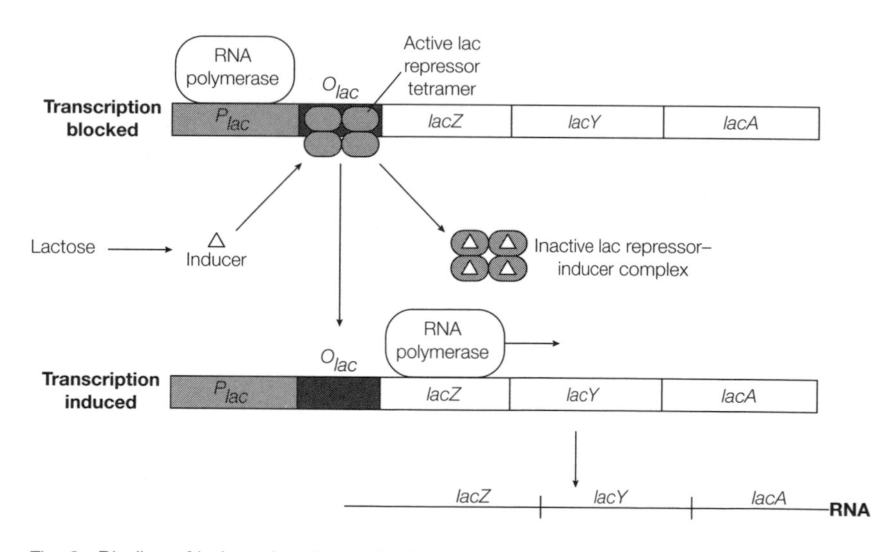

Fig. 3. Binding of inducer inactivates the lac repressor.

cAMP receptor protein

The P_{lac} promoter is not a strong promoter. P_{lac} and related promoters do not have strong –35 sequences and some even have weak –10 consensus sequences. For high level transcription, they require the activity of a specific activator protein called **cAMP receptor protein (CRP)**. CRP may also be called **catabolite activator protein** or **CAP**. When glucose is present, *E. coli* does not require alternative carbon sources such as lactose. Therefore, catabolic operons, such as the lactose operon, are not normally activated. This regulation is mediated by

CRP which exists as a dimer which cannot bind to DNA on its own, nor regulate transcription. Glucose reduces the level of cAMP in the cell. When glucose is absent, the levels of cAMP in *E. coli* increase and CRP binds to cAMP. The CRP–cAMP complex binds to the lactose operon promoter P_{lac} just upstream from the site for RNA polymerase. CRP binding induces a 90° bend in DNA, and this is believed to enhance RNA polymerase binding to the promoter, enhancing transcription by 50-fold.

The CRP-binding site is an inverted repeat and may be adjacent to the promoter (as in the lactose operon), may lie within the promoter itself, or may be much further upstream from the promoter. Differences in the CRP-binding sites of the promoters of different catabolic operons may mediate different levels of response of these operons to cAMP *in vivo*.

L2 THE *TRP* OPERON

Key Notes

The tryptophan operon

The *trp* operon encodes five structural genes involved in tryptophan biosynthesis. One transcript encoding all five enzymes is synthesized using single promoter (P_{trp}) and operator (O_{trp}) sites.

The trp repressor

The trp repressor is the product of a separate operon, the *trpR* operon. The repressor is a dimer which interacts with the *trp* operator only when it is complexed with tryptophan. Repressor binding reduces transcription 70-fold.

The attenuator

A terminator sequence is present in the 162 bp *trp* leader before the start of the *trpE*-coding sequence. It is a rho-independent terminator which terminates transcription at base +140, which is in a run of eight Us just after a hairpin structure. This structure is called the attenuator, because it can cause premature termination of *trp* RNA synthesis.

Leader RNA structure

The *trp* leader RNA contains four regions of complementary sequence which are capable of forming alternative hairpin structures. One of these structures is the attenuator hairpin.

The leader peptide

The leader RNA contains an efficient ribosome-binding site and encodes a 14-amino-acid leader peptide. Codons 10 and 11 of this peptide encode tryptophan. When tryptophan is low the ribosome will pause at these codons.

Attenuation

The RNA polymerase pauses on the DNA template at a site which is at the end of the leader peptide-encoding sequence. When a ribosome initiates translation of the leader peptide, the polymerase continues to transcribe the RNA. If the ribosome pauses at the tryptophan codons (i.e. tryptophan levels are low), it changes the availability of the complementary leader sequences for base pairing so that an alternative RNA hairpin forms instead of the attenuator hairpin. As a result, transcription does not terminate. If the ribosome is not stalled at the tryptophan residues (i.e. tryptophan levels are high), then the attenuator hairpin is able to form and transcription is terminated prematurely.

Importance of attenuation

Attenuation gives rise to 10-fold regulation of transcription by tryptophan. Transcription attenuation occurs in at least six operons involved in amino acid biosynthesis. In some operons (e.g. *His*), it is the only mechanism for feedback regulation of amino acid synthesis.

Related topics

Basic principles of transcription (K1)
Escherichia coli RNA polymerase (K2)
The *E. coli* σ^{70} promoter (K3)
Transcription initiation, elongation and termination (K4)
The *lac* operon (L1)

The tryptophan operon

The *trp* operon encodes five structural genes whose activity is required for tryptophan synthesis (*Fig. 1*). The operon encodes a single transcription unit which produces a 7 kb transcript which is synthesized downstream from the *trp* promoter and *trp* operator sites P_{trp} and O_{trp}. Like many of the operons involved in amino acid biosynthesis, the *trp* operon has evolved systems for co-ordinated expression of these genes when the product of the biosynthetic pathway, tryptophan, is in short supply in the cell. As with the *lac* operon, the RNA product of this transcription unit is very unstable, enabling bacteria to respond rapidly to changing needs for tryptophan.

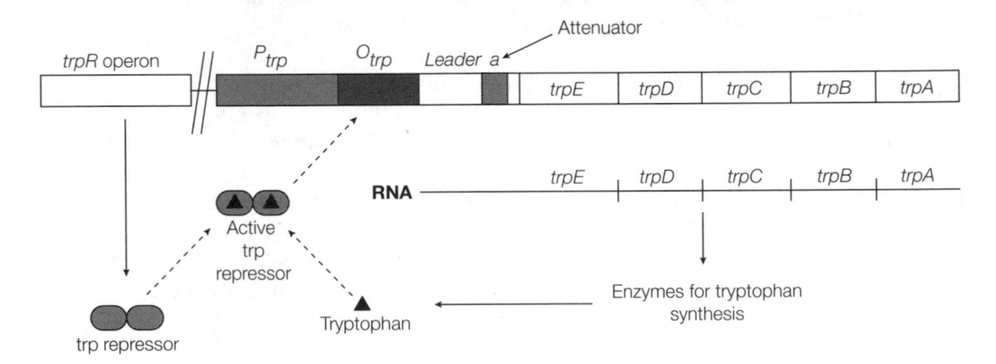

Fig. 1. Structure of the trp *operon and function of the trp repressor.*

The trp repressor

A gene product of the separate *trpR* operon, the trp repressor, specifically interacts with the operator site of the *trp* operon. The symmetrical operator sequence, which forms the trp repressor-binding site, overlaps with the *trp* promoter sequence between bases –21 and +3. The core binding site is a palindrome of 18 bp. The trp repressor binds tryptophan and can only bind to the operator when it is complexed with tryptophan. The repressor is a dimer of two subunits which have structural similarity to the CRP protein and lac repressor (see Topic L1). The repressor dimer has a structure with a central core and two flexible DNA-reading heads each formed from the carboxyl-terminal half of one subunit. Only when tryptophan is bound to the repressor are the reading heads the correct distance apart, and the side chains in the correct conformation, to interact with successive major grooves (see Topic C1) of the DNA at the *trp* operator sequence. Tryptophan, the end-product of the enzymes encoded by the *trp* operon, therefore acts as a **co-repressor** and inhibits its own synthesis through **end-product inhibition**. The repressor reduces transcription initiation by around 70-fold. This is a much smaller transcriptional effect than that mediated by the binding of the lac repressor.

The attenuator

At first, it was thought that the repressor was responsible for all of the transcriptional regulation of the *trp* operon. However, it was observed that the deletion of a sequence between the operator and the *trpE* gene coding region resulted in an increase in both the basal and the activated (derepressed) levels of transcription. This site is termed the attenuator and it lies towards the end of the transcribed leader sequence of 162 nt that precedes the *trpE* initiator codon. The attenuator is a rho-independent terminator site which has a short GC-rich palindrome followed by eight successive U residues (see Topic K4). If this sequence is able to form a hairpin structure in the RNA transcript, then it acts as a highly efficient transcription terminator and only a 140 bp transcript is synthesized.

Leader RNA structure

The leader sequence of the *trp* operon RNA contains four regions of complementary sequence which can form different base-paired RNA structures (*Fig. 2*). These are termed sequences 1, 2, 3 and 4. The attenuator hairpin is the product of the base pairing of sequences 3 and 4 (3:4 structure). Sequences 1 and 2 are also complementary and can form a second 1:2 hairpin. However, sequence 2 is also complementary to sequence 3. If sequences 2 and 3 form a 2:3 hairpin structure, the 3:4 attenuator hairpin cannot be formed and transcription termination will not occur. Under normal conditions, the formation of the 1:2 and 3:4 hairpins is energetically favorable (*Fig. 2a*).

The leader peptide

The leader RNA sequence contains an efficient ribosome-binding site and can form a 14-amino-acid leader peptide encoded by bases 27–68 of the leader RNA. The 10th and 11th codons of this leader peptide encode successive tryptophan residues, the end-product of the synthetic enzymes of the *trp* operon. This leader

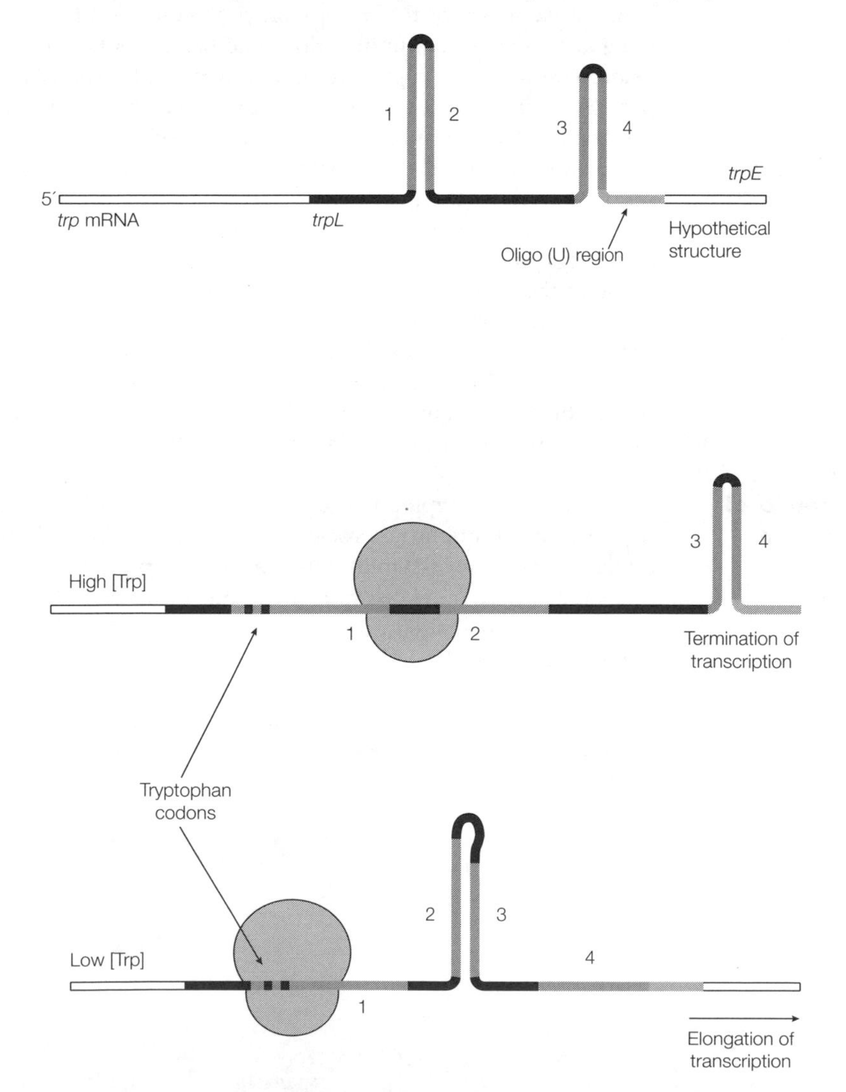

Fig. 2. Transcriptional attenuation in the trp *operon.*

has no obvious function as a polypeptide, and tryptophan is a rare amino acid; therefore, the chances of two tryptophan codons in succession is low and, under conditions of low tryptophan availability, the ribosome would be expected to pause at this site. The function of this leader peptide is to determine tryptophan availability and to regulate transcription termination.

Attenuation

Attenuation depends on the fact that transcription and translation are tightly coupled in *E. coli*; translation can occur as an mRNA is being transcribed. The 3′-end of the *trp* leader peptide coding sequence overlaps complementary sequence 1 (*Fig. 2*); the two trp codons are within sequence 1 and the stop codon is between sequences 1 and 2. The availability of tryptophan (the ultimate product of the enzymes synthesized by the *trp* operon) is sensed through its being required in translation, and determines whether or not the terminator (3:4) hairpin forms in the mRNA.

As transcription of the *trp* operon proceeds, the RNA polymerase pauses at the end of sequence 2 until a ribosome begins to translate the leader peptide. Under conditions of high tryptophan availability, the ribosome rapidly incorporates tryptophan at the two trp codons and thus translates to the end of the leader message. The ribosome is then occluding sequence 2 and, as the RNA polymerase reaches the terminator sequence, the 3:4 hairpin can form, and transcription may be terminated (*Fig. 2b*). This is the process of **attenuation**.

Alternatively, if tryptophan is in scarce supply, it will not be available as an aminoacyl tRNA for translation (see Topic P2), and the ribosome will tend to pause at the two trp codons, occluding sequence 1. This leaves sequence 2 free to form a hairpin with sequence 3 (*Fig. 2c*), known as the **anti-terminator**. The terminator (3:4) hairpin cannot form, and transcription continues into *trpE* and beyond. Thus the level of the end product, tryptophan, determines the probability that transcription will terminate early (attenuation), rather than proceeding through the whole operon.

Importance of attenuation

The presence of tryptophan gives rise to a 10-fold repression of *trp* operon transcription through the process of attenuation alone. Combined with control by the *trp* repressor (70-fold), this means that tryptophan levels exert a 700-fold regulatory effect on expression from the *trp* operon. Attenuation occurs in at least six operons that encode enzymes concerned with amino acid biosynthesis. For example, the *His* operon has a leader which encodes a peptide with seven successive histidine codons. Not all of these other operons have the same combination of regulatory controls that are found in the *trp* operon. The *His* operon has no repressor–operator regulation, and attenuation forms the only mechanism of feedback control.

L3 TRANSCRIPTIONAL REGULATION BY ALTERNATIVE σ FACTORS

Key Notes

Sigma factors

The sigma (σ) factor is responsible for recognition of consensus promoter sequences and is only required for transcription initiation. Many bacteria produce alternative sets of σ factors

Promoter recognition

In *E. coli*, σ^{70} is responsible for recognition of the –10 and –35 consensus sequences. Differing consensus sequences are found in sets of genes which are regulated by the use of alternative σ factors.

Heat shock

Around 17 proteins are specifically expressed in *E. coli* when the temperature is increased above 37°C. These proteins are expressed through transcription by RNA polymerase using an alternative sigma factor σ^{32}. σ^{32} has its own specific promoter consensus sequences.

Sporulation in *Bacillus subtilis*

Under nonoptimal environmental conditions, *B. subtilis* cells form spores through a basic cell differentiation process involving cell partitioning into mother cell and forespore. This process is closely regulated by a set of σ factors which are required to regulate each step in this process.

Bacteriophage σ factors

Many bacteriophages synthesize their own σ factors in order to 'take over' the host cell's own transcription machinery by substituting the normal cellular σ factor and altering the promoter specificity of the RNA polymerase. *B. subtilis* SPO1 phage expresses a cascade of σ factors which allow a defined sequence of expression of early, middle and late phage genes.

Related topics

Cellular classification (A1)
Basic principles of transcription (K1)
Escherichia coli RNA polymerase (K2)
The *E. coli* σ^{70} promoter (K3)
Transcription initiation, elongation and termination (K4)

Sigma factors

The αββ′ω core enzyme of RNA polymerase is unable to start transcription at promoter sites (see Section K). In order to specifically recognize the consensus –35 and –10 elements of general promoters, it requires the σ factor subunit. This subunit is only required for transcription initiation, being released from the core enzyme after initiation and before RNA elongation takes place (see Topic K4). Thus, σ factors appear to be bifunctional proteins that simultaneously can bind to core RNA polymerase and recognize specific promoter sequences in DNA. Many bacteria, including *E. coli*, produce a set of σ factors that recognize different sets of promoters. Transcription initiation from single promoters or small groups of promoters is regulated commonly by single transcriptional repressors (such as the lac repressor) or transcriptional activators (such as the cAMP receptor protein, CRP). However, some environmental conditions require a massive change in the overall pattern of gene expression in the cell. Under such circumstances, bacteria may use a different set of σ factors to direct RNA polymerase binding to different promoter sequences. This process allows the diversion of the cell's basic transcription machinery to the specific transcription of different classes of genes.

Promoter recognition

The binding of an alternative σ factor to RNA polymerase can confer a new promoter specificity on the enzyme responsible for the general RNA synthesis of the cell. Comparisons of promoters activated by polymerase complexed to specific σ factors show that each σ factor recognizes a different combination of sequences centered approximately around the –35 and –10 sites. It seems likely that σ factors themselves contact both of these regions, with the –10 region being most important. The σ^{70} subunit is the most common σ factor in *E. coli* which is responsible for recognition of general promoters which have consensus –35 and –10 elements.

Heat shock

The response to heat shock is one example in *E. coli* where gene expression is altered significantly by the use of different σ factors. When *E. coli* is subjected to an increase in temperature, the synthesis of a set of around 17 proteins, called heat-shock proteins, is induced. If *E. coli* is transferred from 37 to 42°C, this burst of heat-shock protein synthesis is transient. However if the increase in temperature is more extreme, such as to 50°C, where growth of *E. coli* is not possible, then the heat-shock proteins are the only proteins synthesized. The promoters for *E. coli* heat-shock protein-encoding genes are recognized by a unique form of RNA polymerase holoenzyme containing a variant σ factor, σ^{32}, which is encoded by the *rpoH* gene. σ^{32} is a minor protein which is much less abundant than σ^{70}. Holoenzyme containing σ^{32} acts exclusively on promoters of heat-shock genes and does not recognize the general consensus promoters of most of the other genes (*Fig. 1*). Heat-shock promoters accordingly have different sequences to other general promoters which bind to σ^{70}.

Consensus promoter	*–35 sequence*			*–10 sequence*
Standard (σ^{70})	- - - - - - - - - TTGACA	·······*16–18 bp*·······		TATAAT
Heat shock (σ^{32})	T - - C - C - C TTGAA	·····*13–15 bp*·····	CCC	CAT - T

Fig. 1. Comparison of the heat-shock (σ^{32}) and general (σ^{70}) responsive promoters.

Sporulation in *Bacillus subtilis*

Vegetatively growing *B. subtilis* cells form bacterial spores (see Topic A1) in response to a sub-optimal environment. The formation of a spore (or sporulation) requires drastic changes in gene expression, including the cessation of the synthesis of almost all of the proteins required for vegetative existence as well as the production of proteins which are necessary for the resumption of protein synthesis when the spore germinates under more optimal conditions.

The process of spore formation involves the asymmetrical division of the bacterial cell into two compartments, the forespore, which forms the spore, and the mother cell, which is eventually discarded. This system is considered one of the most fundamental examples of cell differentiation. The RNA polymerase in *B. subtilis* is functionally identical to that in *E. coli*. The vegetatively growing *B. subtilis* contains a diverse set of σ factors. Sporulation is regulated by a further set of σ factors in addition to those of the vegetative cell. Different σ factors are specifically active before cell partition occurs, in the forespore and in the mother cell. Cross-regulation of this compartmentalization permits the forespore and mother cell to tightly co-ordinate the differentiation process.

Bacteriophage σ factors

Some bacteriophages provide new σ subunits to endow the host RNA polymerase with a different promoter specificity and hence to selectively express their own phage genes (e.g. phage T4 in *E. coli* and SPO1 in *B. subtilis*). This strategy is an effective alternative to the need for the phage to encode its own complete polymerase (e.g. bacteriophage T7, see Topic K2). The *B. subtilis* bacteriophage SPO1 expresses a '**cascade**' of σ factors in sequence to allow its own genes to be transcribed at specific stages during virus infection. Initially, **early genes** are expressed by the normal bacterial holoenzyme. Among these early genes is the gene encoding σ^{28}, which then displaces the bacterial σ factor from the RNA polymerase. The σ^{28}-containing holoenzyme is then responsible for expression of the **middle genes**. The phage middle genes include genes 33 and 34 which specificy a further σ factor that is responsible for the specific transcription of **late genes**. In this way, the bacteriophage uses the host's RNA polymerase machinery and expresses its genes in a defined sequential order.

M1 The three RNA polymerases: characterization and function

Key Notes

Eukaryotic RNA polymerases

Three eukaryotic polymerases transcribe different sets of genes. Their activities are distinguished by their different sensitivities to the fungal toxin α-amanitin.

- RNA polymerase I is located in the nucleoli. It is responsible for the synthesis of the precursors of most rRNAs.
- RNA polymerase II is located in the nucleoplasm and is responsible for the synthesis of mRNA precursors and some small nuclear RNAs.
- RNA polymerase III is located in the nucleoplasm. It is responsible for the synthesis of the precursors of 5S rRNA, tRNAs and other small nuclear and cytosolic RNAs.

RNA polymerase subunits

Each RNA polymerase has 12 or more different subunits. The largest two subunits are similar to each other and to the β′ and β subunits of *E. coli* RNA polymerase. Other subunits in each enzyme have homology to the α subunit of the *E. coli* enzyme. Five additional subunits are common to all three polymerases, and others are polymerase specific.

Eukaryotic RNA polymerase activities

Like prokaryotic RNA polymerases, the eukaryotic enzymes do not require a primer and synthesize RNA in a 5′ to 3′ direction. Unlike bacterial polymerases, they require accessory factors for DNA binding.

The CTD of RNA Pol II

The largest subunit of RNA polymerase II has a seven amino acid repeat at the C terminus called the carboxyl-terminal domain (CTD). This sequence, Tyr-Ser-Pro-Thr-Ser-Pro-Ser, is repeated 52 times in the mouse RNA polymerase II and is subject to phosphorylation.

Related topics

Protein analysis (B3)
Escherichia coli RNA polymerase (K2)
RNA Pol I genes: the ribosomal repeat (M2)
RNA Pol III genes: 5S and tRNA transcription (M3)
RNA Pol II genes: promoters and enhancers (M4)
General transcription factors and RNA Pol II initiation (M5)
Examples of transcriptional regulation (N2)

Eukaryotic RNA polymerases

The mechanism of eukaryotic transcription is similar to that in prokaryotes. However, the large number of polypeptides associated with the eukaryotic transcription machinery makes it far more complex. Three different RNA polymerase complexes are responsible for the transcription of different types of eukaryotic genes. The different RNA polymerases were identified by chromatographic purification of the enzymes and elution at different salt concentrations (Topic B4). Each RNA polymerase has a different sensitivity to the fungal toxin **α-amanitin** and this can be used to distinguish their activities.

- **RNA polymerase I** (RNA Pol I) transcribes most rRNA genes. It is located in the nucleoli and is insensitive to α-amanitin
- **RNA polymerase II** (RNA Pol II) transcribes all protein-coding genes and some small nuclear RNA (snRNA) genes. It is located in the nucleoplasm and is very sensitive to α-amanitin.
- **RNA polymerase III** (RNA Pol III) transcribes the genes for tRNA, 5S rRNA, U6 snRNA and certain other small RNAs. It is located in the nucleoplasm and is moderately sensitive to α-amanitin.

In addition to these nuclear enzymes, eukaryotic cells contain additional polymerases in mitochondria and chloroplasts.

RNA polymerase subunits

All three polymerases are large enzymes containing 12 or more subunits. The genes encoding the two largest subunits of each RNA polymerase have homology (related DNA coding sequences) to each other. All of the three eukaryotic polymerases contain subunits which have homology to subunits within the *E. coli* core RNA polymerase $\alpha_2\beta\beta'$ (see Topic K2). The largest subunit of each eukaryotic RNA polymerase is similar to the β′ subunit of the *E. coli* polymerase, and the second largest subunit is similar to the β subunit which contains the active site of the *E. coli* enzyme. The functional significance of this homology is supported by the observation that the second largest subunits of the eukaryotic RNA polymerases also contain the active sites. Two subunits which are common to RNA Pol I and RNA Pol III, and a further subunit which is specific to RNA Pol II, have homology to the *E. coli* RNA polymerase α subunit. At least five other smaller subunits are common to the three different polymerases. Each polymerase also contains an additional four to seven subunits which are only present in one type.

Eukaryotic RNA polymerase activities

Like bacterial RNA polymerases, each of the eukaryotic enzymes catalyzes transcription in a 5′ to 3′ direction and synthesizes RNA complementary to the antisense template strand. The reaction requires the precursor nucleotides ATP, GTP, CTP and UTP and does not require a primer for transcription initiation. The purified eukaryotic RNA polymerases, unlike the purified bacterial enzymes, require the presence of additional initiation proteins before they are able to bind to promoters and initiate transcription.

The CTD of RNA Pol II

The carboxyl end of RNA Pol II contains a stretch of seven amino acids that is repeated 52 times in the mouse enzyme and 26 times in yeast. This heptapeptide has the sequence **Tyr-Ser-Pro-Thr-Ser-Pro-Ser** and is known as the **carboxyl-terminal domain** or **CTD**. These repeats are essential for viability. The CTD sequence may be phosphorylated at the serines and some tyrosines. *In vitro* studies have shown that the CTD is unphosphorylated at transcription initiation, but phosphorylation occurs during transcription elongation as the RNA polymerase leaves the promoter. Since RNA Pol II catalyzes the synthesis of all of the eukaryotic protein-coding genes, it is the most important RNA polymerase for the study of differential gene expression. The CTD has been shown to be an important target for differential activation of transcription elongation (see Topics M5 and N2).

M2 RNA Pol I genes: the ribosomal repeat

Key Notes

Ribosomal RNA genes
The pre-rRNA transcription units contain three sequences that encode the 18S, 5.8S and 28S rRNAs. Pre-rRNA transcription units are arranged in clusters in the genome as long tandem arrays separated by nontranscribed spacer sequences.

Role of the nucleolus
Pre-rRNA is synthesized by RNA polymerase I (RNA Pol I) in the nucleolus. The arrays of rRNA genes loop together to form the nucleolus and are known as nucleolar organizer regions.

RNA Pol I promoters
The pre-rRNA promoters consist of two transcription control regions. The core element includes the transcription start site. The upstream control element (UCE) is approximately 50 bp long and begins at around position –100.

Upstream binding factor
Upstream binding factor (UBF) binds to the UCE. It also binds to a different site in the upstream part of the core element, causing the DNA to loop between the two sites.

Selectivity factor 1
Selectivity factor 1 (SL1) binds to and stabilizes the UBF–DNA complex. SL1 then allows binding of RNA Pol I and initiation of transcription

TBP and TAF_Is
SL1 is made up of four subunits. These include the TATA-binding protein (TBP) which is required for transcription initiation by all three RNA polymerases. The other factors are RNA Pol I-specific TBP-associated factors called TAF_Is.

Other rRNA genes
Acanthamoeba has a simple transcription control system. This has a single control element and a single factor TIF-1, which are required for RNA Pol I binding and initiation at the rRNA promoter.

Related topics

Genome complexity (D4)
Escherichia coli RNA polymerase (K2)
The *E. coli* σ^{70} promoter (K3)
Transcription initiation, elongation and termination (K4)
The three RNA polymerases: characterization and function (M1)
rRNA processing and ribosomes (O1)

Ribosomal RNA genes

RNA polymerase I (RNA Pol I) is responsible for the continuous synthesis of rRNAs during interphase. Human cells contain five clusters of around 40 copies of the rRNA gene situated on different chromosomes (see *Fig. 1* and Topic D4). Each rRNA gene produces a 45S rRNA transcript which is about 13 000 nt long (see Topic D4). This transcript is cleaved to give one copy each of the 28S RNA (5000 nt), 18S (2000 nt) and 5.8S (160 nt) rRNAs (see Topic O1). The continuous transcription of multiple gene copies of the RNAs is essential for sufficient production of the processed rRNAs which are packaged into ribosomes.

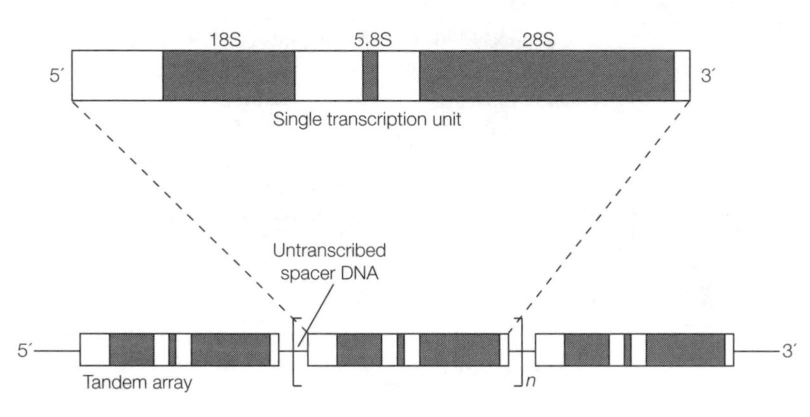

Fig. 1. Ribosomal RNA transcription units.

Role of the nucleolus

Each rRNA cluster is known as a nucleolar organizer region, since the nucleolus contains large loops of DNA corresponding to the gene clusters. After a cell emerges from mitosis, rRNA synthesis restarts and tiny nucleoli appear at the chromosomal locations of the rRNA genes. During active rRNA synthesis, the pre-rRNA transcripts are packed along the rRNA genes and may be visualized in the electron microscope as '**Christmas tree structures**'. In these structures, the RNA transcripts are densely packed along the DNA and stick out perpendicularly from the DNA. Short transcripts can be seen at the start of the gene, which get longer until the end of the transcription unit, which is indicated by the disappearance of the RNA transcripts.

RNA Pol I promoters

Mammalian pre-rRNA gene promoters have a bipartite transcription control region (*Fig. 2*). The core element includes the transcription start site and encompasses bases –31 to +6. This sequence is essential for transcription. An additional element of around 50–80 bp named the **upstream control element** (**UCE**) begins about 100 bp upstream from the start site (–100). The UCE is responsible for an increase in transcription of around 10- to 100-fold compared with that from the core element alone.

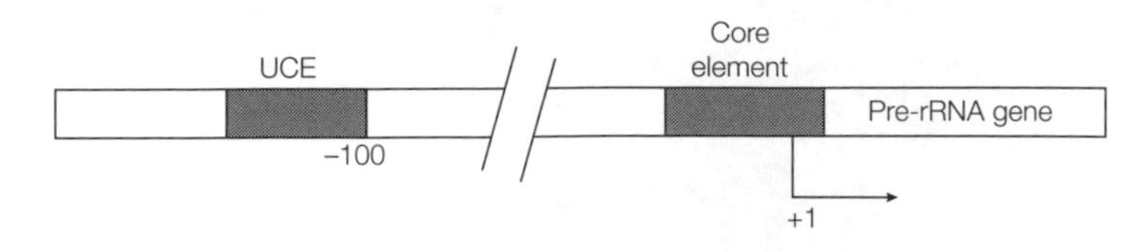

Fig. 2. Structure of a mammalian pre-rRNA promoter.

Upstream binding factor

A specific DNA-binding protein, called **upstream binding factor**, or **UBF**, binds to the UCE. As well as binding to the UCE, UBF binds to a sequence in the upstream part of the core element. The sequences in the two UBF-binding sites have no obvious similarity. One molecule of UBF is thought to bind to each sequence element. The two molecules of UBF may then bind to each other through protein–protein interactions, causing the intervening DNA to form a loop between the two binding sites (*Fig. 3*). A low rate of basal transcription is seen in the absence of UBF, and this is greatly stimulated in the presence of UBF.

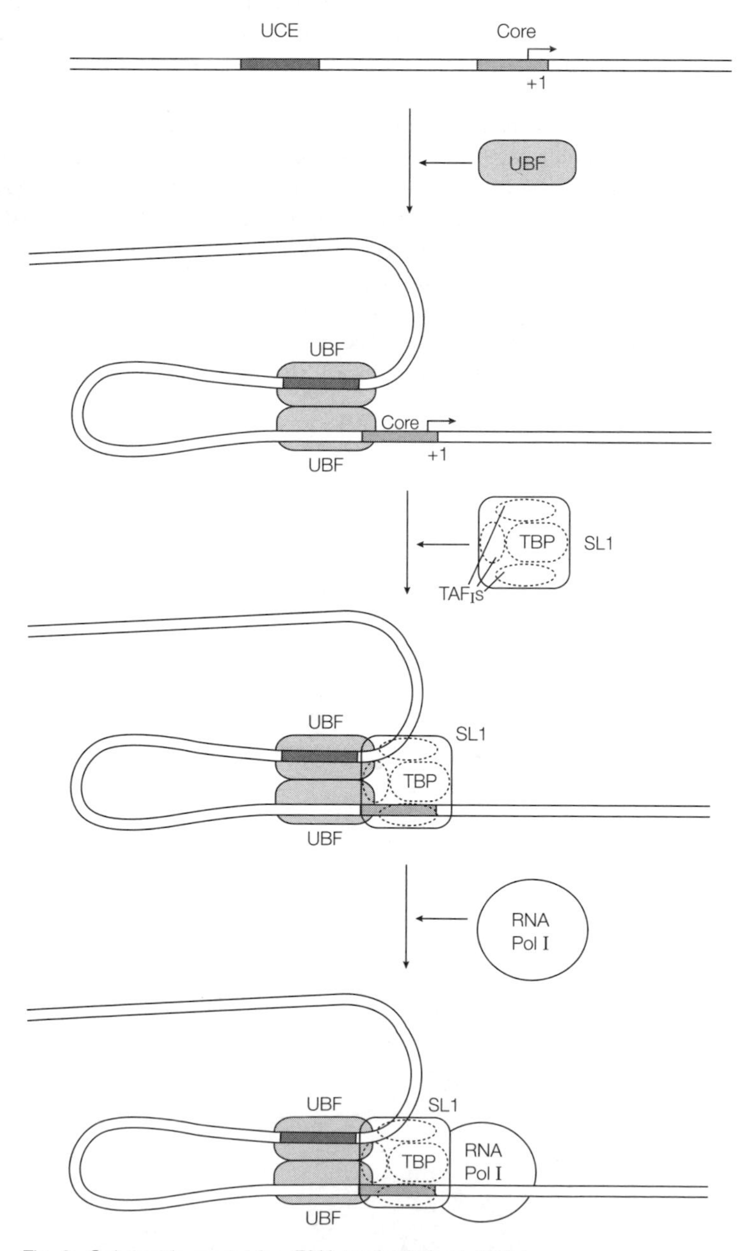

Fig. 3. Schematic model for rRNA transcription initiation.

Selectivity factor 1

An additional factor, called **selectivity factor (SL1)** is essential for RNA Pol I transcription. SL1 binds to and stabilizes the UBF–DNA complex and interacts with the free downstream part of the core element. SL1 binding allows RNA Pol I to bind to the complex and initiate transcription, and is essential for rRNA transcription.

TBP and TAF_Is

SL1 has now been shown to contain several subunits, including a protein called TBP (TATA-binding protein). TBP is required for initiation by all three eukaryotic RNA polymerases (see Topics M1, M3 and M5), and seems to be a critical factor in eukaryotic transcription. The other three subunits of SL1 are referred to as TBP-associated factors or TAFs, and those subunits required for RNA Pol I transcription are referred to as TAF_Is.

Other rRNA genes

In *Acanthamoeba*, a simple eukaryote, there is a single control element in the promoter region of the rRNA genes around 12–72 bp upstream from the transcription start site. A factor named TIF-1, a homolog of SL-1, binds to this site and allows binding of RNA Pol I and transcription initiation. When the polymerase moves along the DNA away from the initiation site, the TIF-1 factor remains bound, permitting initiation of another polymerase and multiple rounds of transcription. This is therefore a very simple transcription control system. It seems that vertebrates have evolved an additional UBF which is responsible for sequence-specific targeting of SL1 to the promoter.

M3 RNA Pol III genes: 5S and tRNA transcription

Key Notes

RNA polymerase III

RNA polymerase III (RNA Pol III) has 16 or more subunits. The enzyme is located in the nucleoplasm and it synthesizes the precursors of 5S rRNA, the tRNAs and other small nuclear and cytosolic RNAs.

tRNA genes

Two transcription control regions, called the A box and the B box, lie downstream from the transcription start site. These sequences are therefore both conserved sequences in tRNAs but also conserved promoter sequences in the DNA. TFIIIC binds to the A and B boxes in the tRNA promoter; TFIIIB binds to the TFIIIC–DNA complex and interacts with DNA upstream from the TFIIIC-binding site. TFIIIB contains three subunits, TBP, BRF and B′′, and is responsible for RNA Pol III recruitment and *hence* transcription initiation.

5S rRNA genes

The genes for 5S rRNA are organized in a tandem cluster. The 5S rRNA promoter contains a conserved C box 81–99 bases downstream from the start site, and a conserved A box at around 50–65 bases downstream. TFIIIA binds strongly to the C box promoter sequence. TFIIIC then binds to the TFIIIA–DNA complex, interacting also with the A box sequence. This complex allows TFIIIB to bind, recruit the polymerase, and initiate transcription.

Alternative RNA Pol III promoters

A number of RNA Pol III promoters are regulated by upstream as well as downstream promoter sequences. Other promoters require only upstream sequences, including the TATA box and other sequences found in RNA Pol II promoters.

RNA Pol III termination

The RNA polymerase can terminate transcription without accessory factors. A cluster of A residues is often sufficient for termination.

Related topics

Escherichia coli RNA polymerase (K2)
The three RNA polymerases: characterization and function (M1)
RNA Pol I genes: the ribosomal repeat (M2)
RNA Pol II genes: promoters and enhancers (M4)
General transcription factors and RNA Pol II initiation (M5)
rRNA processing and ribosomes (O1)
tRNA structure and function (P2)

RNA polymerase III

RNA polymerase III (RNA Pol III) is a complex of at least 16 different subunits. Like RNA Pol II, it is located in the nucleoplasm. RNA Pol III synthesizes the precursors of 5S rRNA, the tRNAs and other snRNAs and cytosolic RNAs.

tRNA genes

The initial transcripts produced from tRNA genes are precursor molecules which are processed into mature RNAs (see Topic O2). The transcription control regions of tRNA genes lie after the transcription start site within the transcription unit. There are two highly conserved sequences within the DNA encoding the tRNA, called the A box (5′-TGGCNNAGTGG-3′) and the B box (5′-GGTTCGANNCC-3′). These sequences also encode important sequences in the tRNA itself, called the D-loop and the TΨC loop (see Topic P2). This means that highly conserved sequences within the tRNAs are also highly conserved promoter DNA sequences.

Two complex DNA-binding factors have been identified which are required for tRNA transcription initiation by RNA Pol III (*Fig. 1*). **TFIIIC** binds to both the A and B boxes in the tRNA promoter. **TFIIIB** binds 50 bp upstream from the A box. TFIIIB consists of three subunits, one of which is TBP, the general

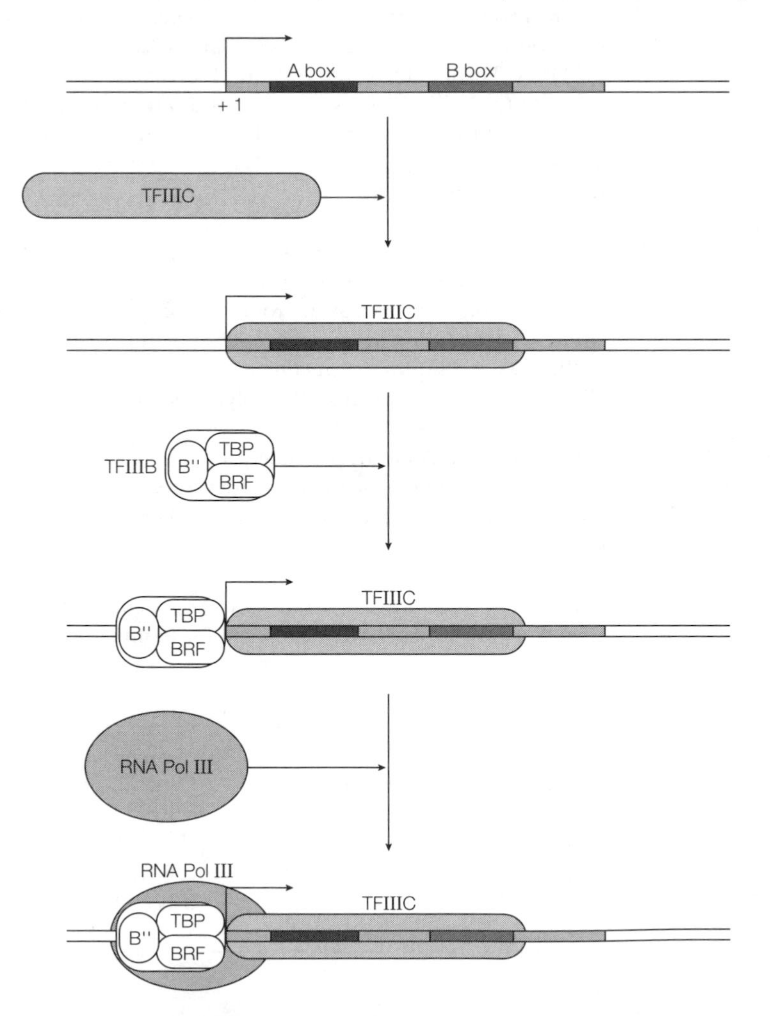

Fig. 1. Initiation of transcription at a eukaryotic tRNA promoter.

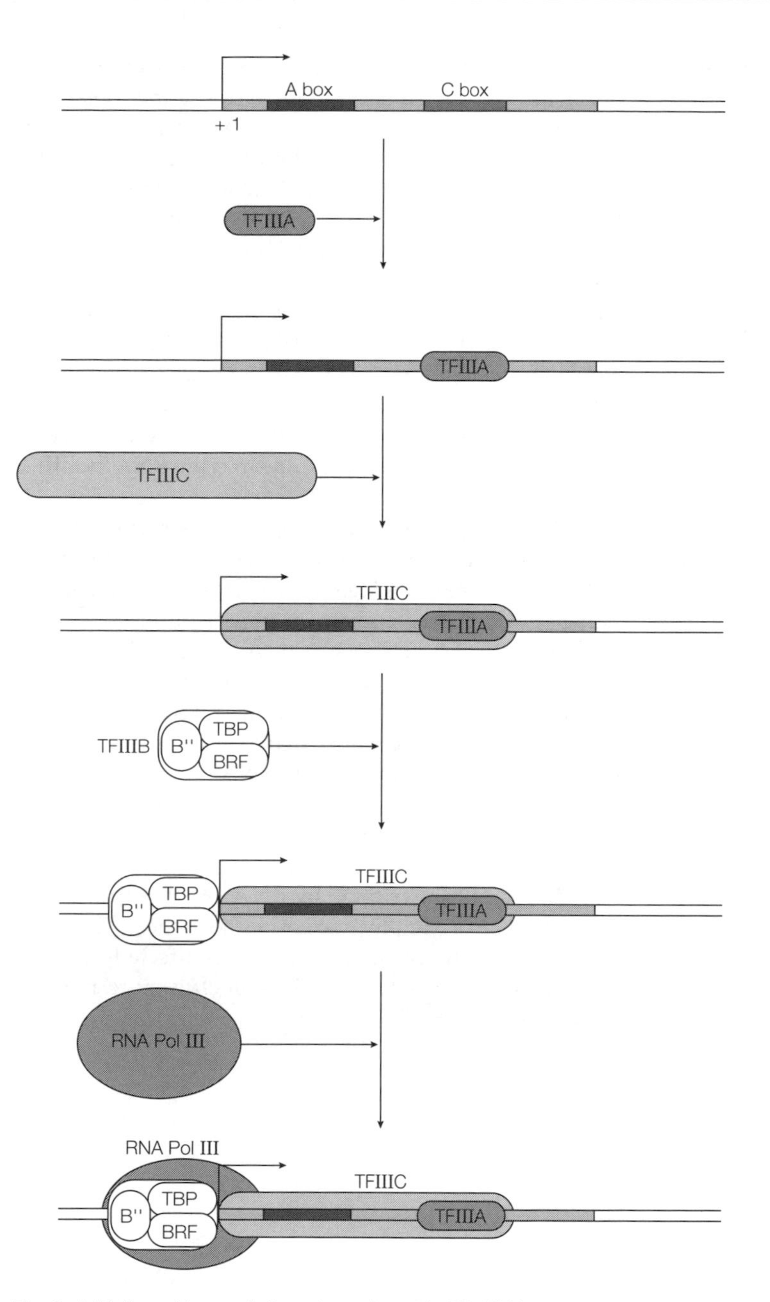

Fig. 2. Initiation of transcription at a eukaryotic 5S rRNA promoter.

initiation factor required by all three RNA polymerases (see Topics M2 and M5). The second is called BRF (TFIIB-related factor, since it has homology to TFIIB, the RNA Pol II initiation factor, see Topic M5). The third subunit is called B''. TFIIIB has no sequence specificity and therefore its binding site appears to be determined by the position of TFIIIC binding to the DNA. TFIIIB allows RNA Pol III to bind and initiate transcription. Once TFIIIB has bound, TFIIIC can be removed without affecting transcription. TFIIIC is therefore an assembly factor for the positioning of the initiation factor TFIIIB.

5S rRNA genes

RNA Pol III transcribes the 5S rRNA component of the large ribosomal subunit. This is the only rRNA subunit to be transcribed separately. Like the other rRNA genes which are transcribed by RNA Pol I, the 5S rRNA genes are tandemly arranged in a gene cluster. In humans, there is a single cluster of around 2000 genes. The promoters of 5S rRNA genes contain an internal control region called the C box which is located 81–99 bp downstream from the transcription start site. A second sequence termed the A box around bases +50 to +65 is also important.

The C box of the 5S rRNA promoter acts as the binding site for a specific DNA-binding protein, **TFIIIA** (*Fig. 2*). TFIIIA acts as an assembly factor which allows TFIIIC to interact with the 5S rRNA promoter. The A box may also stabilize TFIIIC binding. TFIIIC is then bound to the DNA at an equivalent position relative to the start site as in the tRNA promoter. Once TFIIIC has bound, TFIIIB can interact with the complex and recruit RNA Pol III to initiate transcription.

Alternative RNA Pol III promoters

Many RNA Pol III genes also rely on upstream sequences for the regulation of their transcription. Some promoters such as the U6 small nuclear RNA (**U6 snRNA**) and small RNA genes from the Epstein–Barr virus use only regulatory sequences upstream from their transcription start sites. The coding region of the U6 snRNA has a characteristic A box. However, this sequence is not required for transcription. The U6 snRNA upstream sequence contains sequences typical of RNA Pol II promoters, including a TATA box (see Topic M4) at bases –30 to –23. These promoters also share several other upstream transcription factor-binding sequences with many U RNA genes which are transcribed by RNA Pol II. These observations suggest that common transcription factors can regulate both RNA Pol II and RNA Pol III genes.

RNA Pol III termination

Termination of transcription by RNA Pol III appears only to require polymerase recognition of a simple nucleotide sequence. This consists of clusters of dA residues whose termination efficiency is affected by surrounding sequence. Thus the sequence 5′-GCAAAAGC-3′ is an efficient termination signal in the *Xenopus borealis* somatic 5S rRNA gene.

M4 RNA Pol II genes: promoters and enhancers

Key Notes

RNA polymerase II RNA polymerase II (RNA Pol II) catalyzes the synthesis of the mRNA precursors for all protein-coding genes. RNA Pol II-transcribed pre-mRNAs are processed through cap addition, poly(A) tail addition and splicing.

Promoters Many RNA Pol II promoters contain a sequence called a TATA box which is situated 25–30 bp upstream from the start site. Other genes contain an initiator element which overlaps the start site. These elements are required for basal transcription complex formation and transcription initiation.

Upstream regulatory elements Elements within the 100–200 bp upstream from the promoter are generally required for efficient transcription. Examples include the SP1 and CCAAT boxes.

Enhancers These are sequence elements which can activate transcription from thousands of base pairs upstream or downstream. They may be tissue-specific or ubiquitous in their activity and contain a variety of sequence motifs. There is a continuous spectrum of regulatory sequence elements which span from the extreme long-range enhancer elements to the short-range promoter elements.

Related topics

The *E. coli* σ^{70} promoter (K3)
RNA Pol I genes: the ribosomal repeat (M2)
RNA Pol III genes: 5S and tRNA transcription (M3)
General transcription factors and RNA Pol II initiation (M5)
Eukaryotic transcription factors (N1)
mRNA processing, hnRNPs and snRNPs (O3)

RNA polymerase II

RNA polymerase II (RNA Pol II) is located in the nucleoplasm. It is responsible for the transcription of all protein-coding genes and some small nuclear RNA genes. The pre-mRNAs must be processed after synthesis by cap formation at the 5′-end of the RNA and poly(A) addition at the 3′-end, as well as removal of introns by splicing (see Topic O3).

Promoters

Many eukaryotic promoters contain a sequence called the **TATA box** around 25–35 bp upstream from the start site of transcription (*Fig. 1*). It has the 7 bp consensus sequence 5′-TATA(A/T)A(A/T)-3′ although it is now known that the protein which binds to the TATA box, TBP, binds to an 8 bp sequence that includes an additional downstream base pair, whose identity is not important (see Topic M5). The TATA box acts in a similar way to an *E. coli* promoter –10 sequence to position the RNA Pol II for correct transcription initiation (see Topic K3). While the sequence around the TATA box is critical, the sequence between the TATA box and the transcription start site is not critical. However, the

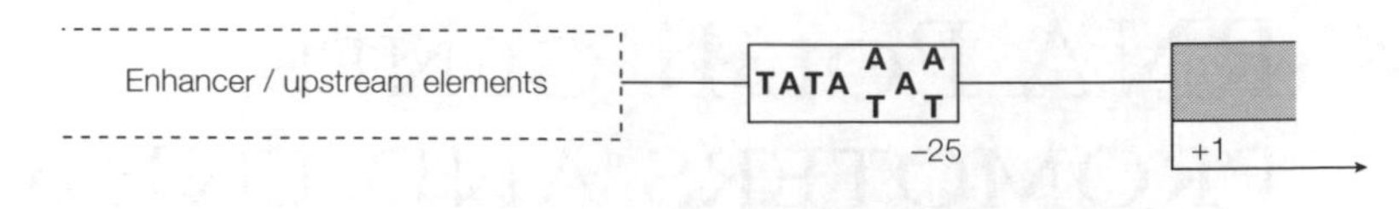

Fig. 1. RNA Pol II promoter containing TATA box.

spacing between the TATA box and the start site is important. Around 50% of the time, the start site of transcription is an adenine residue.

Some eukaryotic genes contain an **initiator element** instead of a TATA box. The initiator element is located around the transcription start site. Many initiator elements have a C at position –1 and an A at +1. Other promoters have neither a TATA box nor an initiator element. These genes are generally transcribed at low rates, and initiation of transcription may occur at different start sites over a length of up to 200 bp. These genes often contain a GC-rich 20–50 bp region within the first 100–200 bp upstream from the start site.

Upstream regulatory elements

The low activity of basal promoters is greatly increased by the presence of other elements located upstream of the promoter. These elements are found in many genes which vary widely in their levels of expression in different tissues. Two common examples are the **SP1 box**, which is found upstream of many genes both with and without TATA boxes, and the **CCAAT box**. Promoters may have one, both or multiple copies of these sequences. These sequences which are often located within 100–200 bp upstream from the promoter are referred to as **upstream regulatory elements** (**UREs**) and play an important role in ensuring efficient transcription from the promoter.

Enhancers

Transcription from many eukaryotic promoters can be stimulated by control elements that are located many thousands of base pairs away from the transcription start site. This was first observed in the genome of the DNA virus SV40. A sequence of around 100 bp from SV40 DNA can significantly increase transcription from a basal promoter even when it is placed far upstream. Enhancer sequences are characteristically 100–200 bp long and contain multiple sequence elements which contribute to the total activity of the enhancer. They may be ubiquitous or cell type-specific. Classically, enhancers have the following general characteristics:

- they exert strong activation of transcription of a linked gene from the correct start site.
- They activate transcription when placed in either orientation with respect to linked genes.
- They are able to function over long distances of more than 1 kb whether from an upstream or downstream position relative to the start site.
- They exert preferential stimulation of the closest of two tandem promoters.

However, as more enhancers and promoters have been identified, it has been shown that the upstream promoter and enhancer motifs overlap physically and functionally. There seems to be a continuum between classic enhancer elements and those promoter elements which are orientation specific and must be placed close to the promoter to have an effect on transcriptional activity.

M5 General transcription factors and RNA Pol II initiation

Key Notes

RNA Pol II basal transcription factors

A complex series of basal transcription factors have been characterized which bind to RNA Pol II promoters and together initiate transcription. These factors and their component subunits are still being identified. They were originally named TFIIA, B, C, etc.

TFIID

TFIID binds to the TATA box. It is a multiprotein complex of the TATA-binding protein (TBP) and multiple accessory factors which are called TBP-associated factors or TAF_{II}s.

TBP

TBP is a transcription factor required for transcription initiation by all three RNA polymerases. It has a saddle structure which binds to the minor groove of the DNA at the TATA box, unwinding the DNA and introducing a 45° bend.

TFIIA

TFIID binding to the TATA box is enhanced by TFIIA. TFIIA appears to stop inhibitory factors binding to TFIID. These inhibitory factors would otherwise block further assembly of the transcription complex.

TFIIB and RNA polymerase binding

TFIIB binds to TFIID and acts as a bridge factor for RNA polymerase binding. The RNA polymerase binds to the complex associated with TFIIF.

Factors binding after RNA polymerase

After RNA polymerase binding, TFIIE, TFIIH and TFIIJ associate with the transcription complex in a defined binding sequence. Each of these proteins is required for transcription *in vitro*.

CTD phosphorylation by TFIIH

TFIIH phosphorylates the carboxyl-terminal domain (CTD) of RNA Pol II. This results in formation of a processive polymerase complex.

The initiator transcription complex

TBP is recruited to initiator-containing promoters by a further DNA-binding protein. The TBP is then able to initiate transcription initiation by a similar mechanism to that in TATA box-containing promoters.

Related topics

Escherichia coli RNA polymerase (K2)
Transcription initiation, elongation and termination (K4)
The three RNA polymerases: characterization and function (M1)
RNA Pol I genes: the ribosomal repeat (M2)
RNA Pol III genes: 5S and tRNA transcription (M3)
RNA Pol II genes: promoters and enhancers (M4)
Eukaryotic transcription factors (N1)
Examples of transcriptional regulation (N2)

RNA Pol II basal transcription factors

A series of nuclear transcription factors have been identified, purified and cloned. These are required for basal transcription initiation from RNA Pol II promoter sequences *in vitro*. These multisubunit factors are named transcription factor IIA, IIB, etc. (TFIIA, etc.). They have been shown to assemble on basal promoters in a specific order (*Fig. 1*) and they may be subject to multiple levels of regulation (see Topic N1).

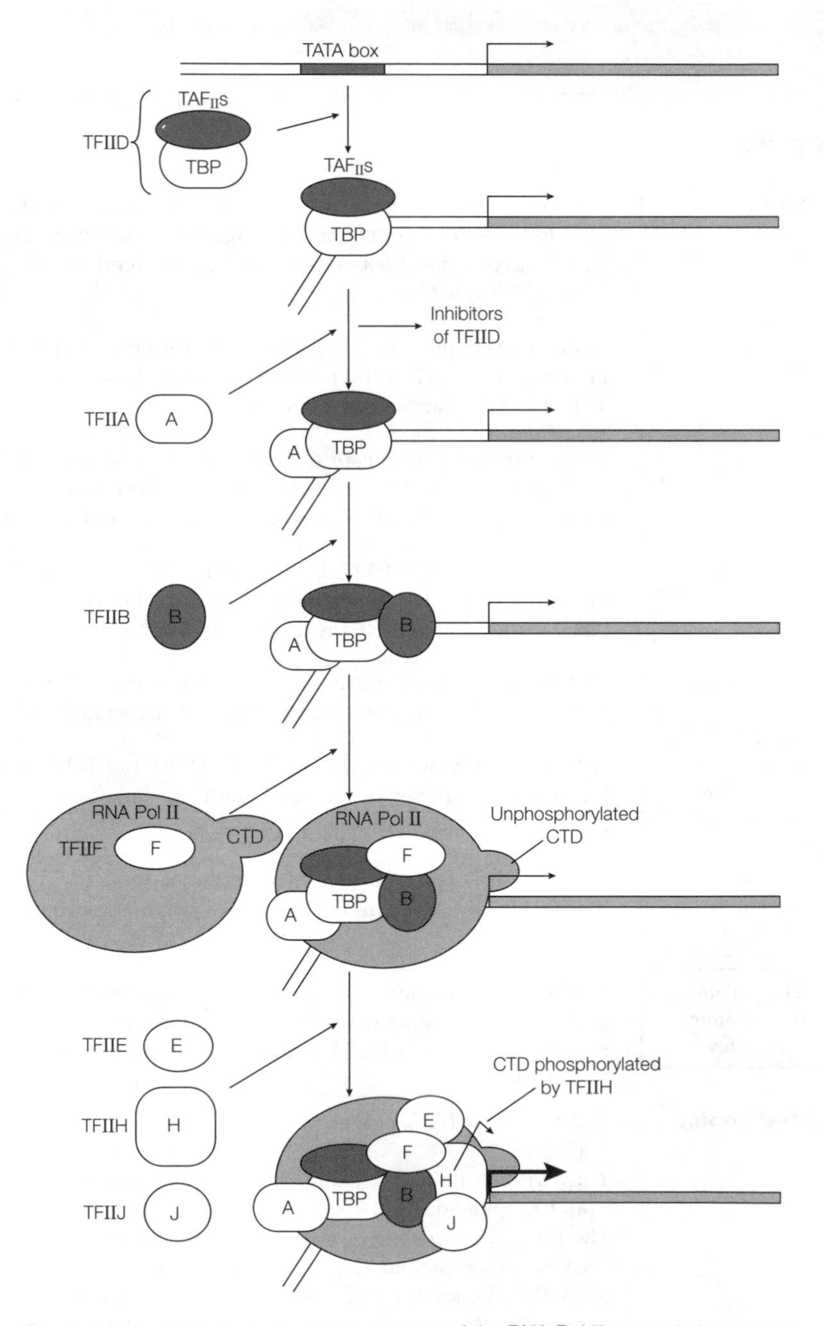

Fig. 1. A schematic diagram of the assembly of the RNA Pol II transcription initiation complex at a TATA box promoter.

TFIID

In promoters containing a TATA box, the RNA Pol II transcription factor **TFIID** is responsible for binding to this key promoter element. The binding of TFIID to the TATA box is the earliest stage in the formation of the RNA Pol II transcription initiation complex. TFIID is a multiprotein complex in which only one polypeptide, **TATA-binding protein (TBP)** binds to the TATA box. The complex also contains other polypeptides known as **TBP-associated factors (TAF_{II}s)**. It seems that in mammalian cells, TBP binds to the TATA box and is then joined by at least eight TAF_{II}s to form TFIID.

TBP

TBP is present in all three eukaryotic transcription complexes (in SL1, TFIIIB and TFIID) and clearly plays a major role in transcription initiation (see Topics M2 and M3). TBP is a monomeric protein. All eukaryotic TBPs analyzed have very highly conserved C-terminal domains of 180 residues and this conserved domain functions as well as the full-length protein in *in vivo* transcription. The function of the less conserved N-terminal domain is therefore not fully understood. TBP has been shown to have a saddle structure with an overall dyad symmetry, but the two halves of the molecule are not identical. TBP interacts with DNA in the minor groove so that the inside of the saddle binds to DNA at the TATA box and the outside surface of the protein is available for interactions with other protein factors. Binding of TBP deforms the DNA so that it is bent into the inside of the saddle and unwound. This results in a kink of about 45° between the first two and last two base pairs of the 8 bp TATA element. A TBP with a mutation in its TATA box-binding domain retains its function for transcription by RNA Pol I and Pol III (see Topics M2 and M3), but it inhibits transcription initiation by RNA Pol II. This indicates that the other two polymerases use TBP to initiate transcription, but the precise role of TBP in these complexes remains unclear.

TFIIA

TFIIA binds to TFIID and enhances TFIID binding to the TATA box, stabilizing the TFIID–DNA complex. TFIIA is made up of at least three subunits. In *in vitro* transcription studies, as TFIID is purified, the requirement for TFIIA is lost. In the intact cell, TFIIA appears to counteract the effects of inhibitory factors such as DR1 and DR2 with which TFIID is associated. It seems likely that TFIIA binding to TFIID prevents binding of these inhibitors and allows the assembly process to continue.

TFIIB and RNA polymerase binding

Once TFIID has bound to the DNA, another transcription factor, **TFIIB**, binds to TFIID. TFIIB can also bind to the RNA polymerase. This seems to be an important step in transcription initiation since TFIIB acts as a bridging factor allowing recruitment of the polymerase to the complex together with a further factor, **TFIIF**.

Factors binding after RNA polymerase

After RNA polymerase binding, three other transcription factors, **TFIIE**, **TFIIH** and **TFIIJ**, rapidly associate with the complex. These proteins are necessary for transcription *in vitro* and associate with the complex in a defined order. TFIIH is a large complex which is made up of at least five subunits. TFIIJ remains to be fully characterized.

CTD phosphorylation by TFIIH

TFIIH is a large multicomponent protein complex which contains both **kinase** and **helicase** activity. Activation of TFIIH results in **phosphorylation** of the carboxyl-terminal domain (CTD) of the RNA polymerase (see Topic M1). This phosphorylation results in formation of a processive RNA polymerase complex and allows the RNA polymerase to leave the promoter region. TFIIH therefore seems to have a very important function in control of transcription elongation (see Tat protein function in Topic N2). Components of TFIIH are also important in DNA repair and in phosphorylation of the cyclin-dependent kinase complexes which regulate the cell cycle.

The initiator transcription complex

Many RNA Pol II promoters which do not contain a TATA box have an **initiator element** overlapping their start site. It seems that at these promoters TBP is recruited to the promoter by a further DNA-binding protein which binds to the initiator element. TBP then recruits the other transcription factors and RNA polymerase in a manner similar to that which occurs in TATA box promoters.

N1 EUKARYOTIC TRANSCRIPTION FACTORS

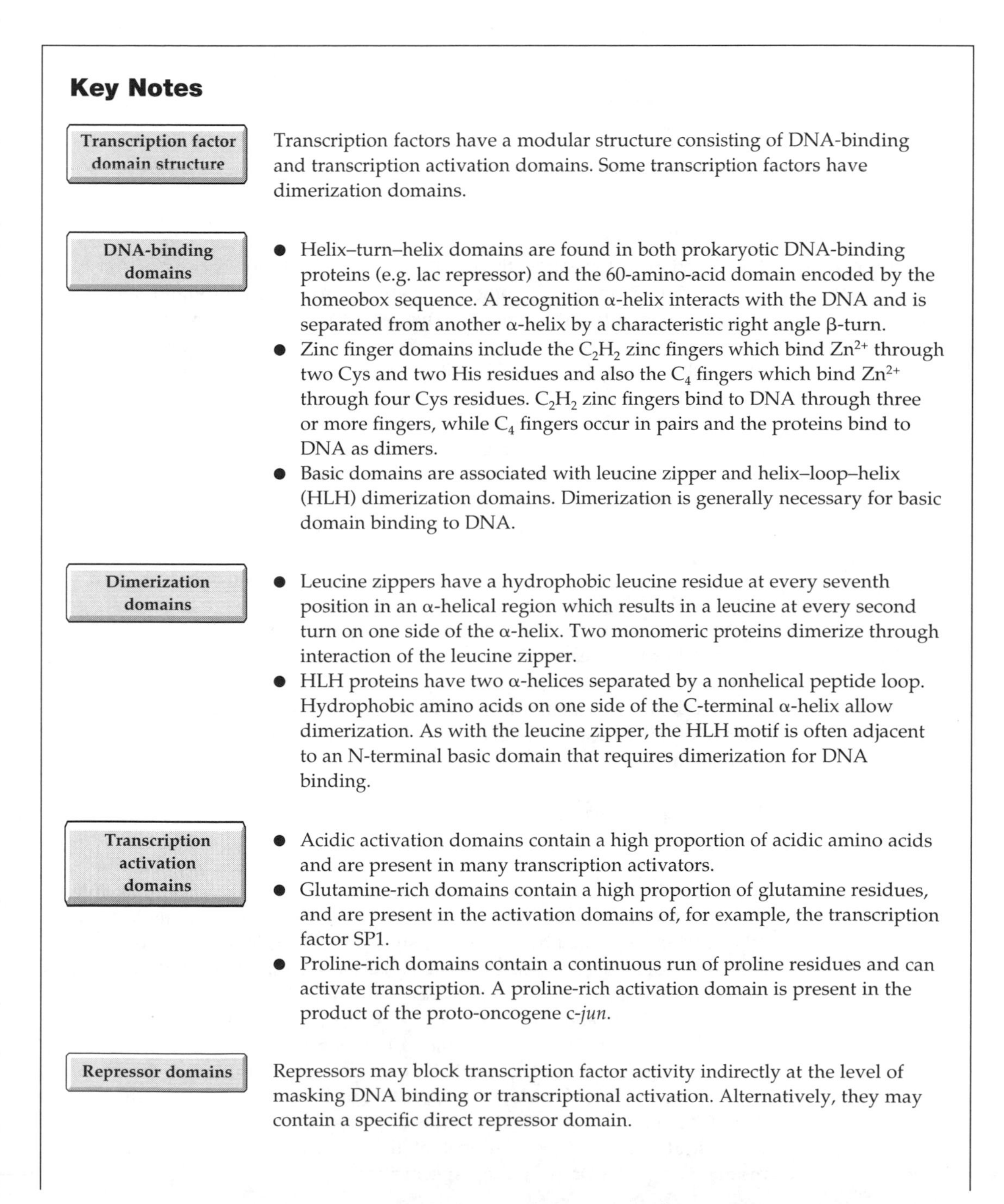

Key Notes

Transcription factor domain structure

Transcription factors have a modular structure consisting of DNA-binding and transcription activation domains. Some transcription factors have dimerization domains.

DNA-binding domains

- Helix–turn–helix domains are found in both prokaryotic DNA-binding proteins (e.g. lac repressor) and the 60-amino-acid domain encoded by the homeobox sequence. A recognition α-helix interacts with the DNA and is separated from another α-helix by a characteristic right angle β-turn.
- Zinc finger domains include the C_2H_2 zinc fingers which bind Zn^{2+} through two Cys and two His residues and also the C_4 fingers which bind Zn^{2+} through four Cys residues. C_2H_2 zinc fingers bind to DNA through three or more fingers, while C_4 fingers occur in pairs and the proteins bind to DNA as dimers.
- Basic domains are associated with leucine zipper and helix–loop–helix (HLH) dimerization domains. Dimerization is generally necessary for basic domain binding to DNA.

Dimerization domains

- Leucine zippers have a hydrophobic leucine residue at every seventh position in an α-helical region which results in a leucine at every second turn on one side of the α-helix. Two monomeric proteins dimerize through interaction of the leucine zipper.
- HLH proteins have two α-helices separated by a nonhelical peptide loop. Hydrophobic amino acids on one side of the C-terminal α-helix allow dimerization. As with the leucine zipper, the HLH motif is often adjacent to an N-terminal basic domain that requires dimerization for DNA binding.

Transcription activation domains

- Acidic activation domains contain a high proportion of acidic amino acids and are present in many transcription activators.
- Glutamine-rich domains contain a high proportion of glutamine residues, and are present in the activation domains of, for example, the transcription factor SP1.
- Proline-rich domains contain a continuous run of proline residues and can activate transcription. A proline-rich activation domain is present in the product of the proto-oncogene c-*jun*.

Repressor domains

Repressors may block transcription factor activity indirectly at the level of masking DNA binding or transcriptional activation. Alternatively, they may contain a specific direct repressor domain.

Targets for transcriptional regulation

Different activation domains may have multiple different targets in the basal transcription complex. Proposed targets include TAF_{II}s in TFIID, TFIIB and the phosphorylation of the C-terminal domain by TFIIH.

Related topics

Protein structure and function (B2)
RNA Pol II genes: promoters and enhancers (M4)
General transcription factors and RNA Pol II initiation (M5)
Examples of transcriptional regulation (N2)
Bacteriophages (R2)

Transcription factor domain structure

Transcription factors other than the general transcription factors of the basal transcription complex (see Topic M5) were first identified through their affinity for specific motifs in promoters, upstream regulatory elements (UREs) or enhancer regions. These factors have two distinct activities. Firstly, they bind specifically to their DNA-binding site and, secondly, they activate transcription. These activities can be assigned to separate protein domains called **activation domains** and **DNA-binding domains**. In addition, many transcription factors occur as homo- or heterodimers, held together by **dimerization domains**. A few transcription factors have **ligand-binding domains** which allow regulation of transcription factor activity by binding of an accessory small molecule. The steroid hormone receptors (see Topic N2) are an example containing all four of these types of domain.

Mutagenesis of the yeast transcription factors **Gal4** and **Gcn4** showed that their DNA-binding and transcription activation domains were in separate parts of the proteins. Experimentally, these activation domains were fused to the bacterial **LexA repressor**. These hybrid fusion proteins activated transcription from a promoter containing the ***lexA*** **operator** sequence, indicating that the transcriptional activation function of the yeast proteins was separable from their DNA-binding activity. These type of experiments are called **domain swap experiments**.

DNA-binding domains

The helix–turn–helix domain

This domain is characteristic of DNA-binding proteins containing a 60-amino-acid **homeodomain** which is encoded by a sequence called the **homeobox** (see Topic N2). In the Antennapedia transcription factor of *Drosophila*, this domain consists of four α-helices in which helices II and III are at right angles to each other and are separated by a characteristic β-turn. The characteristic helix–turn–helix structure (*Fig. 1*) is also found in bacteriophage DNA-binding proteins such as the phage λ cro repressor (see Topic R2), lac and trp repressors (see Topics L1 and L2), and cAMP receptor protein, CRP (see Topic L1). The domain binds so that one helix, known as the recognition helix, lies partly in the major groove and interacts with the DNA. The recognition helices of two homeodomain factors Bicoid and Antennapedia can be exchanged, and this swaps their DNA-binding specificities. Indeed, the specificity of this interaction is demonstrated by the observation that the exchange of only one amino acid residue swaps the DNA-binding specificities.

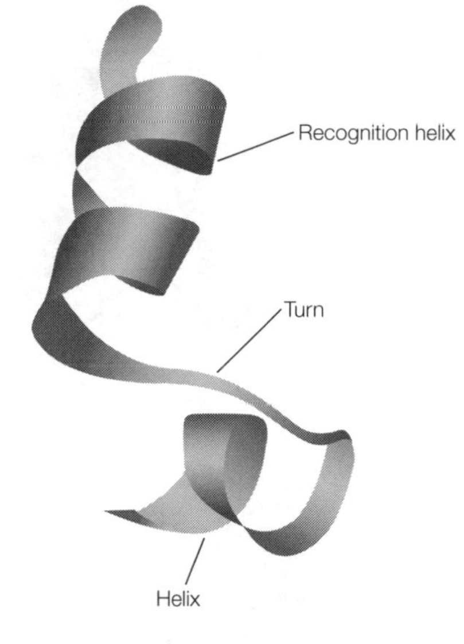

Fig. 1. The helix–turn–helix core structure.

The zinc finger domain

This domain exists in two forms. The **C_2H_2 zinc finger** has a loop of 12 amino acids anchored by two cysteine and two histidine residues that tetrahedrally co-ordinate a zinc ion (*Fig. 2a*). This motif folds into a compact structure comprising two β-strands and one α-helix, the latter binding in the major groove of DNA (*Fig. 2b*) The α-helical region contains conserved basic amino acids which are responsible for interacting with the DNA. This structure is repeated nine times in TFIIIA, the RNA Pol III transcription factor (see Topic M3). It is also present in transcription factor SP1 (three copies). Usually, three or more C_2H_2 zinc fingers are required for DNA binding. A related motif, in which the zinc ion is co-ordinated by four cysteine residues, occurs in over 100 steroid hormone receptor transcription factors (see Topic N2). These factors consist of homo- or hetero-dimers, in which each monomer contains two **C_4** 'zinc finger' motifs (*Fig. 2c*). The two motifs are now known to fold together into a more complex conformation stabilized by zinc, which binds to DNA by the insertion of one α-helix from each monomer into successive major grooves, in a manner reminiscent of the helix–turn–helix proteins.

The basic domain

A **basic domain** is found in a number of DNA-binding proteins and is generally associated with one or other of two dimerization domains, the **leucine zipper** or the **helix–loop–helix** (**HLH**) motif (see below). These are referred to as basic leucine zipper (**bZIP**) or **basic HLH proteins**. Dimerization of the proteins brings together two basic domains which can then interact with DNA.

Dimerization domains

Leucine zippers

Leucine zipper proteins contain a hydrophobic leucine residue at every seventh position in a region that is often at the C-terminal part of the DNA-binding domain. These leucines lie in an α-helical region and the regular repeat of these residues forms a hydrophobic surface on one side of the α-helix with a leucine

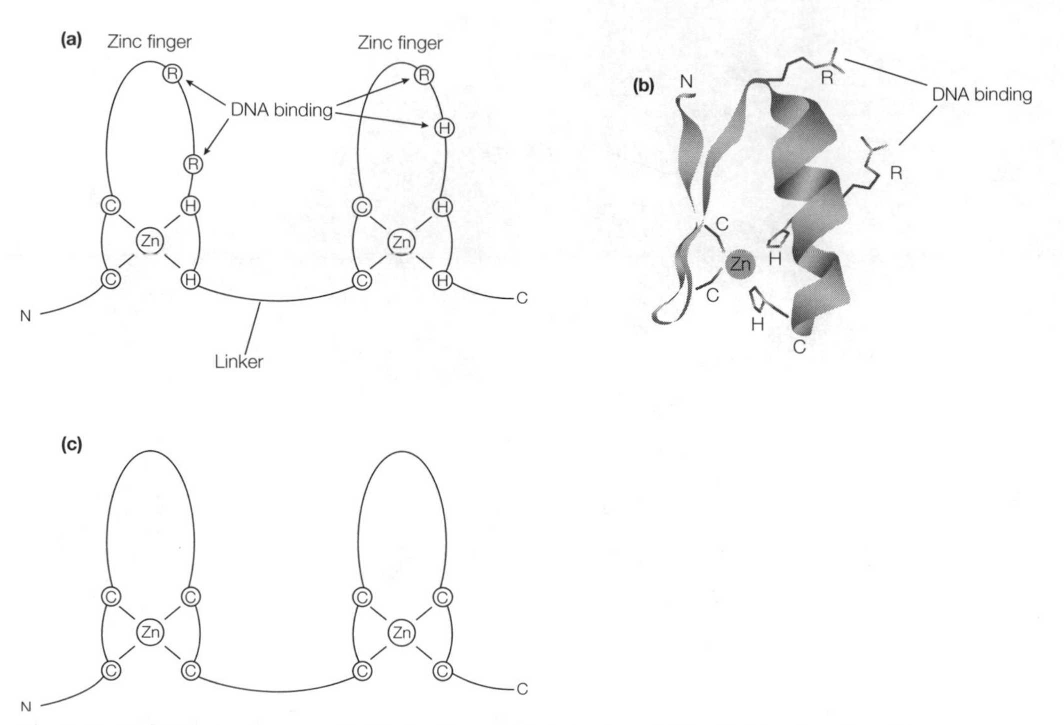

Fig. 2. (a) The C_2H_2 zinc finger motif; (b) zinc finger folded structure; (c) the C_4 'zinc finger' motif.

every second turn of the helix. These leucines are responsible for dimerization through interactions between the hydrophobic faces of the α-helices (see *Fig. 3*). This interaction forms a **coiled-coil structure**. bZIP transcription factors contain a basic DNA-binding domain N-terminal to the leucine zipper. This is present on an α-helix which is a continuation from the leucine zipper α-helical C-terminal domain. The N-terminal basic domains of each helix form a symmetrical structure in which each basic domain lies along the DNA in opposite directions, interacting with a symmetrical DNA recognition site so that the

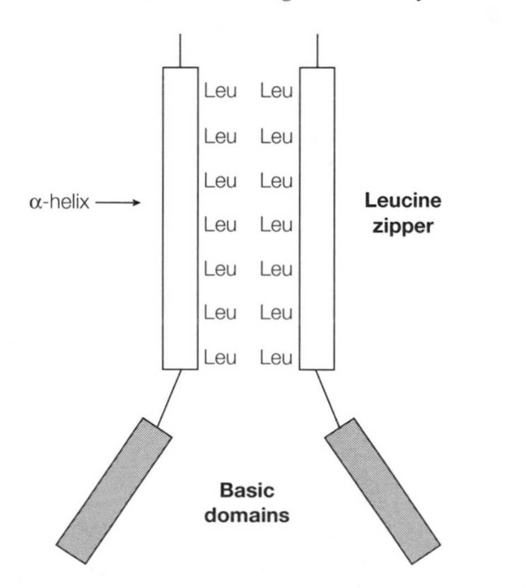

Fig. 3. The leucine zipper and basic domain dimer of a bZIP protein.

protein in effect forms a clamp around the DNA. The leucine zipper is also used as a dimerization domain in proteins that use DNA-binding domains other than the basic domain, including some homeodomain proteins.

The helix–loop–helix domain
The overall structure of this domain is similar to the leucine zipper, except that a nonhelical loop of polypeptide chain separates two α-helices in each monomeric protein. Hydrophobic residues on one side of the C-terminal α-helix allow dimerization. This structure is found in the **MyoD** family of proteins (see Topic N2). As with the leucine zipper, the HLH motif is often found adjacent to a basic domain that requires dimerization for DNA binding. With both basic HLH proteins and bZIP proteins the formation of heterodimers allows much greater diversity and complexity in the transcription factor repertoire.

Transcription activation domains

Acidic activation domains
Comparison of the transactivation domains of yeast Gcn4 and Gal4, mammalian glucocorticoid receptor and herpes virus activator VP16 shows that they have a very high proportion of acidic amino acids. These have been called **acidic activation domains** or 'acid blobs' or 'negative noodles' and are characteristic of many transcription activation domains. It is still uncertain what other features are required for these regions to function as efficient transcription activation domains.

Glutamine-rich domains
Glutamine-rich domains were first identified in two activation regions of the transcription factor SP1. As with acidic domains, the proportion of glutamine residues seems to be more important than overall structure. Domain swap experiments between glutamine-rich transactivation regions from the diverse transcription factors SP1 and the *Drosophila* protein Antennapedia showed that these domains could substitute for each other.

Proline-rich domains
Proline-rich domains have been identified in several transcription factors. As with glutamine, a continuous run of proline residues can activate transcription. This domain is found, for example, in the c-Jun, AP2 and Oct-2 transcription factors (see Topic S2).

Repressor domains

Repression of transcription may occur by indirect interference with the function of an activator. This may occur by:

- **Blocking** the activator DNA-binding site (as with prokaryotic repressors; see Section L).
- Formation of a **non-DNA-binding complex** (e.g. the repressors of steroid hormone receptors, or the Id protein which blocks HLH protein–DNA interactions, since it lacks a DNA-binding domain; see Topic N2).
- **Masking** of the activation domain without preventing DNA binding (e.g. Gal80 masks the activation domain of the yeast transcription factor Gal4).

In other cases, a specific domain of the repressor is directly responsible for inhibition of transcription. For example, a domain of the mammalian **thyroid**

hormone receptor can repress transcription in the absence of thyroid hormone and activates transcription when bound to its ligand (see Topic N2). The product of the **Wilms tumor gene**, *WT1*, is a tumor-suppressor protein having a specific proline-rich repressor domain that lacks charged residues.

Targets for transcriptional regulation

The presence of diverse activation domains raises the question of whether they each have the same target in the basal transcription complex or different targets for the activation of transcription. They are distinguishable from each other since the acidic activation domain can activate transcription from a downstream enhancer site while the proline domain only activates weakly and the glutamine domain not at all. While proline and acidic domains are active in yeast, glutamine domains have no activity, implying that they have a different transcription target which is not present in the yeast transcription complex. Proposed targets of different transcriptional activators include:

- chromatin structure;
- interaction with TFIID through specific TAF_{II}s;
- interaction with TFIIB;
- interaction or modulation of the TFIIH complex activity leading to differential phosphorylation of the CTD of RNA Pol II.

It seems likely that different activation domains may have different targets, and almost any component or stage in initiation and transcription elongation could be a target for regulation resulting in multistage regulation of transcription.

N2 EXAMPLES OF TRANSCRIPTIONAL REGULATION

Key Notes

Constitutive transcription factors: SP1

SP1 is a ubiquitous transcription factor which contains three zinc finger motifs and two glutamine-rich transactivation domains.

Hormonal regulation: steroid hormone receptors

Steroid hormones enter the cell and bind to a steroid hormone receptor. The receptor dissociates from a bound inhibitor protein, dimerizes and translocates to the nucleus. The DNA-binding domain of the steroid hormone receptor binds to response elements, giving rise to activation of target genes. Thyroid hormone receptors act as transcription repressors until they are converted to transcriptional activators by thyroid hormone binding.

Regulation by phosphorylation: STAT proteins

Interferon-γ activates JAK kinase which phosphorylates STAT1α. STAT1α dimerizes and translocates to the nucleus, where it activates expression of target genes.

Transcription elongation: HIV Tat

Tat protein binds to an RNA sequence present at the 5′-end of all human immunodeficiency virus (HIV) RNAs called TAR. In the absence of Tat, HIV transcription terminates prematurely. The Tat–TAR complex activates TFIIH in the transcription initiation complex at the promoter, leading to phosphorylation of the RNA Pol II carboxyl-terminal domain (CTD). This permits full-length transcription by the polymerase.

Cell determination: *myoD*

The expression of *myoD* and related genes (*myf5*, *mrf4* and *myogenin*) can convert nonmuscle cells into muscle cells. Their expression activates muscle-specific gene expression and blocks cell division. Each gene encodes a helix–loop–helix (HLH) transcription factor. HLH heterodimer formation and the non-DNA binding inhibitor, Id, give rise to diversity and regulation of these transcription factors.

Embryonic development: homeodomain proteins

The homeobox, which encodes a DNA-binding domain, was originally found in *Drosophila melanogaster* homeotic genes that encode transcription factors which specify the development of body parts. The conservation of both function and organization of the homeotic gene clusters between *Drosophila* and mammals suggests that these proteins have important common roles in development.

Related topics

Cellular classification (A1)
Eukaryotic DNA replication (E3)
The three RNA polymerases: characterization and function (M1)
General transcription factors and RNA Pol II initiation (M5)
Eukaryotic transcription factors (N1)

Constitutive transcription factors: SP1

SP1 binds to a GC-rich sequence with the consensus sequence GGGCGG. It is a constitutive transcription factor whose binding site is found in the promoter of many housekeeping genes. SP1 is present in all cell types. It contains three zinc finger motifs and has been shown to contain two glutamine-rich transactivation domains. The glutamine-rich domains of SP1 have been shown to interact specifically with $TAF_{II}110$, one of the $TAF_{II}s$ which bind to the TATA-binding protein (TBP) to make up TFIID. This represents one target through which SP1 may interact with and regulate the basal transcription complex.

Hormonal regulation: steroid hormone receptors

Many transcription factors are activated by **hormones** which are secreted by one cell type and transmit a signal to a different cell type. One class of hormones, the **steroid hormones**, are lipid soluble and can diffuse through cell membranes to interact with transcription factors called **steroid hormone receptors** (see Topic N1). In the absence of the steroid hormone, the receptor is bound to an inhibitor, and located in the cytoplasm (*Fig. 1*). The steroid hormone binds to the receptor and releases the receptor from the inhibitor, allowing the receptor to dimerize and translocate to the nucleus. The DNA-binding domain of the steroid hormone receptor then interacts with its specific DNA-binding sequence,

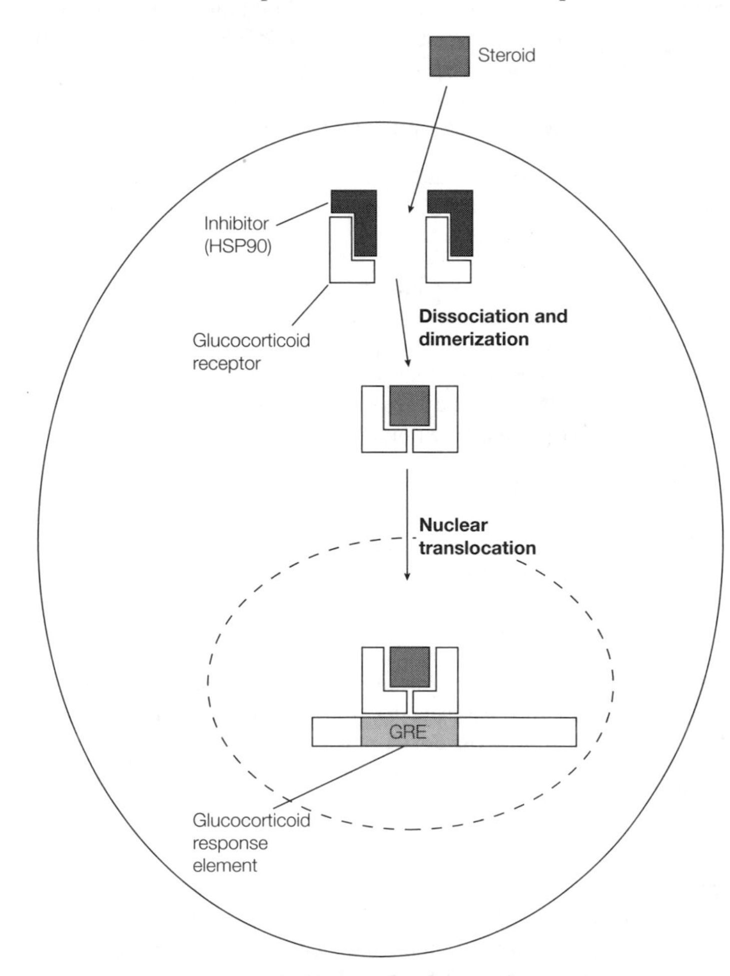

Fig. 1. Steroid hormone activation of the glucocorticoid receptor.

or **response element**, and this gives rise to activation of the target gene. Important classes of related receptors include the glucocorticoid, estrogen, retinoic acid and thyroid hormone receptors. The general model described above is not true for all of these. For example, the thyroid hormone receptors act as DNA-bound repressors in the absence of hormone. In the presence of the hormone, the receptor is converted from a transcriptional repressor to a transcriptional activator.

Regulation by phosphorylation: STAT proteins

Many hormones do not diffuse into the cell. Instead, they bind to cell-surface receptors and pass a signal to proteins within the cell through a process called **signal transduction**. This process often involves protein phosphorylation. Interferon-γ induces phosphorylation of a transcription factor called **STAT1α** through activation of the intracellular kinase called Janus activated kinase (JAK). When STAT1α protein is unphosphorylated, it exists as a monomer in the cell cytoplasm and has no transcriptional activity. However, when STAT1α becomes phosphorylated at a specific tyrosine residue, it is able to form a homodimer which moves from the cytoplasm into the nucleus. In the nucleus, STAT1α is able to activate the expression of target genes whose promoter regions contain a consensus DNA-binding motif (*Fig. 2*).

Transcription elongation: HIV Tat

Human immunodeficiency virus (HIV) encodes an activator protein called **Tat**, which is required for productive HIV gene expression. Tat binds to an RNA stem–loop structure called **TAR**, which is present in the 5′-untranslated region of all HIV RNAs, just after the HIV transcription start site. The predominant effect of Tat in mammalian cells lies at the level of transcription elongation. In the absence of Tat, the HIV transcripts terminate prematurely due to poor processivity of the RNA Pol II transcription complex. Tat is thought to bind to TAR on one transcript in a complex together with cellular RNA-binding factors. This protein–RNA complex may loop backwards and interact with the new transcription initiation complex which is assembled at the promoter. This interaction is thought to result in the activation of the kinase activity of TFIIH. This leads to phosphorylation of the carboxyl-terminal domain (CTD) of RNA Pol II, making the RNA polymerase a processive enzyme (see Topics M1 and M5). As a result, the polymerase is able to read through the HIV transcription unit, leading to the productive synthesis of HIV proteins (*Fig. 3*).

Cell determination: *myoD*

Muscle cells arise from mesodermal embryonic cells called **somites**. Somites become committed to forming muscle cells (**myoblasts**) before there is an appreciable sign of cell differentiation to form skeletal muscle cells (called **myotomes**). This process is called **cell determination**. *myoD* was identified originally as a gene that was expressed in undifferentiated cells which were committed to form muscle and had therefore undergone cell determination. Overexpression of ***myoD*** can turn fibroblasts (cells that lay down the basement matrix in many tissues) into muscle-like cells which express muscle-specific genes and resemble myotomes. MyoD protein has been shown to activate muscle-specific gene expression directly. MyoD also activates expression of ***p21* waf1/cip1** expression. p21 waf1/cip1 is a small molecule inhibitor of the **cyclin-dependent kinases (CDKs)**. Inhibition of CDK activity by *p21* waf1/cip1 causes arrest at the G1-phase of the cell cycle (see Topic E3). *myoD* expression is therefore responsible for the withdrawal from the cell cycle which is characteristic of differentiated muscle cells. There have now been shown to be four genes,

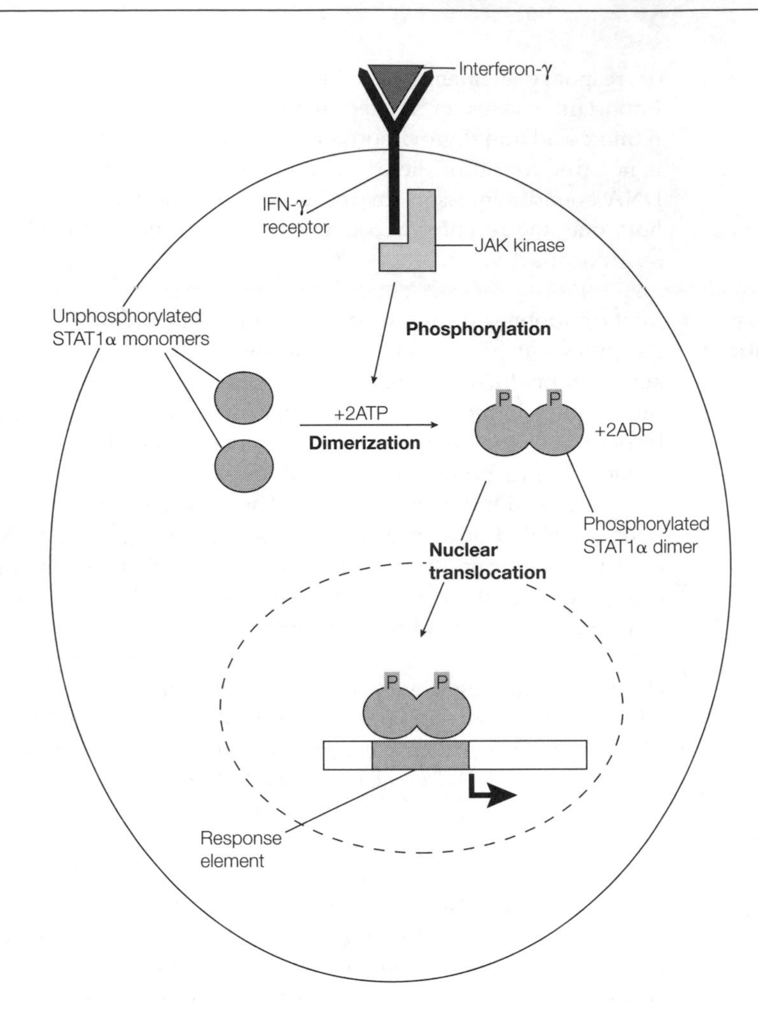

Fig. 2. Interferon-γ-mediated transcription activation caused by phosphorylation and dimerization of the STAT1α transcription factor.

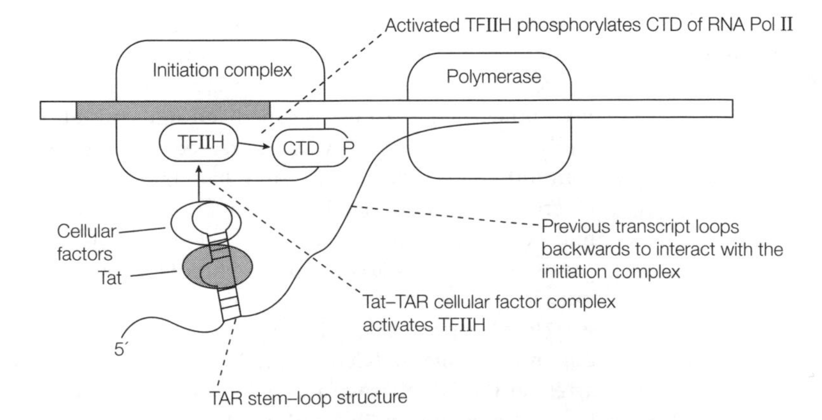

Fig. 3. Mechanism of activation of transcriptional elongation by the HIV Tat protein.

myoD, ***myogenin***, ***myf5*** and ***mrf4***, the expression of each of which has the ability to convert fibroblasts into muscle. The encoded proteins are all members of the helix–loop–helix (HLH) transcription factor family. MyoD is most active as a heterodimer with constitutive HLH transcription factors E12 and E47. The HLH group of proteins therefore produce a diverse range of hetero- and homodimeric transcription factors that may each have different activities and roles. These proteins are regulated by an inhibitor called **Id** that lacks a DNA-binding domain, but contains the HLH dimerization domain. Therefore, Id protein can bind to MyoD and related proteins, but the resulting heterodimers cannot bind DNA, and hence cannot regulate transcription.

Embryonic development: homeodomain proteins

The **homeobox** is a conserved DNA sequence which encodes the helix–turn–helix DNA binding protein structure called the homeodomain (see topic N1). The homeodomain was first discovered in the transcription factors encoded by **homeotic genes** of *Drosophila*. Homeotic genes are responsible for the correct specification of body parts (see Topic A1). For example, mutation of one of these genes, *Antennapedia*, causes the fly to form a leg where the antenna should be. These genes are very important in spatial pattern formation in the embryo. The homeobox sequence has been conserved between a wide range of eukaryotes and homeobox-containing genes have been shown to be important in mammalian development. In *Drosophila* and mammals, the homeobox genes are arranged in gene clusters in which homologous genes are in the same order. The gene homologs are also expressed in a similar order in the embryo on the anterior to posterior axis. This suggests that the conserved homeobox-encoded DNA-binding domain is characteristic of transcription factors which have a conserved function in embryonic development.

O1 rRNA PROCESSING AND RIBOSOMES

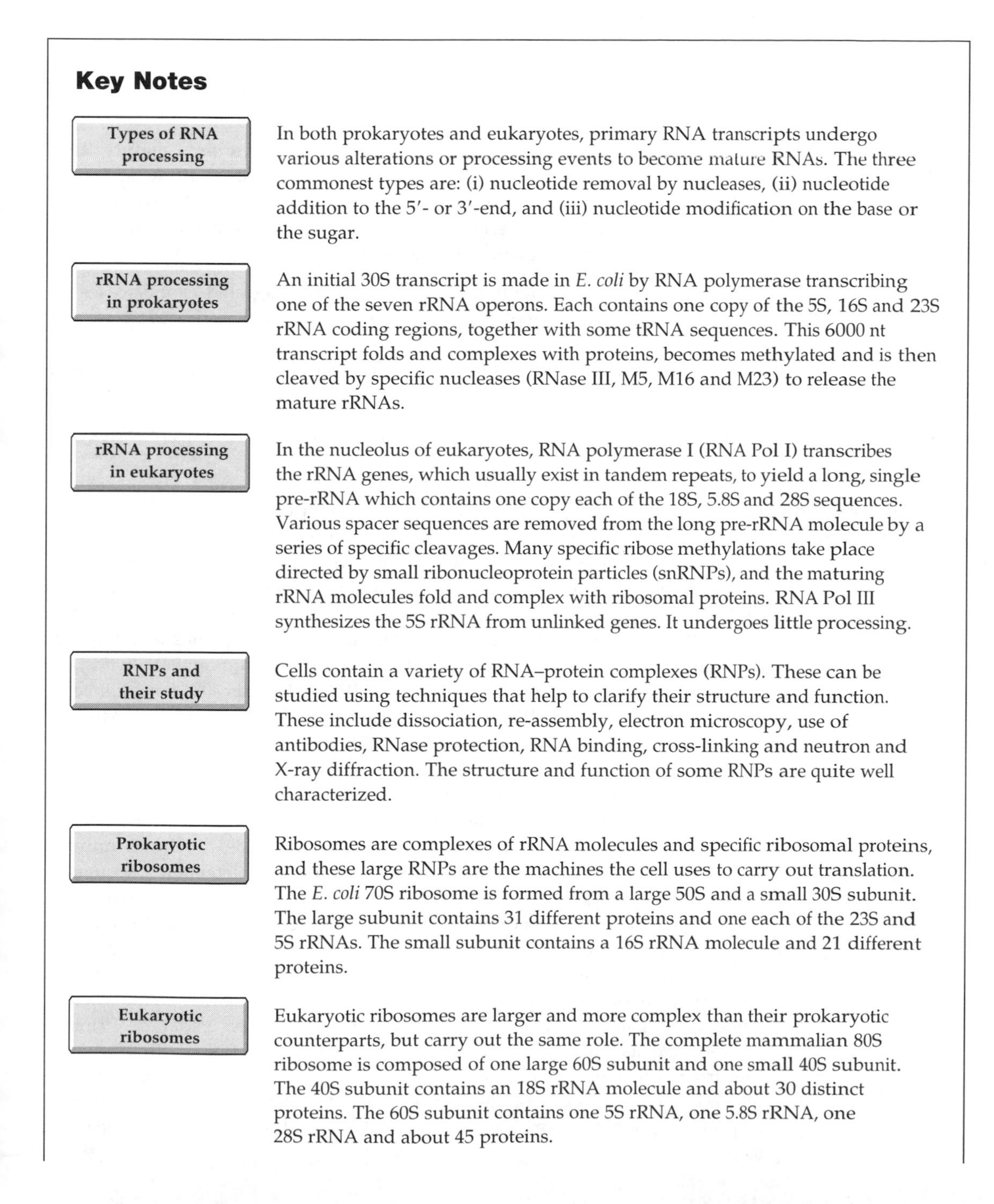

Key Notes

Types of RNA processing

In both prokaryotes and eukaryotes, primary RNA transcripts undergo various alterations or processing events to become mature RNAs. The three commonest types are: (i) nucleotide removal by nucleases, (ii) nucleotide addition to the 5′- or 3′-end, and (iii) nucleotide modification on the base or the sugar.

rRNA processing in prokaryotes

An initial 30S transcript is made in *E. coli* by RNA polymerase transcribing one of the seven rRNA operons. Each contains one copy of the 5S, 16S and 23S rRNA coding regions, together with some tRNA sequences. This 6000 nt transcript folds and complexes with proteins, becomes methylated and is then cleaved by specific nucleases (RNase III, M5, M16 and M23) to release the mature rRNAs.

rRNA processing in eukaryotes

In the nucleolus of eukaryotes, RNA polymerase I (RNA Pol I) transcribes the rRNA genes, which usually exist in tandem repeats, to yield a long, single pre-rRNA which contains one copy each of the 18S, 5.8S and 28S sequences. Various spacer sequences are removed from the long pre-rRNA molecule by a series of specific cleavages. Many specific ribose methylations take place directed by small ribonucleoprotein particles (snRNPs), and the maturing rRNA molecules fold and complex with ribosomal proteins. RNA Pol III synthesizes the 5S rRNA from unlinked genes. It undergoes little processing.

RNPs and their study

Cells contain a variety of RNA–protein complexes (RNPs). These can be studied using techniques that help to clarify their structure and function. These include dissociation, re-assembly, electron microscopy, use of antibodies, RNase protection, RNA binding, cross-linking and neutron and X-ray diffraction. The structure and function of some RNPs are quite well characterized.

Prokaryotic ribosomes

Ribosomes are complexes of rRNA molecules and specific ribosomal proteins, and these large RNPs are the machines the cell uses to carry out translation. The *E. coli* 70S ribosome is formed from a large 50S and a small 30S subunit. The large subunit contains 31 different proteins and one each of the 23S and 5S rRNAs. The small subunit contains a 16S rRNA molecule and 21 different proteins.

Eukaryotic ribosomes

Eukaryotic ribosomes are larger and more complex than their prokaryotic counterparts, but carry out the same role. The complete mammalian 80S ribosome is composed of one large 60S subunit and one small 40S subunit. The 40S subunit contains an 18S rRNA molecule and about 30 distinct proteins. The 60S subunit contains one 5S rRNA, one 5.8S rRNA, one 28S rRNA and about 45 proteins.

Related topics	Basic principles of transcription (K1)	RNA Pol III genes: 5S and tRNA transcription (M3)
	RNA Pol I genes: the ribosomal repeat (M2)	tRNA processing, RNase P and ribozymes (O2)

Types of RNA processing

Very few RNA molecules are transcribed directly into the final **mature RNA** product (see Sections K and M). Most newly transcribed RNA molecules undergo various alterations to yield the mature product. **RNA processing** is the collective term used to describe these alterations to the **primary transcript**. The commonest types of alterations include:

(i) the **removal of nucleotides** by both endonucleases and exonucleases;
(ii) the **addition of nucleotides** to the 5′- or 3′-ends of the primary transcripts or their cleavage products;
(iii) the **modification of certain nucleotides** on either the base or the sugar moiety.

These processing events take place in both prokaryotes and eukaryotes on the major classes of RNA.

rRNA processing in prokaryotes

In the prokaryote, *E. coli*, there are seven different operons for rRNA that are dispersed throughout the genome and which are called rrnH, rrnE, etc. Each operon contains one copy of each of the **5S,** the **16S** and the **23S rRNA** sequences. Between one and four coding sequences for tRNA molecules are also present in these rRNA operons, and these primary transcripts are processed to give both rRNA and tRNA molecules. The initial transcript has a sedimentation coefficient of 30S (approx. 6000 nt) and is normally quite short-lived (*Fig. 1a*). However, in *E. coli* mutants defective for **RNase III**, this 30S transcript accumulates, indicating that RNase III is involved in rRNA processing. Mutants defective in other RNases such as **M5, M16** and **M23** have also shown the involvement of these RNases in *E. coli* rRNA processing.

The post-transcriptional processing of *E. coli* rRNA takes place in a defined series of steps (*Fig. 1a*). Following, and to some extent during, transcription of the 6000 nt primary transcript, the RNA folds up into a number of stem–loop structures by base pairing between complementary sequences in the transcript. The formation of this secondary structure of stems and loops allows some proteins to bind to form a **ribonucleoprotein (RNP)** complex. Many of these proteins remain attached to the RNA and become part of the ribosome. After the binding of proteins, modifications such as 24 specific base methylations take place. *S*-Adenosylmethionine (SAM) is the methylating agent, and usually a methyl group is added to the base adenine. Primary cleavage events then take place, mainly carried out by RNase III, to release precursors of the 5S, 16S and 23S molecules. Further cleavages at the 5′- and 3′-ends of each of these precursors by RNases M5, M16 and M23, respectively, release the mature length rRNA molecules in a secondary cleavage step.

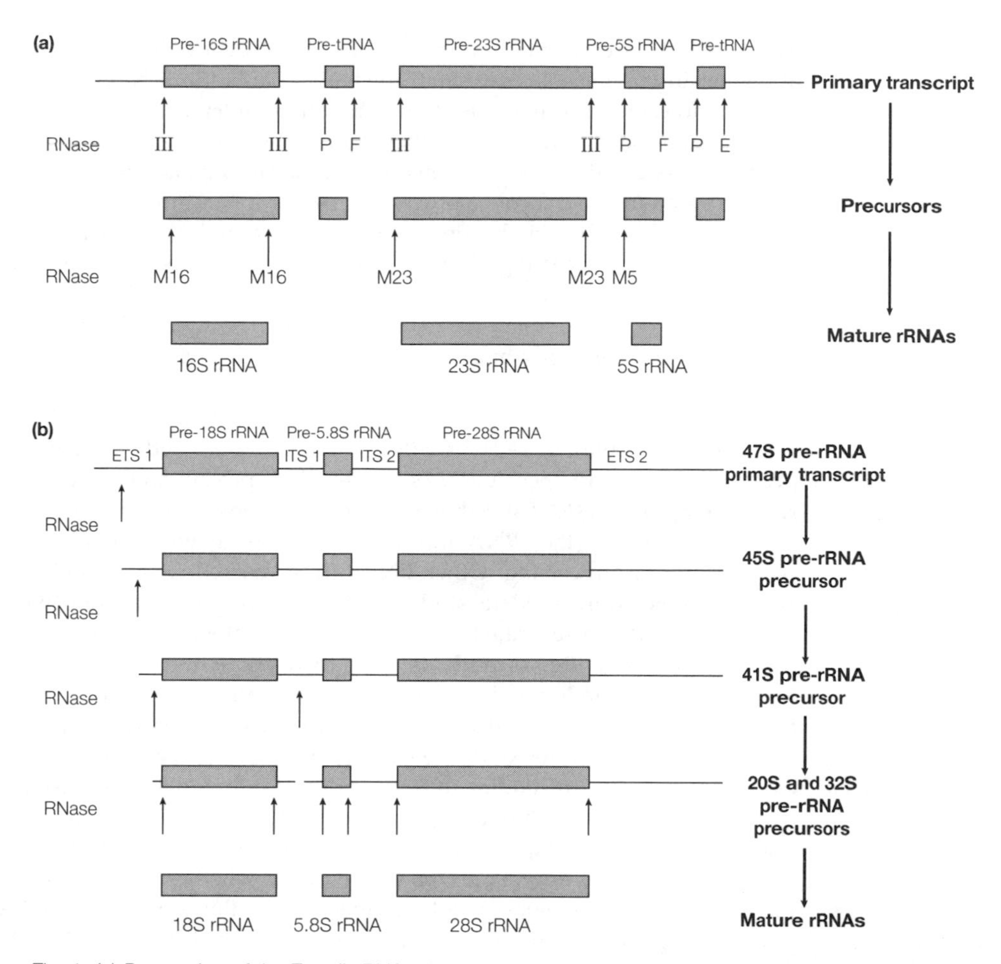

Fig. 1. (a) Processing of the E. coli *rRNA primary transcript; (b) mammalian pre-rRNA processing.*

rRNA processing in eukaryotes

rRNA in eukaryotes is also generated from a single, long precursor molecule by specific modification and cleavage steps, although these are not so well understood. In many eukaryotes, the rRNA genes are present in a tandemly repeated cluster containing 100 or more copies of the transcription unit and, as described in Topic M2, they are transcribed in the nucleolus (see Topic D4) by **RNA Pol I**. The precursor has a characteristic size in each organism, being about 7000 nt in yeast and 13 500 nt in mammals (*Fig. 1b*). It contains one copy of the **18S** coding region and one copy each of the **5.8S** and **28S** coding regions, which together are the equivalent of the 23S rRNA in prokaryotes. The eukaryotic **5S** rRNA is transcribed by **RNA Pol III** from unlinked genes to give a 121 nt transcript which undergoes little or no processing.

For mammalian pre-rRNA, the 13 500 nt precursor (47S) undergoes a number of cleavages (*Fig. 1b*), firstly in the **external transcribed spacers (ETSs) 1 and 2**. Cleavages in the **internal transcribed spacers (ITSs)** then release the 20S pre-rRNA from the 32S pre-rRNA. Both of these precursors must be trimmed further and the 5.8S region must base-pair to the 28S rRNA before the mature molecules are produced. As with prokaryotic pre-rRNA, the precursor folds and complexes with proteins as it is being transcribed. This takes place in the nucleolus. **Methylation** takes place at over 100 sites to give 2′-*O*-methylribose and this is now known to be carried out by a subset of small nuclear RNP particles which are abundant in the nucleolus i.e. **small nucleolar RNPs (SnoRNPs)**. They contain snRNA molecules that have short stretches of complementarity to parts of the rRNA and, by base pairing with it, they define where methylation takes place.

At least one eukaryote, *Tetrahymena thermophila*, makes a pre-rRNA that undergoes an unusual form of processing before it can function. It contains an intron (see Topic O3) in the precursor for the largest rRNA which must be removed during processing. Although this process occurs *in vivo* in the presence of protein, it has been shown that the intron can actually excise itself in the test tube in the complete absence of protein. The RNA folds into an enzymatically active form or **ribozyme** (see Topic O2) to perform self-cleavage and ligation.

RNPs and their study

The RNA molecules in cells usually exist complexed with proteins, specific proteins attaching to specific RNAs. These RNA–protein complexes are called **ribonucleoproteins (RNPs)**. **Ribosomes** are the largest and most complex RNPs and are formed by the rRNA molecules complexing with specific ribosomal proteins during processing. Other RNPs are discussed in Topics O2 and O3. Several methods are used to study RNPs, including **dissociation**, where the RNP is purified and separated into its RNA and protein components which are then characterized. **Re-assembly** is used to discover the order in which the components fit together and, if the components can be modified, it is possible to gain clues as to their individual functions. **Electron microscopy** can allow direct visualization if the RNPs are large enough, otherwise it can roughly indicate overall shape. **Antibodies to RNPs** or their individual components can be used for purification, inhibition of function and, in combination with electron microscopy, they can show the crude positions of the components in the overall structure. **RNA binding experiments** can show whether a particular protein binds to an RNA, and subsequent treatment of the RNA–protein complex with RNase (**RNase protection experiment**) can show which parts of the RNA are protected by bound protein (i.e. the site of binding). **Cross-linking experiments** using UV light with or without chemical cross-linking agents can show which parts of the RNA and protein molecules are in close contact in the complex. Physical methods such as **neutron** and **X-ray diffraction** can ultimately give the complete 3-D structure (see Topic B3). Collectively, these methods have provided much information on the structure of the RNPs described in this and the subsequent topics.

Prokaryotic ribosomes

The importance of ribosomes to a cell is well illustrated by the fact that in *E. coli* ribosomes account for 25% of the dry weight (10% of total protein and 80% of total RNA). *Figure 2* shows the components present in the *E. coli* ribosome. The **70S ribosome** of molecular mass 2.75×10^6 Da is made up of a large subunit of **50S** and a small subunit of **30S**. The latter is composed of one copy of the 16S rRNA molecule and 21 different proteins denoted S_1 to S_{21}. The large subunit contains one 23S and one 5S rRNA molecule and 31 different proteins. These were named L_1 to L_{34} after fractionation on two-dimensional gels. However, the L_{26} spot was later found to be S_{20}, L_7 is the acetylated version of L_{12}, and L_8 is a complex of L_{10} and L_7, hence there are only 31 different large subunit proteins. The sizes of these ribosomal proteins vary widely, from L_{34} which is only 46 amino acids to S_1 which is 557. Mostly, these relatively small proteins are basic, which might be expected since they bind to RNA. It is

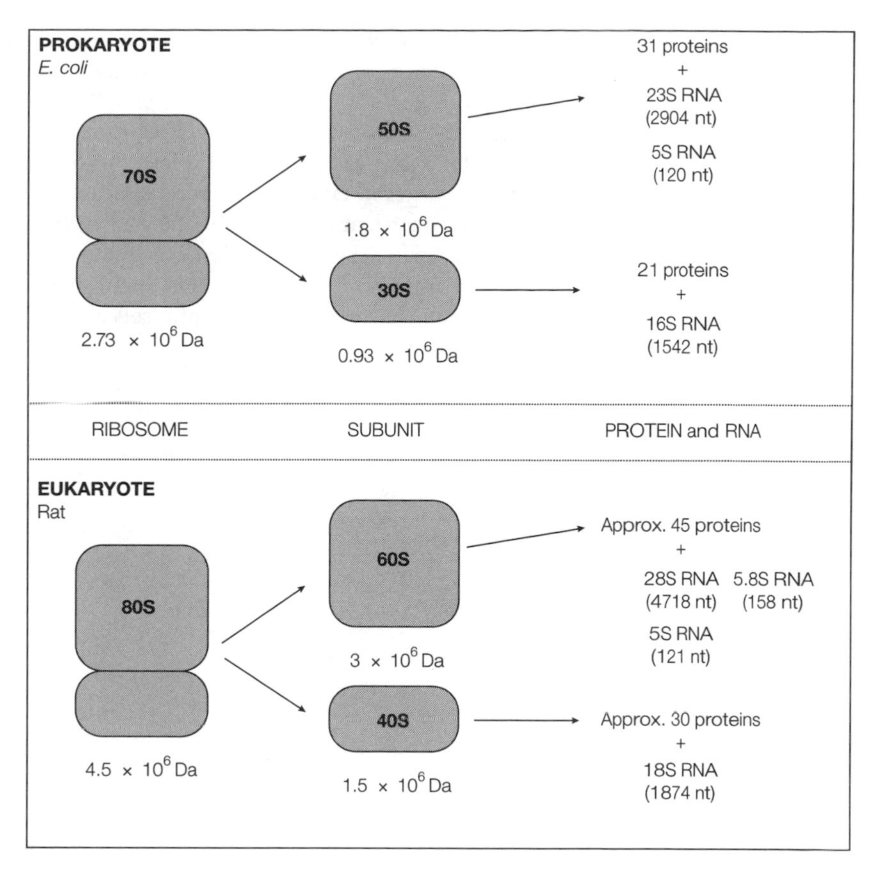

Fig. 2. Composition of typical prokaryotic and eukaryotic ribosomes.

possible to re-assemble functional *E. coli* ribosomes from the RNA and protein components, and there is a defined pathway of assembly. The various methods of studying RNPs have led to the structures shown in *Fig. 3* for the *E. coli* ribosomal subunits.

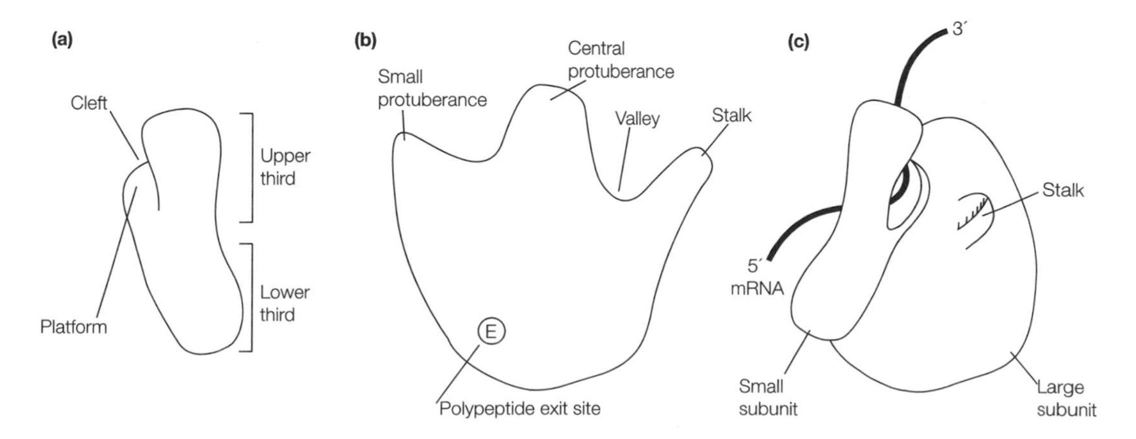

Fig. 3. Features of the E. coli *ribosome. (a) The 30S subunit; (b) the 50S subunit; (c) the complete 70S ribosome. After Dr James Lake [see* Sci. Amer. *(1981), Vol. 2245, p. 86].*

Eukaryotic ribosomes

The corresponding sizes of, and components in, the 80S eukaryotic (rat) ribosome are shown in *Fig. 2*. In the large 60S subunit, which contains about 45 proteins and one 5S rRNA molecule, the 5.8S rRNA and the 28S rRNA molecules together are the equivalent of the prokaryotic 23S molecule. The small 40S subunit contains the 18S rRNA and about 30 different proteins. Although all the rRNAs are larger in eukaryotes, there is a considerable degree of conservation of secondary structure in each of the corresponding molecules. Due to their greater complexity, the eukaryotic ribosomal subunits have not yet been re-assembled into functional complexes and their structure is less well understood. The ribosomes in a typical eukaryotic cell can collectively make about a million peptide bonds per second.

O2 tRNA PROCESSING, RNASE P AND RIBOZYMES

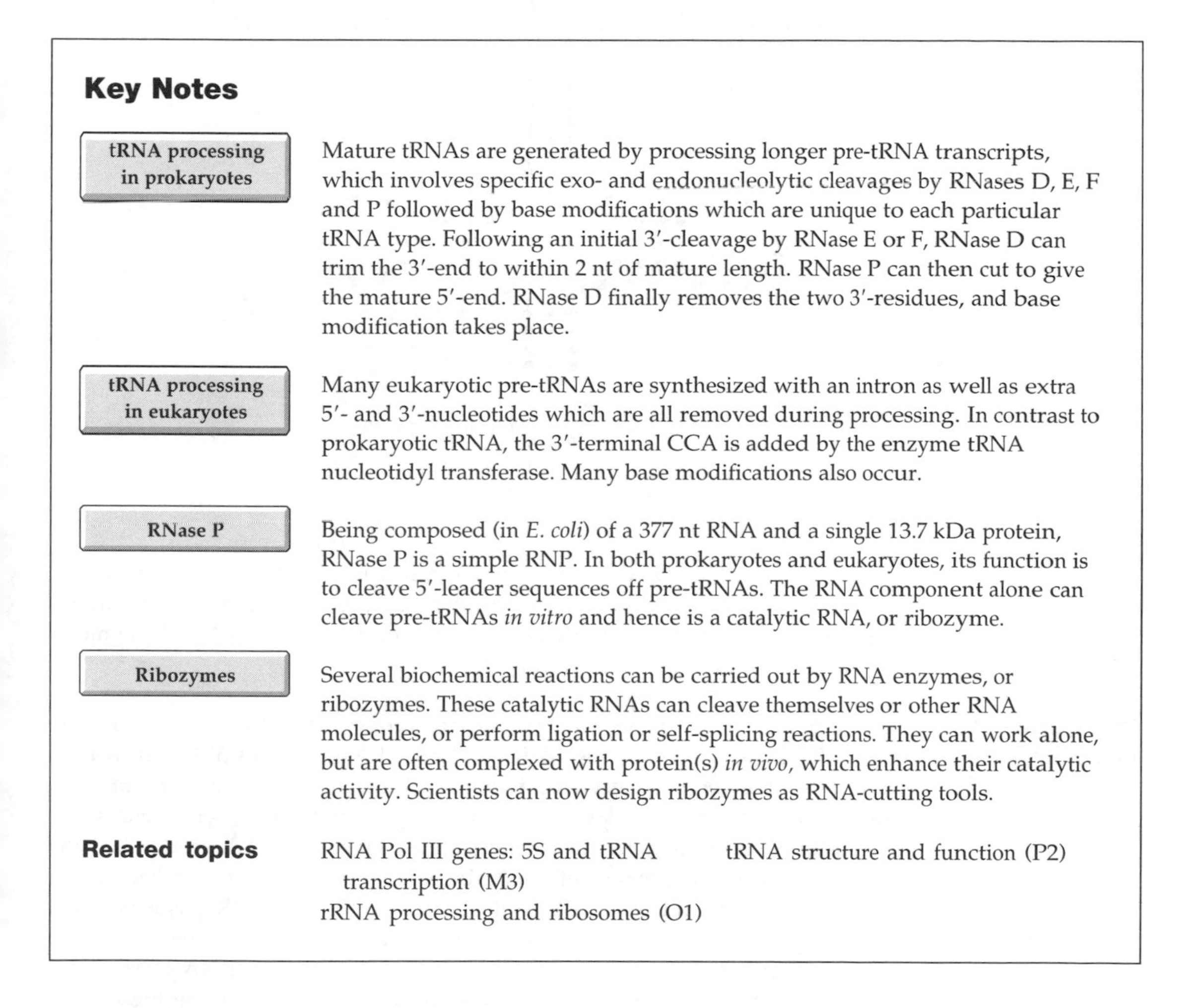

Key Notes

tRNA processing in prokaryotes

Mature tRNAs are generated by processing longer pre-tRNA transcripts, which involves specific exo- and endonucleolytic cleavages by RNases D, E, F and P followed by base modifications which are unique to each particular tRNA type. Following an initial 3′-cleavage by RNase E or F, RNase D can trim the 3′-end to within 2 nt of mature length. RNase P can then cut to give the mature 5′-end. RNase D finally removes the two 3′-residues, and base modification takes place.

tRNA processing in eukaryotes

Many eukaryotic pre-tRNAs are synthesized with an intron as well as extra 5′- and 3′-nucleotides which are all removed during processing. In contrast to prokaryotic tRNA, the 3′-terminal CCA is added by the enzyme tRNA nucleotidyl transferase. Many base modifications also occur.

RNase P

Being composed (in *E. coli*) of a 377 nt RNA and a single 13.7 kDa protein, RNase P is a simple RNP. In both prokaryotes and eukaryotes, its function is to cleave 5′-leader sequences off pre-tRNAs. The RNA component alone can cleave pre-tRNAs *in vitro* and hence is a catalytic RNA, or ribozyme.

Ribozymes

Several biochemical reactions can be carried out by RNA enzymes, or ribozymes. These catalytic RNAs can cleave themselves or other RNA molecules, or perform ligation or self-splicing reactions. They can work alone, but are often complexed with protein(s) *in vivo*, which enhance their catalytic activity. Scientists can now design ribozymes as RNA-cutting tools.

Related topics

RNA Pol III genes: 5S and tRNA transcription (M3)
rRNA processing and ribosomes (O1)
tRNA structure and function (P2)

tRNA processing in prokaryotes

In Topic O1, *Fig. 1*, it was seen that the rRNA operons of *E. coli* contain coding sequences for tRNAs. In addition, there are other operons in *E. coli* that contain up to seven tRNA genes separated by spacer sequences. Mature tRNA molecules are processed from precursor transcripts of both of these types of operon by **RNases D, E, F and P** in an ordered series of steps, illustrated for *E. coli* $tRNA^{Tyr}$ in *Fig. 1*. Once the primary transcript has folded and formed characteristic stems and loops (see Topic P2), an endonuclease (see Topic D2) (RNase E or F) cuts off a flanking sequence at the 3′-end, at the base of a stem, to leave a precursor with nine extra nucleotides. The exonuclease RNase D then removes seven of these 3′-nucleotides one at a time. RNase P can then make an endonucleolytic cut to produce the mature 5′-end of the tRNA. In turn, this allows RNase D to trim the remaining 2 nt from the 3′-end, giving the molecule the

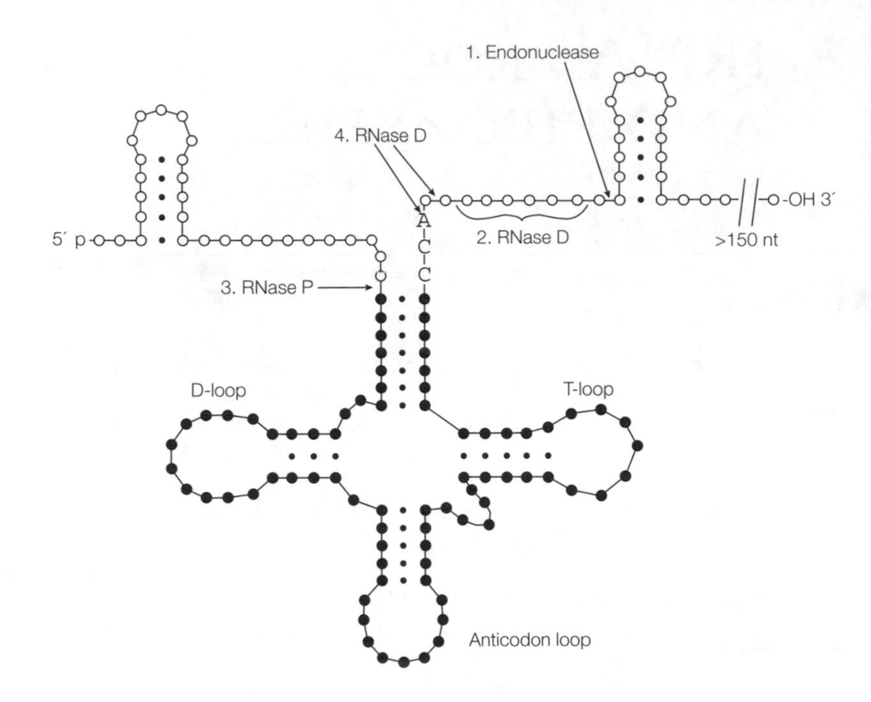

Fig. 1. Pre-tRNA processing in E. coli.

mature 3′-end. Finally, the tRNA undergoes a series of **base modifications**. Different pre-tRNAs are processed in a similar way, but the base modifications are unique to each particular tRNA type. The more common tRNA base modifications are shown in Topic P2, *Fig. 1*.

tRNA processing in eukaryotes

For comparison, the processing of the eukaryotic yeast $tRNA^{Tyr}$ is shown in *Fig. 2*. In this case, the **pre-tRNA** is synthesized with a 16 nt 5′-leader, a 14 nt **intron** (intervening sequence) and two extra 3′-nucleotides. Again, the primary transcript forms a secondary structure with characteristic stems and loops which allow endonucleases to recognize and cleave off the 5′-leader and the two 3′-nucleotides. A major difference between prokaryotes and eukaryotes is that, in the former, the 5′-CCA-3′ at the 3′-end of the mature tRNAs is encoded by the genes. In eukaryotic nuclear-encoded tRNAs this is not the case. After the two 3′-nucleotides have been cleaved off, the enzyme **tRNA nucleotidyl transferase** adds the sequence 5′-CCA-3′ to the 3′-end to generate the mature 3′-end of the tRNA. The next step is the removal of the intron, which occurs by endonucleolytic cleavage at each end of the intron followed by ligation of the half molecules of tRNA. The introns of yeast pre-tRNAs can be processed in vertebrates and therefore the eukaryotic tRNA processing machinery seems to have been highly conserved during evolution.

RNase P

RNase P is an **endonuclease** composed of one RNA molecule and one protein molecule. It is therefore a very simple **RNP**. Its role in cells is to generate the mature 5′-end of tRNAs from their precursors. RNase P enzymes are found in both prokaryotes and eukaryotes, being located in the nucleus of the latter where they are therefore small nuclear RNPs (snRNPs). In *E. coli*, the endonuclease is composed of a 377 nt RNA and a small basic protein of 13.7 kDa. The secondary

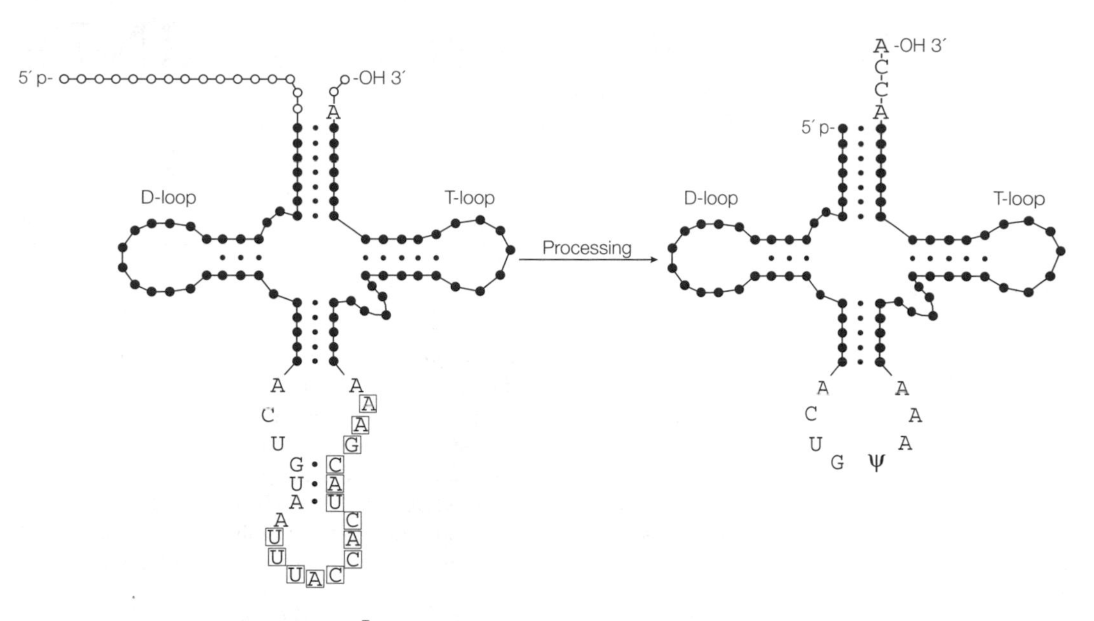

Fig. 2. Processing of yeast pre-tRNA^{Tyr}. Intron nucleotides are boxed.

structure of the RNA has been highly conserved during evolution. Surprisingly, it has been found that the RNA component alone will work as an endonuclease if given pre-tRNA in the test tube. This RNA is therefore a **catalytic RNA**, or **ribozyme**, capable of catalyzing a chemical reaction in the absence of protein. There are several kinds of ribozymes now known. The *in vitro* RNase P ribozyme reaction requires a higher Mg^{2+} concentration than occurs *in vivo*, so the protein component probably helps to catalyze the reaction in cells.

Ribozymes

Ribozymes are **catalytic RNA** molecules that can catalyze particular biochemical reactions. Although only relatively recently discovered, there are several types known to occur naturally, and researchers are now able to create new ones using *in vitro* selection techniques. RNase P is a ubiquitous enzyme that matures tRNA, and its RNA component is a ribozyme that acts as an endonuclease. There is an intron in the large subunit rRNA of *Tetrahymena* that can remove itself from the transcript *in vitro* in the absence of protein (see Topic O1). The process is called **self-splicing** and requires guanosine, or a phosphorylated derivative, as co-factor. The *in vitro* reaction is about 50 times less efficient than the *in vivo* reaction, so it is probable that cellular proteins may assist the reaction *in vivo*. During the replication of some plant viruses, concatameric molecules of the genomic RNA are produced. These are caused by the polymerase continuing to synthesize RNA after it has completed one circle of template. These molecules are able to fold up in such a way as to self-cleave themselves into monomeric, genome-sized lengths. Studies of these self-cleaving molecules have identified the minimum sequences needed, and researchers have managed to develop ribozymes that can cleave other target RNA molecules in *cis* or in *trans*. Currently, there is much interest in using ribozymes to inhibit gene expression by cleaving mRNA molecules *in vivo*, as it may be possible to prevent virus replication, kill cancer cells and discover the function of new genes by inactivating them.

O3 mRNA processing, hnRNPs and snRNPs

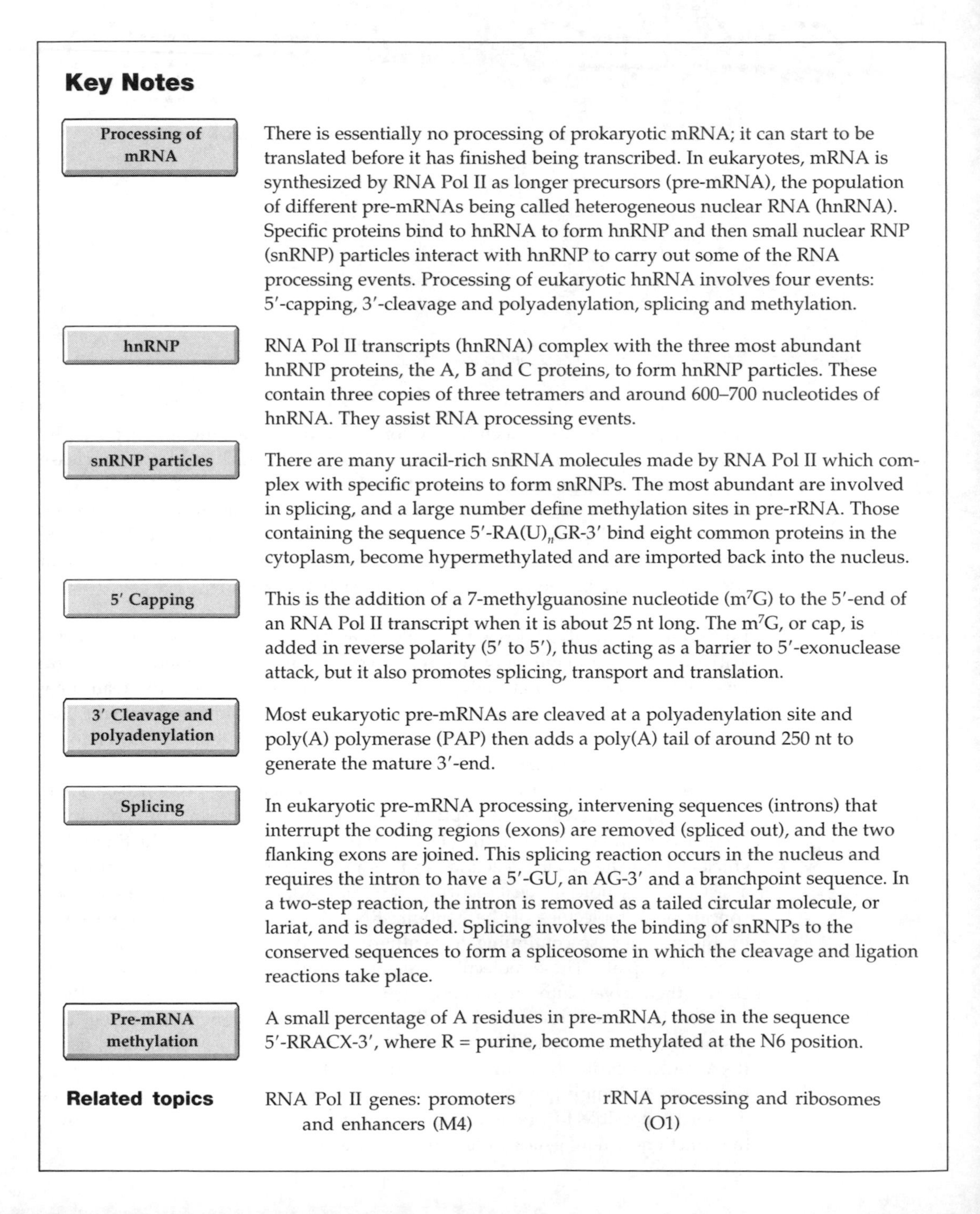

Key Notes

Processing of mRNA

There is essentially no processing of prokaryotic mRNA; it can start to be translated before it has finished being transcribed. In eukaryotes, mRNA is synthesized by RNA Pol II as longer precursors (pre-mRNA), the population of different pre-mRNAs being called heterogeneous nuclear RNA (hnRNA). Specific proteins bind to hnRNA to form hnRNP and then small nuclear RNP (snRNP) particles interact with hnRNP to carry out some of the RNA processing events. Processing of eukaryotic hnRNA involves four events: 5′-capping, 3′-cleavage and polyadenylation, splicing and methylation.

hnRNP

RNA Pol II transcripts (hnRNA) complex with the three most abundant hnRNP proteins, the A, B and C proteins, to form hnRNP particles. These contain three copies of three tetramers and around 600–700 nucleotides of hnRNA. They assist RNA processing events.

snRNP particles

There are many uracil-rich snRNA molecules made by RNA Pol II which complex with specific proteins to form snRNPs. The most abundant are involved in splicing, and a large number define methylation sites in pre-rRNA. Those containing the sequence 5′-RA(U)$_n$GR-3′ bind eight common proteins in the cytoplasm, become hypermethylated and are imported back into the nucleus.

5′ Capping

This is the addition of a 7-methylguanosine nucleotide (m^{7}G) to the 5′-end of an RNA Pol II transcript when it is about 25 nt long. The m^{7}G, or cap, is added in reverse polarity (5′ to 5′), thus acting as a barrier to 5′-exonuclease attack, but it also promotes splicing, transport and translation.

3′ Cleavage and polyadenylation

Most eukaryotic pre-mRNAs are cleaved at a polyadenylation site and poly(A) polymerase (PAP) then adds a poly(A) tail of around 250 nt to generate the mature 3′-end.

Splicing

In eukaryotic pre-mRNA processing, intervening sequences (introns) that interrupt the coding regions (exons) are removed (spliced out), and the two flanking exons are joined. This splicing reaction occurs in the nucleus and requires the intron to have a 5′-GU, an AG-3′ and a branchpoint sequence. In a two-step reaction, the intron is removed as a tailed circular molecule, or lariat, and is degraded. Splicing involves the binding of snRNPs to the conserved sequences to form a spliceosome in which the cleavage and ligation reactions take place.

Pre-mRNA methylation

A small percentage of A residues in pre-mRNA, those in the sequence 5′-RRACX-3′, where R = purine, become methylated at the N6 position.

Related topics

RNA Pol II genes: promoters and enhancers (M4)

rRNA processing and ribosomes (O1)

Processing of mRNA

There appears to be little or no processing (see Topic O1) of mRNA transcripts in prokaryotes. In fact, ribosomes can assemble on, and begin to translate, mRNA molecules that have not yet been completely synthesized. Prokaryotic mRNA is degraded rapidly from the 5′-end and the first **cistron** (protein-coding region) can therefore only be translated for a limited amount of time. Some internal cistrons are partially protected by stem–loop structures that form at the 5′- and 3′-ends and provide a temporary barrier to exonucleases and can thus be translated more often before they are eventually degraded.

Because eukaryotic RNA Pol II transcribes such a wide variety of different genes, from the snRNA genes of 60–300 nt to the large *Antennapedia* gene, whose transcript can be over 100 kb in length, the collection of products made by this enzyme is referred to as **heterogeneous nuclear RNA (hnRNA)**. Those transcripts that will be processed to give mRNAs are called **pre-mRNAs**. Pre-mRNA molecules are processed to mature mRNA by 5′-capping, 3′-cleavage and polyadenylation, splicing and methylation.

hnRNP

The hnRNA synthesized by RNA Pol II is mainly pre-mRNA and rapidly becomes covered in proteins to form **heterogeneous nuclear ribonucleoprotein (hnRNP)**. The proteins involved have been classified as hnRNP proteins A–U. There are two forms of each of the three more abundant hnRNP proteins, the A, B and C proteins. Purification of this material from nuclei gives a fairly homogeneous preparation of 30–40S particles called hnRNP particles. These particles are about 20 nm in diameter and each contains about 600–700 nt of RNA complexed with three copies of three different tetramers. These tetramers are $(A_1)_3B_2$, $(A_2)_3B_1$ and $(C_1)_3C_2$. The hnRNP proteins are thought to help keep the hnRNA in a single-stranded form and to assist in the various RNA processing reactions.

snRNP particles

RNA Pol II also transcribes most **snRNAs** which complex with specific proteins to form **snRNPs**. These RNAs are rich in the base uracil and are thus denoted U1, U2, etc. The most abundant are those involved in **pre-mRNA splicing – U1, U2, U4, U5 and U6**. However, the list of snRNAs is growing, and the majority seem to be involved in determining the sites of methylation of pre-rRNA and are thus located in the nucleolus (see Topic O1). The major nucleoplasmic snRNPs are formed by the individual snRNAs complexing with a common set of eight proteins, which are small and basic, and a variable number of snRNP-specific proteins. These core proteins, known as the **Sm proteins** (after an antibody which recognizes them), require the sequence 5′-RA(U)$_n$GR-3′ to be present in a single-stranded region of the RNA. U6 does not have this sequence but it is usually base-paired to U4 which does. The snRNPs are formed as follows. They are synthesized in the nucleus by RNA Pol II and have a normal 5′-cap (see below). They are exported to the cytoplasm where they associate with the common core proteins and with other specific proteins. Their 5′-cap gains two methyl groups and they are then imported back into the nucleus where they function in splicing.

5′ Capping

Very soon after RNA Pol II starts making a transcript, and before the RNA chain is more than 20–30 nt long, the 5′-end is chemically modified by the addition of a **7-methylguanosine** (m^{7}G) residue (*Fig. 1*). This 5′ modification is called a **cap** and occurs by addition of a GMP nucleotide to the new RNA transcript in the reverse orientation compared with the normal 3′–5′ linkage, giving a

5′–5′ triphosphate bridge. The reaction is carried out by an enzyme called mRNA guanyltransferase and there can be subsequent methylations of the sugars on the first and second transcribed nucleotides, particularly in vertebrates. The cap structure forms a barrier to 5′-exonucleases and thus stabilizes the transcript, but the cap is also important in other reactions undergone by pre-mRNA and mRNA, such as splicing, nuclear transport and translation.

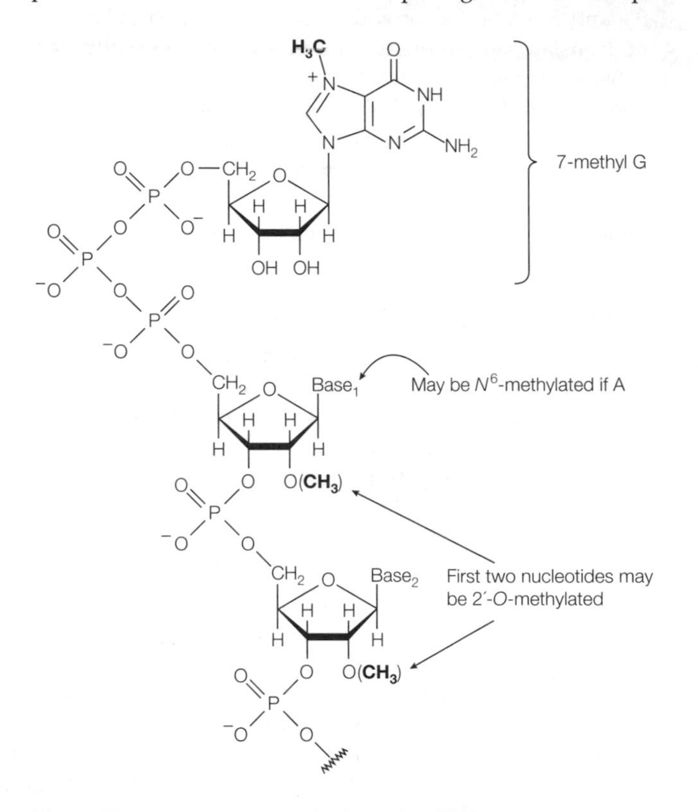

Fig. 1. The 5′ cap structure of eukaryotic mRNA.

3′ Cleavage and polyadenylation

In most pre-mRNAs, the mature 3′-end of the molecule is generated by cleavage followed by the addition of a run, or tail, of A residues which is called the **poly(A) tail**. This feature has allowed the purification of mRNA molecules from the other types of cellular RNAs, permitting the construction of cDNA libraries as described in Topics I1 and I2, from which specific genes have been isolated and their functions analyzed.

The cleavage and polyadenylation reaction requires that specific sequences be present in the DNA and its pre-mRNA transcript. These consist of a 5′-AAUAAA-3′, the **polyadenylation signal**, followed by a 5′-YA-3′, where Y = pyrimidine, in the next 11–20 nt (*Fig. 2a*). Downstream, a GU-rich sequence is often present. Collectively, these sequence elements make up the requirements of a **polyadenylation site**.

A number of specific protein factors recognize these sequence elements and bind to the pre-mRNA. When the complex has assembled, cleavage takes place and then one of the factors, **poly(A) polymerase (PAP)**, adds up to 250 A residues to the 3′-end of the cleaved pre-mRNA. The poly(A) tail on pre-mRNA is thought

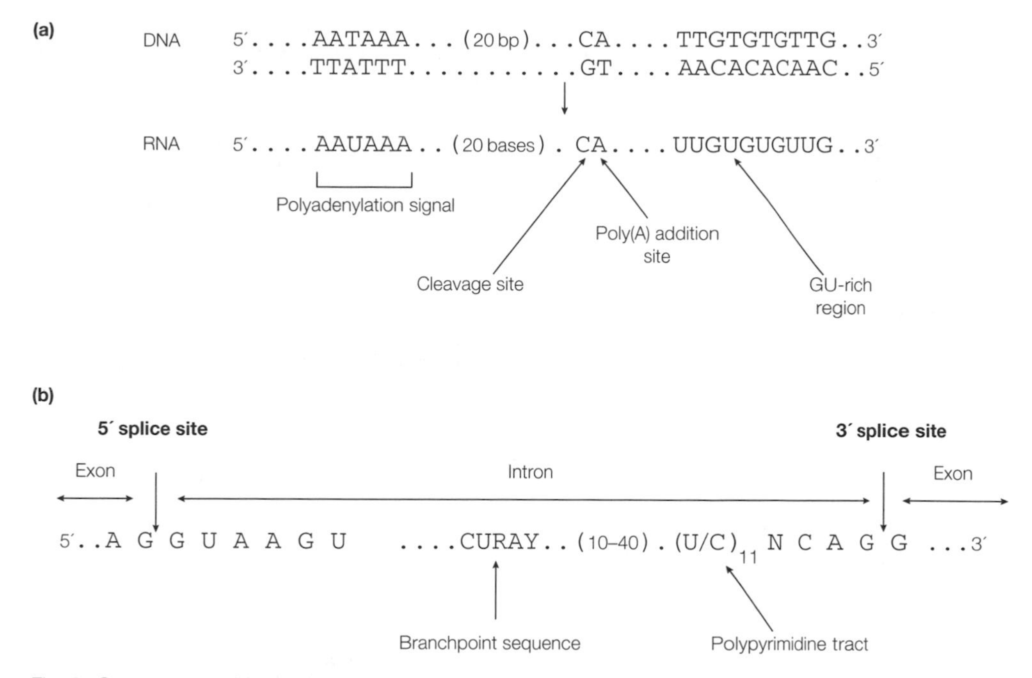

Fig. 2. Sequences of (a) a typical polyadenylation site and (b) the splice site consensus.

to help stabilize the molecule since a **poly(A)-binding protein** binds to it which should act to resist 3′-exonuclease action. In addition, the poly(A) tail may help in the translation of the mature mRNA in the cytoplasm. Histone pre-mRNAs do not get polyadenylated, but are cleaved at a special sequence to generate their mature 3′-ends.

Splicing

During the processing of pre-mRNA in eukaryotes, some sequences that are transcribed and which are upstream of the polyadenylation site are also eventually removed to create the mature mRNA. These sequences are cut out from central regions of the pre-mRNA and the outer portions joined. These intervening sequences, or **introns**, interrupt those sequences that will become adjacent regions in the mature mRNA, the **exons** which are usually the protein-coding regions of the mRNA. The process of cutting the pre-mRNA to remove the introns and joining together of the exons is called **splicing**. Like the polyadenylation process, it takes place in the nucleus before the mature mRNA can be exported to the cytoplasm.

Splicing also requires a set of specific sequences to be present (*Fig. 2b*). The 5′-end of almost all introns has the sequence 5′-GU-3′ and the 3′-end is usually 5′-AG-3′. The AG at the 3′-end is preceded by a pyrimidine-rich sequence called the **polypyrimidine tract**. About 10–40 residues upstream of the polypyrimidine tract is a sequence called the **branchpoint sequence** which is 5′-CURAY-3′ in vertebrates, where R = purine and Y = pyrimidine, but in yeast is the more specific sequence 5′-UACUAAC-3′.

Splicing has been shown to take place in a **two-step reaction** (*Fig. 3a*). First, the bond in front of the G at the 5′-end of the intron at the so-called **5′-splice site** is attacked by the 2′-hydroxyl group of the A residue of the branchpoint

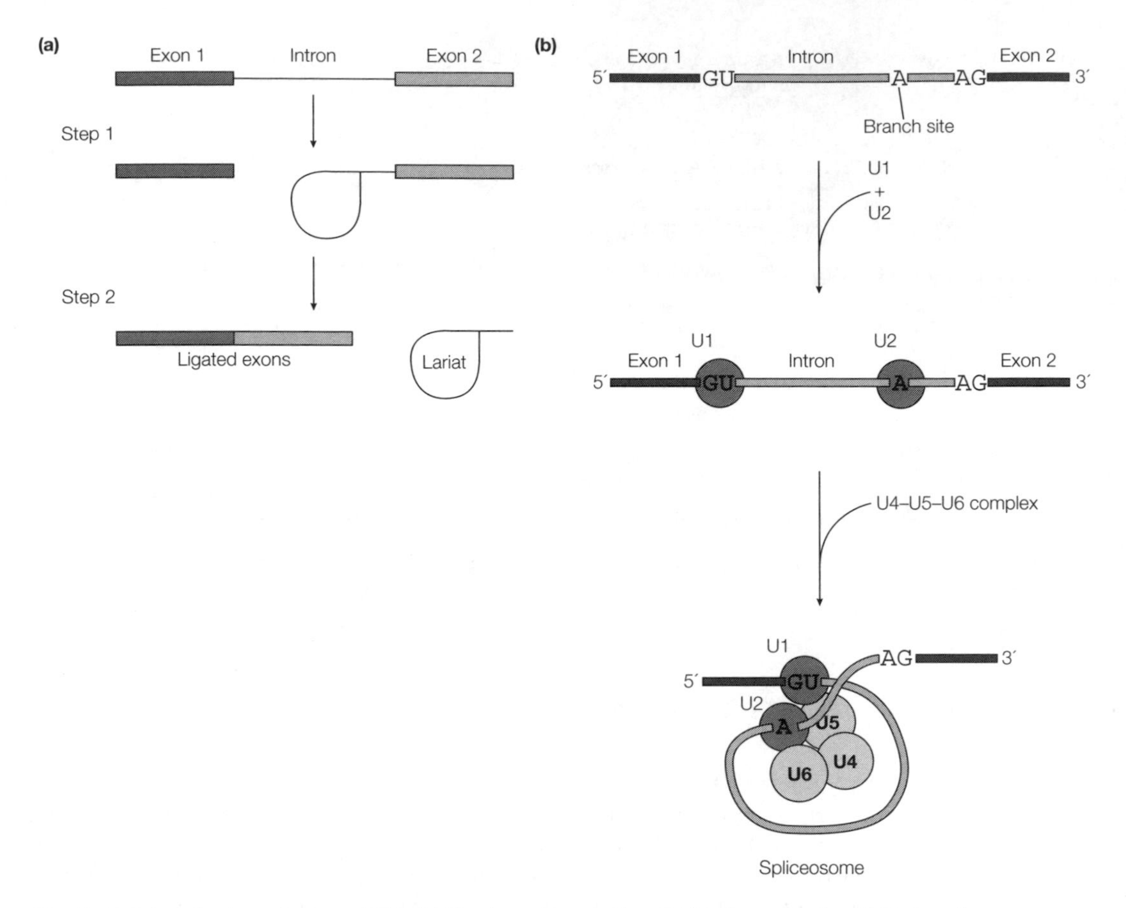

Fig. 3. Splicing of eukaryotic pre-mRNA. (a) The two-step reaction; (b) involvement of snRNPs in spliceosome formation.

sequence to create a tailed circular molecule called a **lariat** and free exon 1. In the second step, cleavage at the **3′-splice site** occurs after the G of the AG, as the two exon sequences are joined together. The intron is released in the lariat form and is eventually degraded.

The splicing process is catalyzed by the U1, U2, U4, U5 and U6 **snRNPs**, as well as other splicing factors. The RNA components of these snRNPs form base pairs with the various conserved sequences at the 5′- and 3′-splice sites and the branchpoint (*Fig. 3b*). Early in splicing, the 5′-end of the U1 snRNP binds to the 5′-splice site and then U2 binds to the branchpoint. The tri-snRNP complex of U4, U5 and U6 can then bind, and in so doing the intron is looped out and the 5′- and 3′-exons are brought into close proximity. The snRNPs interact with one another forming a complex which folds the pre-mRNA into the correct conformation for splicing. This complex of snRNPs and pre-mRNA which forms to hold the upstream and downstream exons close together while looping out the intron is called a **spliceosome**. After the spliceosome forms, a rearrangement takes place before the two-step splicing reaction can occur with release of the intron as a lariat.

Pre-mRNA methylation

The final modification or processing event that many pre-mRNAs undergo is specific methylation of certain bases. In vertebrates, the most common methylation event is on the N6 position of A residues, particularly when these A residues occur in the sequence 5′-RRACX-3′, where X is rarely G. Up to 0.1% of pre-mRNA A residues are methylated, and the methylations seem to be largely conserved in the mature mRNA, though their function is unknown.

O4 ALTERNATIVE mRNA PROCESSING

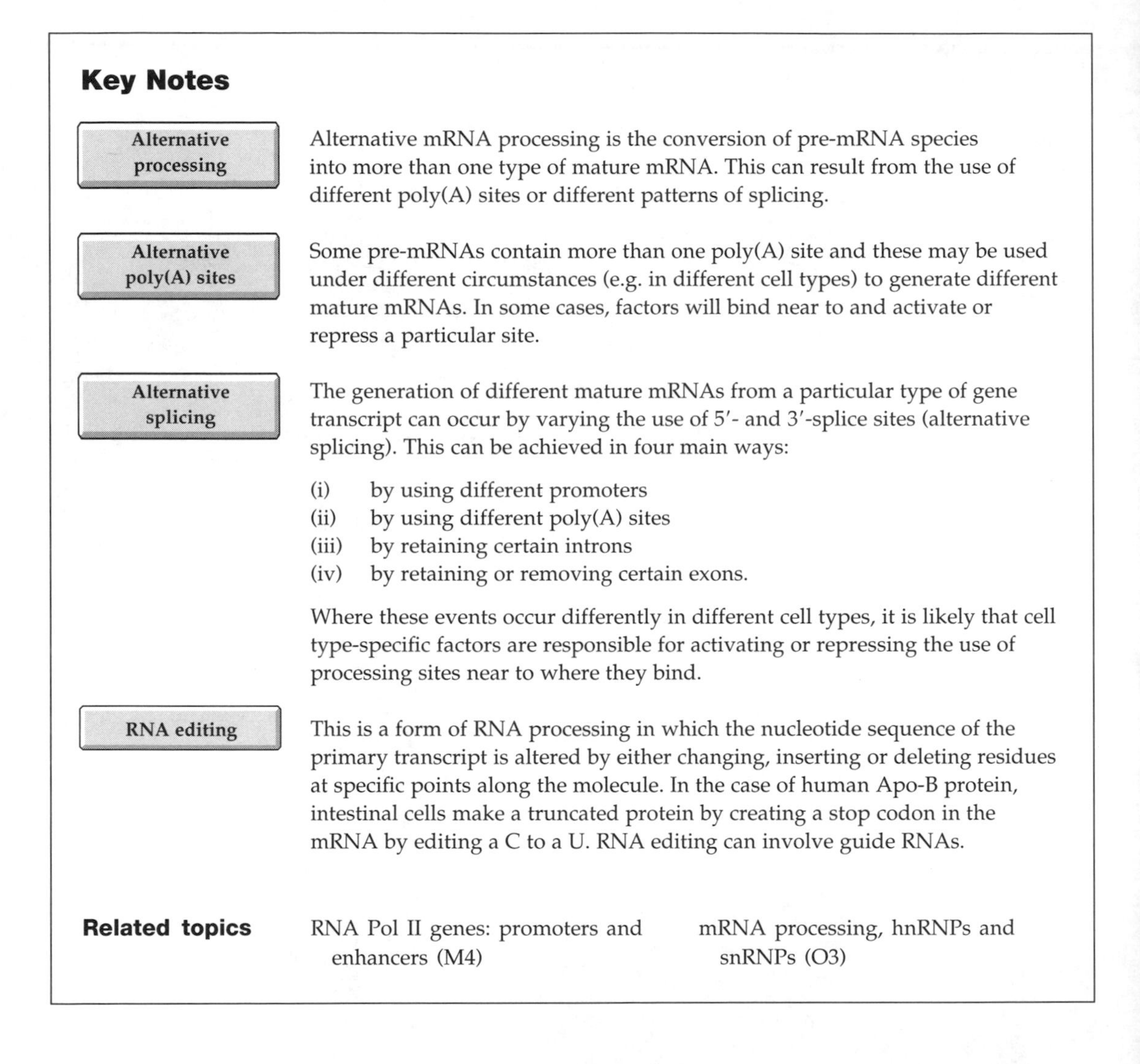

Key Notes

Alternative processing

Alternative mRNA processing is the conversion of pre-mRNA species into more than one type of mature mRNA. This can result from the use of different poly(A) sites or different patterns of splicing.

Alternative poly(A) sites

Some pre-mRNAs contain more than one poly(A) site and these may be used under different circumstances (e.g. in different cell types) to generate different mature mRNAs. In some cases, factors will bind near to and activate or repress a particular site.

Alternative splicing

The generation of different mature mRNAs from a particular type of gene transcript can occur by varying the use of 5′- and 3′-splice sites (alternative splicing). This can be achieved in four main ways:

(i) by using different promoters
(ii) by using different poly(A) sites
(iii) by retaining certain introns
(iv) by retaining or removing certain exons.

Where these events occur differently in different cell types, it is likely that cell type-specific factors are responsible for activating or repressing the use of processing sites near to where they bind.

RNA editing

This is a form of RNA processing in which the nucleotide sequence of the primary transcript is altered by either changing, inserting or deleting residues at specific points along the molecule. In the case of human Apo-B protein, intestinal cells make a truncated protein by creating a stop codon in the mRNA by editing a C to a U. RNA editing can involve guide RNAs.

Related topics

RNA Pol II genes: promoters and enhancers (M4)

mRNA processing, hnRNPs and snRNPs (O3)

Alternative processing

It has become clear that in many cases in eukaryotes a particular pre-mRNA species can give rise to more than one type of mRNA. This can occur when certain exons (**alternative exons**) are removed by splicing and so are not retained in the mature mRNA product. Additionally, if there are alternative possible poly(A) sites that can be used, different 3′-ends can be present in the mature mRNAs. Types of **alternative RNA processing** include **alternative** (or differential) **splicing** and **alternative** (or differential) **poly(A) processing**.

Alternative poly(A) sites

Some pre-mRNAs contain more than one set of the sequences required for cleavage and polyadenylation (see Topic O3). The cell, or organism, has a choice of which one to use. It is possible that if the upstream site is used then sequences that control mRNA stability or location are removed in the portion that is cleaved off. Thus mature mRNAs with the same coding region, but differing stabilities or locations, could be produced from one gene. In some situations, both sites could be used in the same cell at a frequency that reflects their relative efficiencies (strengths) and the cell would contain both types of mRNA. The efficiency of a poly(A) site may reflect how well it matches the consensus sequences (see Topic O3). In other situations, one cell may exclusively use one poly(A) site, while a different cell uses another. The most likely explanation is that in one cell the stronger site is used by default, but in the other cell a factor is present that activates the weaker site so it is used exclusively, or that prevents the stronger site from being used. In some cases, the use of alternative poly(A) sites can cause different patterns of splicing to occur (see below).

Alternative splicing

Four common types of alternative splicing are summarized in *Fig. 1*. In *Fig. 1a*, it is the choice of promoter (see Topic M4) that forces the pattern of splicing, as happens in the α-amylase and myosin light chain genes. The exon transcribed from the **upstream promoter** has the stronger 5′-splice site which out-competes the downstream one for use of the the first 3′-splice site. This happens in salivary gland for the α-amylase gene when specific transcription factors cause transcription from the upstream promoter. In the liver, the **downstream promoter** is used and the weaker (second) 5′-splice site is used by default. Alternative splicing caused by differential use of poly(A) sites is shown in *Fig. 1b*. The stronger 3′-splice site is only present if the downstream poly(A) site is used and thus the penultimate 'exon' will be removed. When the upstream poly(A) site is used (such as in a different cell or at a different stage of development), splicing occurs by default using the weaker (upstream) 3′-splice site. In the case of immunoglobulins, use of a downstream poly(A) site includes exons encoding membrane-anchoring regions whereas when the upstream site is used these regions are not present and the secreted form of immunoglobulin is produced. In some situations, introns can be retained, as shown in *Fig. 1c*. If the intron contains a stop codon then a truncated protein will be produced on translation. This can give rise to an inactive protein, as in the case of the P element transposase in *Drosophila* somatic cells. In germ cells, a specific factor (or the lack of one present in somatic cells) causes the correct splicing of the intron and a longer mRNA is made which is translated into a functional enzyme in these cells only. The final type of alternative splicing (*Fig. 1d*) illustrates that some exons can be retained or removed in different circumstances. A likely reason is the existence of a factor in one cell type that either promotes the use of a particular splice site or prevents the use of another. The rat troponin-T pre-mRNA can be differentially spliced in this way.

RNA editing

An unusual form of RNA processing in which the sequence of the primary transcript is altered is called **RNA editing**. Several examples exist, and they seem to be more common in nonvertebrates. In man, editing causes a single base change from C to U in the apolipoprotein B pre-mRNA, creating a stop codon in the mRNA in intestinal cells at position 6666 in the 14 500 nt molecule. The unedited RNA in the liver makes apolipoprotein B100, a 512 kDa protein, but in the intestine editing causes the truncated apolipoprotein B48

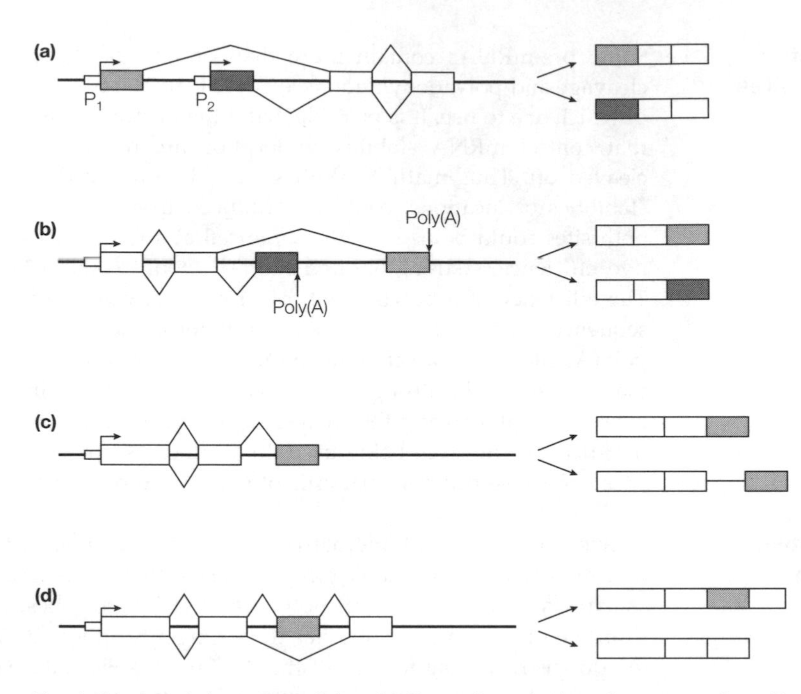

Fig. 1. Modes of alternative splicing. (a) Alternative selection of promoters P_1 or P_2; (b) alternative selection of cleavage/polyadenylation sites; (c) retention of an intron; (d) exon skipping. Empty boxes are exons, filled boxes are alternative exons and thin lines are introns. Reproduced from D.M. Freifelder (1987) Molecular Biology, *2nd Edn.*

(241 kDa) to be made. Similarly, a single A to G change in the glutamate receptor pre-mRNA gives rise to an altered form of the receptor in neuronal cells. The RNA editing changes in the ciliated protozoan, *Leishmania*, are much more dramatic. When the cDNA for the mitochondrial cytochrome *b* gene was cloned, it had a coding region corresponding to the protein sequence. However, although the gene was known to be encoded by the mitochondrial genome, no corresponding sequence was apparent. Eventually, some cDNA clones were obtained which had sequences corresponding to intermediates between the mature mRNA and the genomic sequence. It seems that the primary transcript is edited successively by introducing U residues at specific points. Many cycles of editing eventually produce the mature mRNA which can be translated. Short RNA molecules called **guide RNAs** seem to be involved. Their sequences are complementary to regions of the genomic DNA and the edited RNA. Several other types of RNA editing are known.

P1 THE GENETIC CODE

Key Notes

Nature

The genetic code is the way in which the nucleotide sequence in nucleic acids specifies the amino acid sequence in proteins. It is a triplet code, where the codons (groups of three nucleotides) are adjacent (nonoverlapping) and are not separated by punctuation (comma-less). Because many of the 64 codons specify the same amino acid, the genetic code is degenerate (has redundancy).

Deciphering

The standard genetic code was deciphered by adding homopolymers, co-polymers or synthetic nucleotide triplets to cell extracts which were capable of limited translation. It was found that 61 codons specify the 20 amino acids and there are three stop codons.

Features

Eighteen of the 20 amino acids are specified by multiple (or synonymous) codons which are grouped together in the genetic code table. Usually they differ only in the third codon position. If this is a pyrimidine, then the codons always specify the same amino acid. If a purine, then this is usually also true.

Effect of mutation

The grouping of synonymous codons means that the effects of mutations are minimized. Transitions in the third position often have no effect, as do transversions more than half the time. Mutations in the first and second position often result in a chemically similar type of amino acid being used.

Universality

Until recently, the standard genetic code was considered universal: however, some deviations are now known to occur in mitochondria and some unicellular organisms.

ORFs

Open reading frames are suspected coding regions usually identified by computer in newly sequenced DNA. They are continuous groups of adjacent codons following a start codon and ending at a stop codon.

Overlapping genes

These occur when the coding region of one gene partially or completely overlaps that of another. Thus one reading frame encodes one protein, and one of the other possible frames encodes part or all of a second protein. Some small viral genomes use this strategy to increase the coding capacity of their genomes.

Related topics

Mutagenesis (F1)
Alternative mRNA processing (O4)
tRNA structure and function (P2)
Mechanism of protein synthesis (Q2)
Initiation in eukaryotes (Q3)

Nature

The **genetic code** is the correspondence between the sequence of the four bases in nucleic acids and the sequence of the 20 amino acids in proteins. It has been shown that the code is a triplet code, where three nucleotides encode one amino acid, and this agrees with mathematical argument as being the minimum necessary $[(4^2 = 16) < 20 < (4^3 = 64)]$. However, since there are only 20 amino acids to be specified and potentially 64 different triplets, most amino acids are specified by more than one triplet and the genetic code is said to be **degenerate**, or to have **redundancy**. From a fixed start point, each group of three bases in the coding region of the mRNA represents a **codon** which is recognized by a complementary triplet, or **anticodon**, on the end of a particular tRNA molecule (see Topic P2). The triplets are read in nonoverlapping groups and there is no punctuation between the codons to separate or delineate them. They are simply decoded as adjacent triplets once the process of decoding has begun at the correct start point (initiation site, see Topic Q3). As more gene and protein sequence information has been obtained, it has become clear that the genetic code is very nearly, but not quite, universal. This supports the hypothesis that all life has evolved from a single common origin.

Deciphering

In the 1960s, Nirenberg developed a **cell-free protein synthesizing system** from *E. coli*. Essentially, it was a centrifuged cell lysate which was DNase-treated to prevent new transcription and which would carry out limited protein synthesis if natural or synthetic mRNA was added. To determine which amino acids were being polymerized into polypeptides, it was necessary to carry out 20 reactions in parallel. Each reaction had 19 nonradioactive amino acids and one amino acid labeled with radioactivity. The enzyme **polynucleotide phosphorylase** was used to make synthetic mRNAs that were composed of only one nucleotide, that is poly(U), poly(C), poly(A) and poly(G). If protein synthesis took place after adding one of these **homopolymeric synthetic mRNAs**, then in one of the 20 reaction tubes radioactivity would be incorporated into polypeptide. In this way, it was found that poly(U) caused the synthesis of polyphenylalanine, poly(C) coded for polyproline and poly(A) for polylysine. Poly(G) did not work because it formed a complex secondary structure.

If polynucleotide phosphorylase is used to polymerize a mixture of two nucleotides, say U and G, at unequal ratios such as 0.76 : 0.24, then the triplet GGG is the rarest and UUU will be most common. Triplets with two Us and one G will be the next most frequent. By using these **random co-polymers** as synthetic mRNAs in the cell-free system and determining the frequency of incorporation of particular amino acids, it was possible to determine the composition of the codon for many amino acids. The precise sequence of the triplet codon can only be worked out if additional information is available.

Towards the end of the 1960s, it was found that **synthetic trinucleotides** could attach to the ribosome and bind their corresponding aminoacyl-tRNAs from a mixture. Upon filtering through a membrane, only the complex of ribosome, synthetic triplet and aminoacyl-tRNA (see Topic P2) was retained on the membrane. If the mixture of aminoacyl-tRNAs was made up 20 times, but each time with a different radioactive amino acid, then in this experiment specific triplets could be assigned unambiguously to specific amino acids. A total of 61 codons were shown to code for amino acids and there were three stop codons (*Table 1*) (see Topic B1, *Fig. 2* for the one-letter and three-letter amino acid codes).

Table 1. The universal genetic code

First position (5′ end)	Second position U		Second position C		Second position A		Second position G		Third position (3′ end)
	U		C		A		G		
U	Phe	UUU	Ser	UCU	Tyr	UAU	Cys	UGU	U
	Phe	UUC	Ser	UCC	Tyr	UAC	Cys	UGC	C
	Leu	UUA	Ser	UCA	**Stop**	UAA	**Stop**	UGA	A
	Leu	UUG	Ser	UCG	**Stop**	UAG	Trp	UGG	G
C	Leu	CUU	Pro	CCU	His	CAU	Arg	CGU	U
	Leu	CUC	Pro	CCC	His	CAC	Arg	CGC	C
	Leu	CUA	Pro	CCA	Gln	CAA	Arg	CGA	A
	Leu	CUG	Pro	CCG	Gln	CAG	Arg	CGG	G
A	Ile	AUU	Thr	ACU	Asn	AAU	Ser	AGU	U
	Ile	AUC	Thr	ACC	Asn	AAC	Ser	AGC	C
	Ile	AUA	Thr	ACA	Lys	AAA	Arg	AGA	A
	Met	AUG	Thr	ACG	Lys	AAG	Arg	AGG	G
G	Val	GUU	Ala	GCU	Asp	GAU	Gly	GGU	U
	Val	GUC	Ala	GCC	Asp	GAC	Gly	GGC	C
	Val	GUA	Ala	GCA	Glu	GAA	Gly	GGA	A
	Val	GUG	Ala	GCG	Glu	GAG	Gly	GGG	G

Features

The genetic code is degenerate (or it shows redundancy). This is because 18 out of 20 amino acids have more than one codon to specify them, called **synonymous codons**. Only methionine and tryptophan have single codons. The synonymous codons are not positioned randomly, but are grouped in the table. Generally they differ only in their third position. In all cases, if the third position is a pyrimidine, then the codons specify the same amino acid (are synonymous). In most cases, if the third position is a purine the codons are also synonymous. If the second position is a pyrimidine then generally the amino acid specified is hydrophilic. If the second position is a purine then generally the amino acid specified is polar.

Effect of mutation

It is generally considered that the genetic code evolved in such a way as to minimize the effect of mutations (see Topic F1). The most common type of mutation is a **transition**, where either a purine is mutated to the other purine or a pyrimidine is changed to the other pyrimidine. **Transversions** are where a pyrimidine changes to a purine or vice versa. In the third position, transitions usually have no effect, but can cause changes between Met and Ile, or Trp and stop. Just over half of transversions in the third position have no effect and the remainder usually result in a similar type of amino acid being specified, for example Asp or Glu. In the second position, transitions will usually result in a similar chemical type of amino acid being used, but transversions will change the type of amino acid. In the first position, mutations (both transition and transversions) usually specify a similar type of amino acid, and in a few cases it is the same amino acid.

Universality

For a long time after the genetic code was deciphered, it was thought to be universal, that is the same in all organisms. However, since 1980, it has been discovered that mitochondria, which have their own small genomes, utilize a genetic code that differs slightly from the standard, or 'universal' code. Indeed, it is now known that some other unicellular organisms also have a variant genetic code. *Table 2* shows the variations in the genetic code.

Table 2. Modifications of the genetic code

Codon	Usual meaning	Alternative	Organelle or organism
AGA AGG	Arg	**Stop**, Ser	Some animal mitochondria
AUA	Ile	Met	Mitochondria
CGG	Arg	Trp	Plant mitochondria
CUN	Leu	Thr	Yeast mitochondria
AUU GUG UUG	Ile Val Leu	Start	Some prokaryotes
UAA UAG	**Stop**	Glu	Some protozoans
UGA	**Stop**	Trp	Mitochondria, mycoplasma

ORFs

Inspection of DNA sequences, such as those obtained by genome sequencing projects, by eye or by computer will identify continuous groups of adjacent codons that start with ATG and end with TGA, TAA or TAG. These are referred to as **open reading frames**, or **ORFs**, when there is no known protein product. When a particular ORF is known to encode a certain protein, the ORF is usually referred to as a coding region. Hence, an ORF is a **suspected coding region**.

Overlapping genes

Although it is generally true that one gene encodes one polypeptide, and the evolutionary constraints on having more than one protein encoded in a given region of sequence are great, there are now known to be several examples of overlapping coding regions (overlapping genes). Generally these occur where the genome size is small and there is a need for greater information storage density. For example, the phage φX174 makes 11 proteins of combined molecular mass 262 kDa from a 5386 bp genome. Without overlapping genes, this genome could encode at most 200 kDa of protein. Three proteins are encoded within the coding regions for longer proteins. In prokaryotes, the ribosomes simply have to find the second start codon to be able to translate the overlapping gene and they may achieve this without detaching from the template. Eukaryotes have a different way of initiating protein synthesis (see Topic Q3) and tend to make use of alternative RNA processing (see Topic O4) to generate variant proteins from one gene.

P2 tRNA STRUCTURE AND FUNCTION

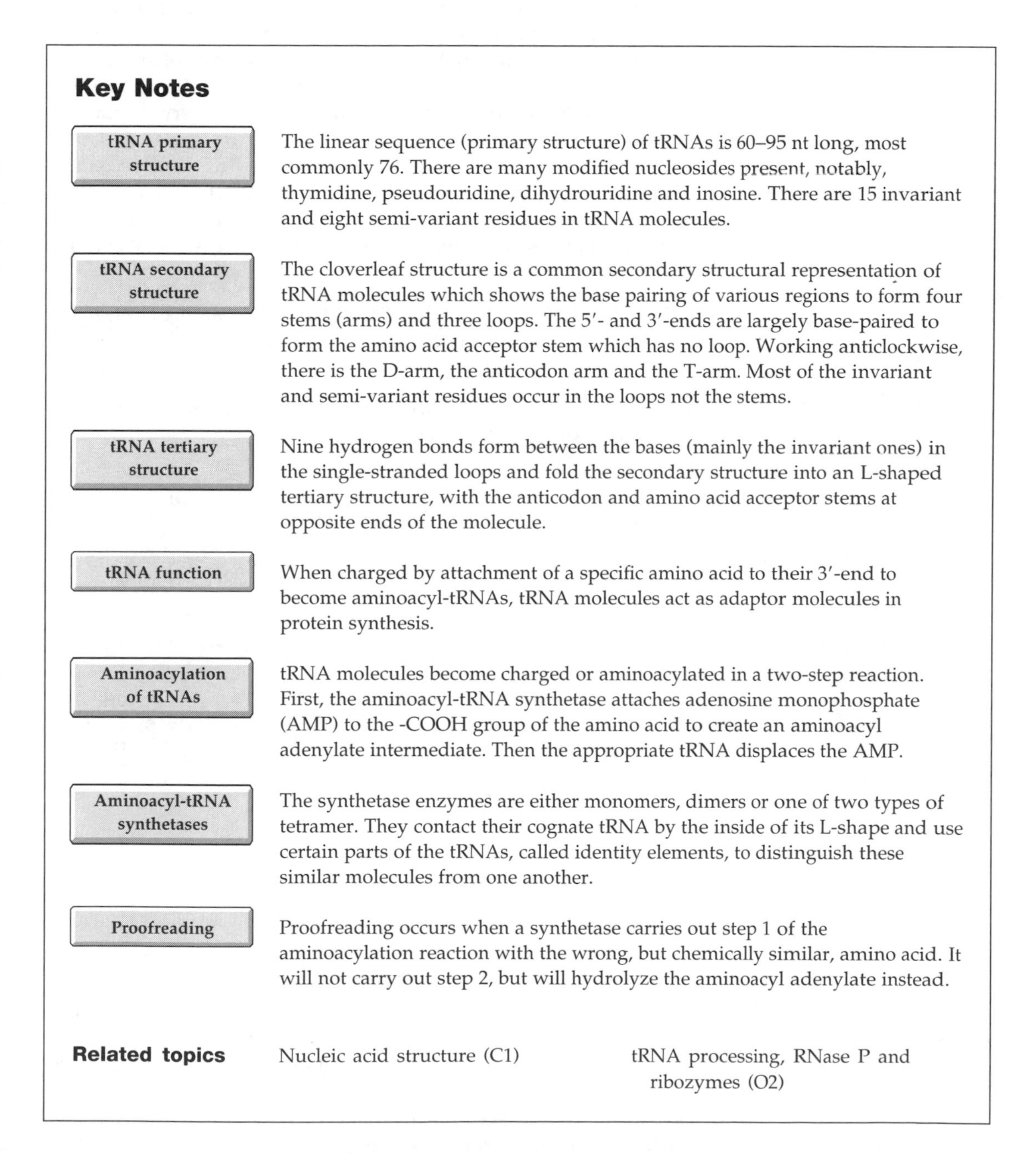

Key Notes

tRNA primary structure	The linear sequence (primary structure) of tRNAs is 60–95 nt long, most commonly 76. There are many modified nucleosides present, notably, thymidine, pseudouridine, dihydrouridine and inosine. There are 15 invariant and eight semi-variant residues in tRNA molecules.
tRNA secondary structure	The cloverleaf structure is a common secondary structural representation of tRNA molecules which shows the base pairing of various regions to form four stems (arms) and three loops. The 5′- and 3′-ends are largely base-paired to form the amino acid acceptor stem which has no loop. Working anticlockwise, there is the D-arm, the anticodon arm and the T-arm. Most of the invariant and semi-variant residues occur in the loops not the stems.
tRNA tertiary structure	Nine hydrogen bonds form between the bases (mainly the invariant ones) in the single-stranded loops and fold the secondary structure into an L-shaped tertiary structure, with the anticodon and amino acid acceptor stems at opposite ends of the molecule.
tRNA function	When charged by attachment of a specific amino acid to their 3′-end to become aminoacyl-tRNAs, tRNA molecules act as adaptor molecules in protein synthesis.
Aminoacylation of tRNAs	tRNA molecules become charged or aminoacylated in a two-step reaction. First, the aminoacyl-tRNA synthetase attaches adenosine monophosphate (AMP) to the -COOH group of the amino acid to create an aminoacyl adenylate intermediate. Then the appropriate tRNA displaces the AMP.
Aminoacyl-tRNA synthetases	The synthetase enzymes are either monomers, dimers or one of two types of tetramer. They contact their cognate tRNA by the inside of its L-shape and use certain parts of the tRNAs, called identity elements, to distinguish these similar molecules from one another.
Proofreading	Proofreading occurs when a synthetase carries out step 1 of the aminoacylation reaction with the wrong, but chemically similar, amino acid. It will not carry out step 2, but will hydrolyze the aminoacyl adenylate instead.

Related topics Nucleic acid structure (C1) tRNA processing, RNase P and ribozymes (O2)

tRNA primary structure

tRNAs are the **adaptor** molecules that deliver amino acids to the ribosome and decode the information in mRNA. Their **primary structure** (i.e. the linear sequence of nucleotides) is 60–95 nt long, but most commonly 76. They have many **modified bases** sometimes accounting for 20% of the total bases in any one tRNA molecule. Indeed, over 50 different types of modified base have been observed in the several hundred tRNA molecules characterized to date, and all of them are created post-transcriptionally. Seven of the most common types are shown in *Fig. 1* as nucleosides. Four of these, **ribothymidine (T)**, which contains the base thymine not usually found in RNA, **pseudouridine (Ψ)**, **dihydrouridine (D)** and **inosine (I)**, are very common in tRNA, all but the last being present in nearly all tRNA molecules in similar positions in the sequences. The letters D and T are used to name secondary structural features (see below). In the tRNA primary structure, there are 15 invariant nucleotides and eight which are either purines (R) or pyrimidines (Y). Using the standard numbering convention where position 1 is the 5′-end and 76 is the 3′-end, these are:

8, 11,	14, 15, 18, 19, 21,	24,	32, 33, 37,	48,	53, 54, 55, 56, 57, 58, 60, 61,	74, 75, 76.
	D-loop		anticodon loop		T-loop	acceptor stem

The positions of invariant and semi-variant nucleotides play a role in either the secondary or tertiary structure (see below).

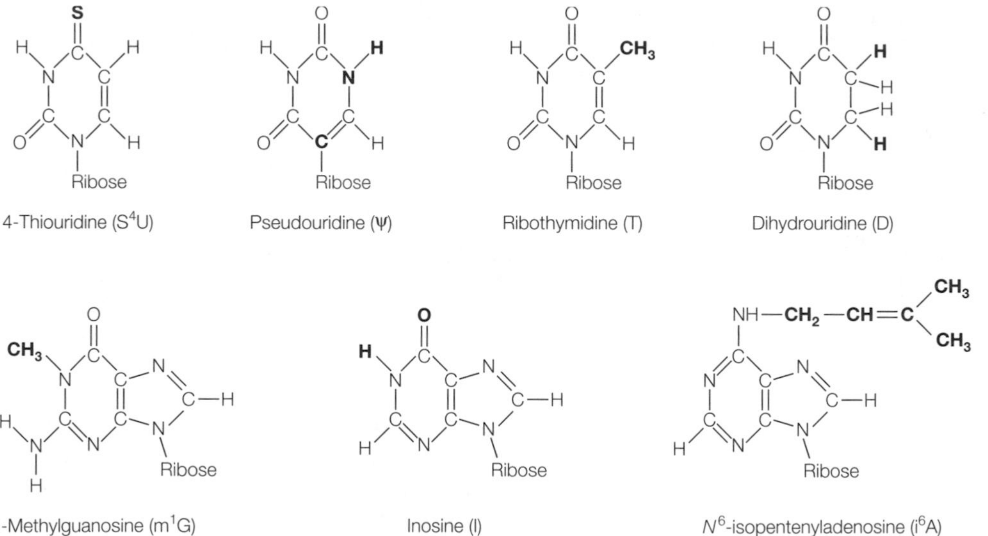

Fig. 1. Modified nucleosides found in tRNA.

tRNA secondary structure

All tRNAs have a common **secondary structure** (i.e. base pairing of different regions to form stems and loops), the **cloverleaf** structure shown in *Fig. 2a*. This structure has a 5′-phosphate formed by RNase P cleavage (see Topic O2), not the usual 5′-triphosphate. It has a 7 bp stem formed by base pairing between the 5′- and 3′-ends of the tRNA; however, the invariant residues 74–76 (i.e. the terminal 5′-CCA-3′) which are added during processing in eukaryotes (see Topic O2) are not included in this base pairing region. This stem is called the **amino acid acceptor stem**. Working 5′ to 3′ (anticlockwise), the next secondary structural feature is called the D-arm which is composed of a 3 or 4 bp stem and a loop called the **D-loop** (DHU-loop) usually containing the modified base dihydrouracil. The next structural feature consists of a 5 bp stem and a seven residue loop in which there are three adjacent nucleotides called the **anticodon** which are complementary to the codon sequence (a triplet in the mRNA) that the tRNA recognizes. The presence of inosine in the anticodon gives a tRNA the ability to base-pair to more than one codon sequence (see Topic Q1). Next there is a **variable arm** which can have between three and 21 residues and may form a stem of up to 7 bp. The other positions of length variation in tRNAs are in the D-loop shown as dashed lines in *Fig. 2a*. The final major feature of secondary structure is the **T-arm** or TΨC-arm which

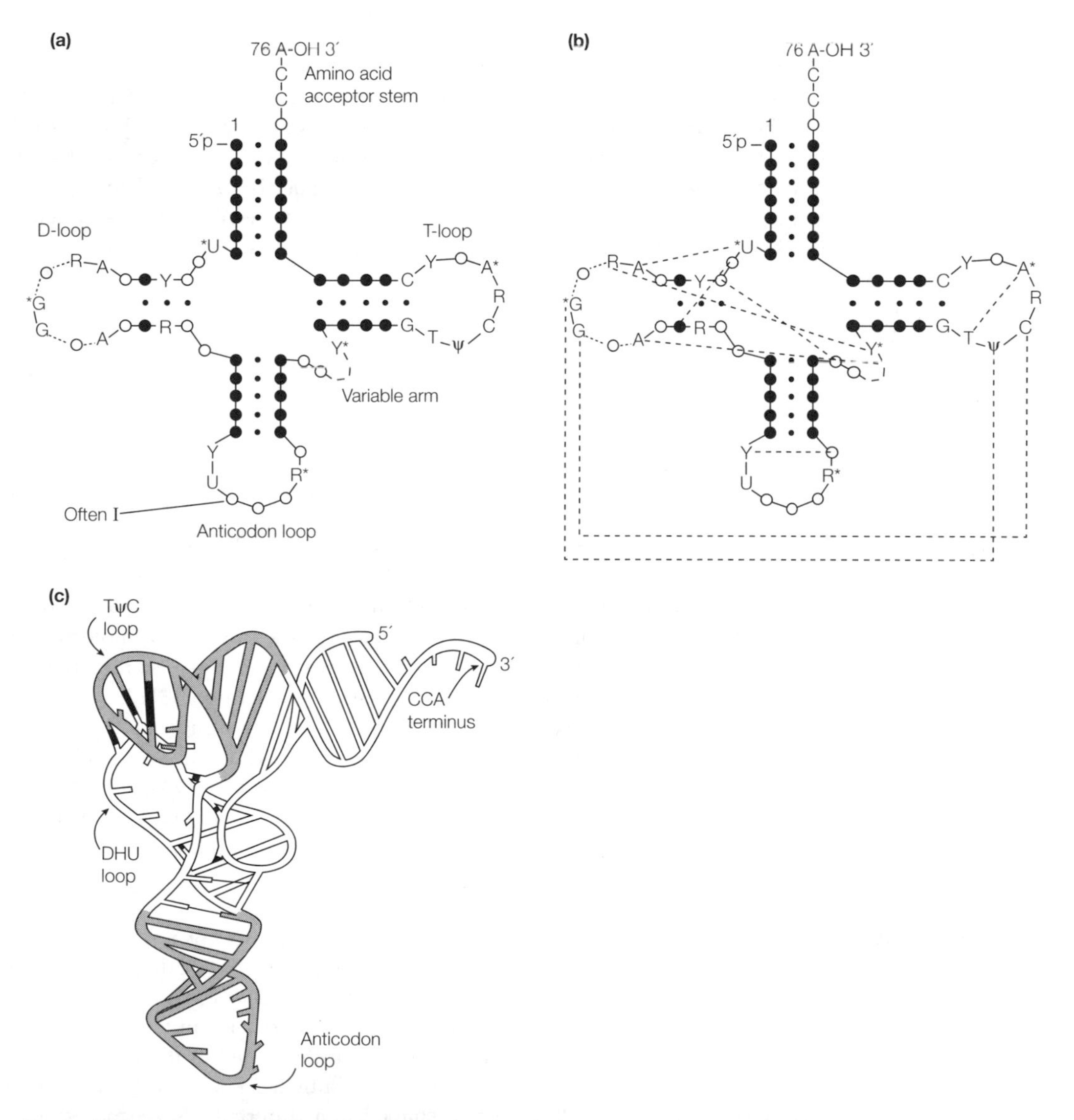

*Fig. 2. tRNA structure. (a) Cloverleaf structure showing the invariant and semi-variant nucleotides, where I = inosine, Ψ = pseudouridine, R = purine, Y = pyrimidine and * indicates a modification. (b) Tertiary hydrogen bonds between the nucleotides in tRNA are shown as dashed lines. (c) The L-shaped tertiary structure of yeast tRNA^{Tyr}. Part (c) reproduced from D.M. Freifelder (1987)* Molecular Biology, *2nd Edn.*

is composed of a 5 bp stem ending in a loop containing the invariant residues GTΨC. Note that the majority of the invariant residues in tRNA molecules are in the loops and do not play a major role in forming the secondary structure. Several of them help to form the tertiary structure.

tRNA tertiary structure

There are nine hydrogen bonds (**tertiary hydrogen bonds**) that help form the 3-D structure of tRNA molecules. They mainly involve base pairing between several invariant bases and are shown in *Fig. 2b*. Base pairing between residues in the D- and T-arms fold the tRNA molecule over into an **L-shape**, with the anticodon at one end and the amino acid acceptor site at the other. The tRNA tertiary structure is strengthened by base stacking interactions (*Fig. 2c*) (see Topic C2).

tRNA function

tRNAs are joined to amino acids to become **aminoacyl-tRNAs** (charged tRNAs) in a reaction called aminoacylation (see below). It is these charged tRNAs that are the adaptor molecules in protein synthesis. Special enzymes called aminoacyl-tRNA synthetases carry out the joining reaction which is extremely specific (i.e. a specific amino acid is joined to a specific tRNA). These pairs of specific amino acids and tRNAs, or tRNAs and aminoacyl-tRNA synthetases are called cognate pairs, and the nomenclature used is shown in *Table 1*.

Table 1. Nomenclature of tRNA-synthetases and charged tRNAs

Amino acid	Cognate tRNA	Cognate aminoacyl-tRNA synthetase	Aminoacyl-tRNA
Serine	$tRNA^{Ser}$	Seryl-tRNA synthetase	Seryl-$tRNA^{Ser}$
Leucine	$tRNA^{Leu}$	Leucyl-tRNA synthetase	Leucyl-$tRNA^{Leu}$
	$tRNA^{Leu}_{UUA}$		Leucyl-$tRNA^{Leu}_{UUA}$

Aminoacylation of tRNAs

The general **aminoacylation reaction** is shown in *Fig. 3*. It is a two-step reaction driven by ATP. In the first step, AMP is linked to the carboxyl group of the amino acid giving a high-energy intermediate called an **aminoacyl adenylate**. The hydrolysis of the pyrophosphate released (to two molecules of inorganic phosphate) drives the reaction forward. In the second step, the aminoacyl adenylate reacts with the **appropriate** uncharged tRNA to give the aminoacyl-tRNA and AMP. Some synthetases join the amino acid to the 2′-hydroxyl of the ribose and some to the 3′-hydroxyl, but once joined the two species can interconvert. The formation of an aminoacyl-tRNA helps to drive protein synthesis as the aminoacyl-tRNA bond is of a higher energy than a peptide bond and thus peptide bond formation is a favorable reaction once this energy-consuming step has been performed.

Aminoacyl-tRNA synthetases

Despite the fact that they all carry out the same reaction of joining an amino acid to a tRNA, the various synthetase enzymes can be quite different. They fall into one of four classes of subunit structure, being either α, α_2, α_4 or $\alpha_2\beta_2$. The polypeptide chains range from 334 to over 1000 amino acids in length, and these enzymes contact the tRNA on the underside (in the angle) of the L-shape. They have a separate amino acid-binding site. The synthetases have to be able to distinguish between about 40 similarly shaped, but different, tRNA molecules in cells, and they use particular parts of the tRNA molecules, called **identity elements**, to be able to do this (*Fig. 4*). These are not always the anticodon sequence (which does differ between tRNA molecules). They often include base pairs in

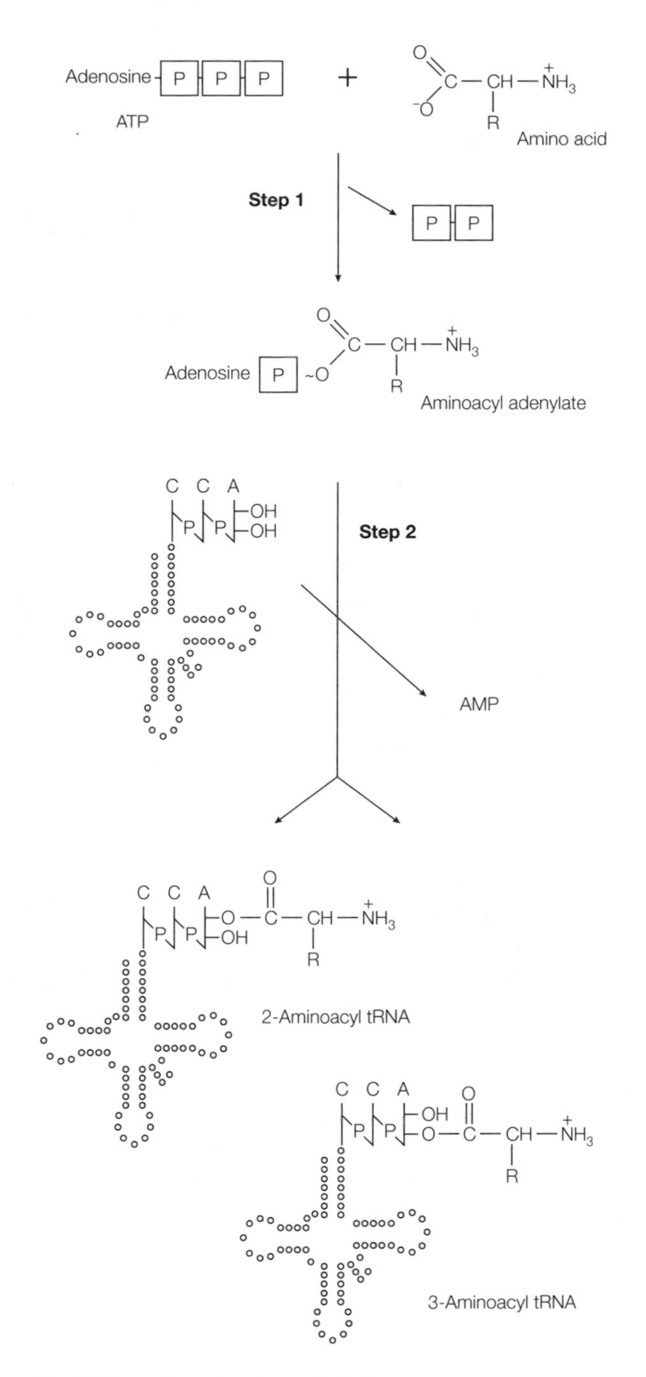

Fig. 3. Formation of aminoacyl-tRNA.

the acceptor stem, and if these are swapped between tRNAs then the synthetase enzymes can be tricked into adding the amino acid to the wrong tRNA. For example, if the G3:U70 identity element of $tRNA^{Ala}$ is used to replace the 3:70 base pair of either $tRNA^{Cys}$ or $tRNA^{Phe}$, then these modified tRNAs are recognized by alanyl-tRNA synthetase and charged with alanine.

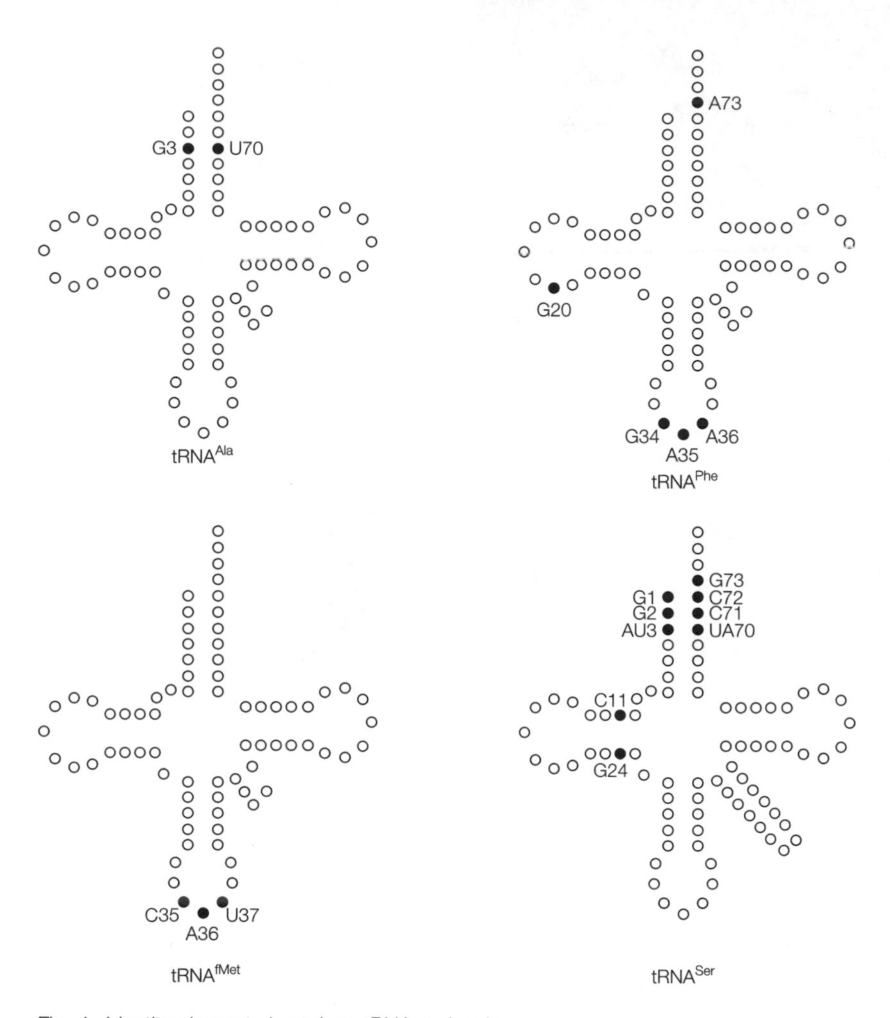

Fig. 4. Identity elements in various tRNA molecules.

Proofreading

Some synthetase enzymes that have to distinguish between two chemically similar amino acids can carry out a proofreading step. If they accidentally carry out step 1 of the aminoacylation reaction with the wrong amino acid, then they will not carry out step 2. Instead they will hydrolyze the amino acid adenylate. This proofreading ability is only necessary when a single recognition step is not sufficiently discriminating. Discrimination between the amino acids Phe and Tyr can be achieved in one step because of the -OH group difference on the benzene ring, so in this case there is no need for proofreading.

Q1 ASPECTS OF PROTEIN SYNTHESIS

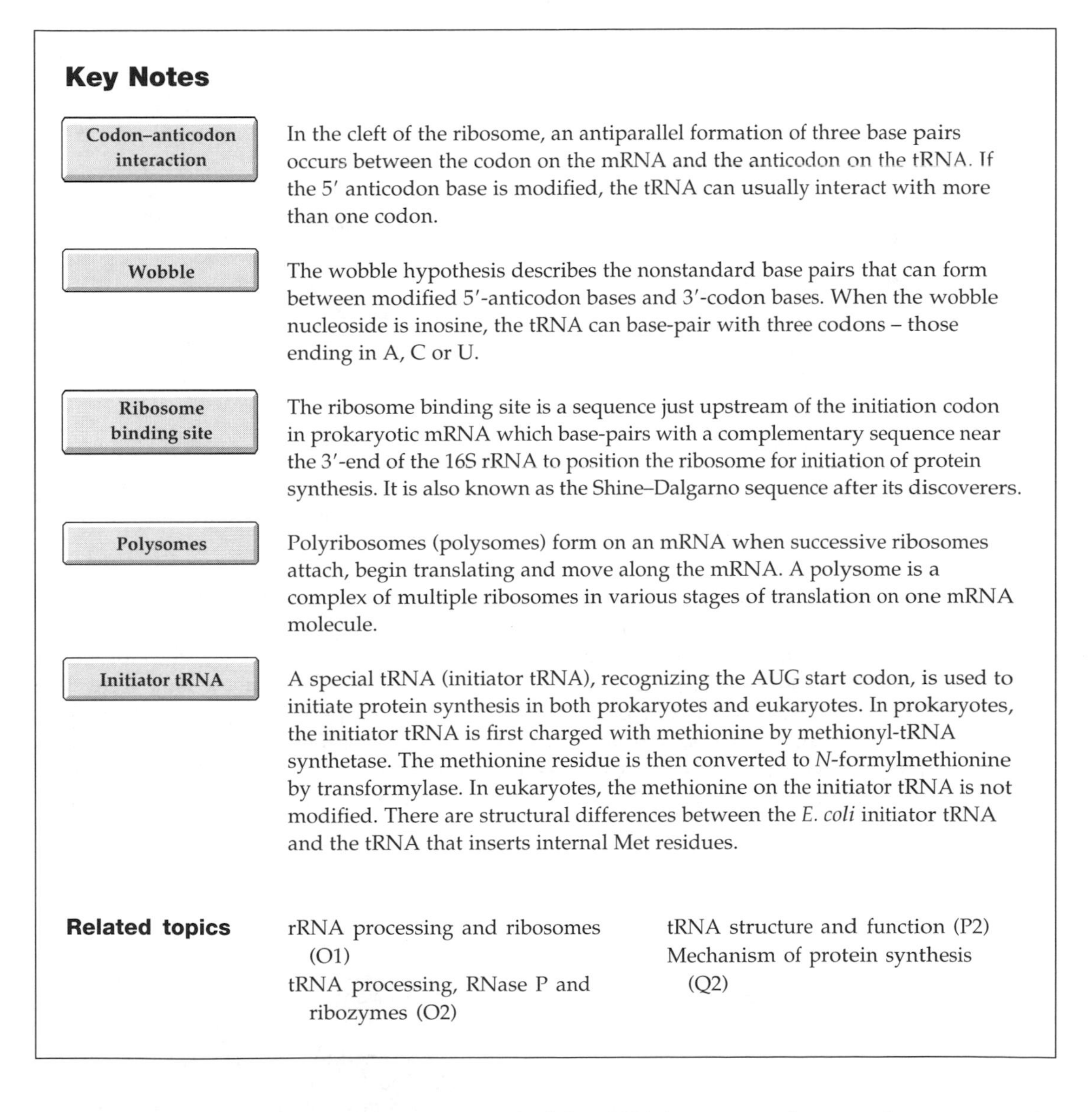

Key Notes

Codon–anticodon interaction

In the cleft of the ribosome, an antiparallel formation of three base pairs occurs between the codon on the mRNA and the anticodon on the tRNA. If the 5′ anticodon base is modified, the tRNA can usually interact with more than one codon.

Wobble

The wobble hypothesis describes the nonstandard base pairs that can form between modified 5′-anticodon bases and 3′-codon bases. When the wobble nucleoside is inosine, the tRNA can base-pair with three codons – those ending in A, C or U.

Ribosome binding site

The ribosome binding site is a sequence just upstream of the initiation codon in prokaryotic mRNA which base-pairs with a complementary sequence near the 3′-end of the 16S rRNA to position the ribosome for initiation of protein synthesis. It is also known as the Shine–Dalgarno sequence after its discoverers.

Polysomes

Polyribosomes (polysomes) form on an mRNA when successive ribosomes attach, begin translating and move along the mRNA. A polysome is a complex of multiple ribosomes in various stages of translation on one mRNA molecule.

Initiator tRNA

A special tRNA (initiator tRNA), recognizing the AUG start codon, is used to initiate protein synthesis in both prokaryotes and eukaryotes. In prokaryotes, the initiator tRNA is first charged with methionine by methionyl-tRNA synthetase. The methionine residue is then converted to *N*-formylmethionine by transformylase. In eukaryotes, the methionine on the initiator tRNA is not modified. There are structural differences between the *E. coli* initiator tRNA and the tRNA that inserts internal Met residues.

Related topics

rRNA processing and ribosomes (O1)
tRNA processing, RNase P and ribozymes (O2)
tRNA structure and function (P2)
Mechanism of protein synthesis (Q2)

Codon–anticodon interaction

The **anticodon** at one end of the tRNA interacts with a complementary triplet of bases on the mRNA, the **codon**, when both are brought together in the cleft of the ribosome (see Topic O1). The interaction is **antiparallel** in nature (*Fig. 1*). Some highly purified tRNA molecules were found to interact with more than one codon, and this ability correlated with the presence of modified nucleosides in the 5′-anticodon position, particularly inosine (see Topic P2). Inosine

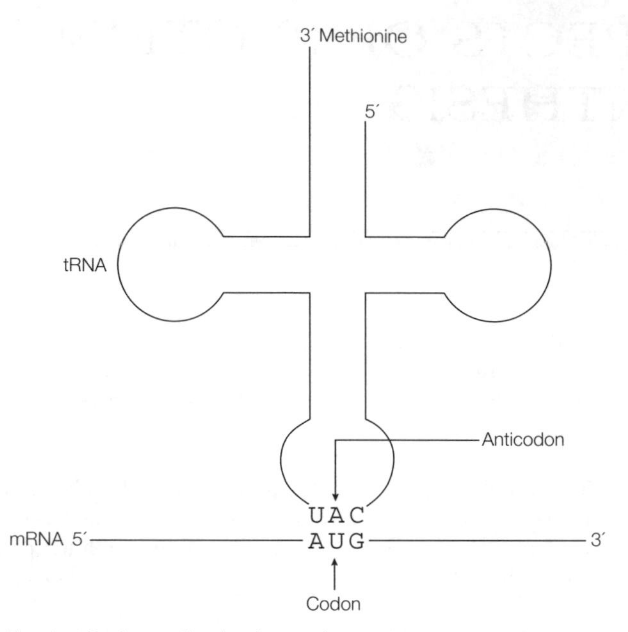

Fig. 1. Codon–anticodon interaction.

is formed by post-transcriptional processing (see Topic O2) of adenosine if it occurs at this position. This is carried out by **anticodon deaminase** which converts the 6-amino group to a keto group.

Wobble

The wobble hypothesis was suggested by Crick to explain the redundancy of the genetic code. He realized, by model building, that the 5′-anticodon base was able to undergo more movement than the other two bases and could thus form **non-standard base pairs** as long as the distances between the ribose units were close to normal. His specific predictions are shown in *Table 1* along with actual observations. No purine–purine or pyrimidine–pyrimidine base pairs are allowed as the ribose distances would be incorrect. No single tRNA could recognize more than three codons, hence, at least 32 tRNAs would be needed to decode the 61 codons, excluding stop codons. tRNAs can recognize either one, two or three codons, depending on their wobble base (the 5′-anticodon base). If it is C it will

Table 1. Original wobble predictions

5′ Anticodon base	Predicted 3′ codon base read	Observations
A	U	A converted to I by anticodon deaminase
C	G	No wobble, normal base pairing
G	C and U	G, and modified G, can pair with C and U
U	A and G	U not found as 5′-anticodon base
I	A and C and U	Wobble as predicted. Inosine (I) can recognize 3′ -A, -C or -U

recognize only the codon ending in G. If it is G, it will recognize the two codons ending in U or C. If U, which is subsequently modified, it will pair with either A or G. The wobble nucleoside is never A, as this is converted to inosine which then pairs with A, C or U.

Ribosome binding site

In prokaryotic mRNAs there is a conserved sequence 8–13 nt upstream of the first codon to be translated (the initiation codon). It was discovered by Shine and Dalgarno and is a purine-rich sequence usually containing all or part of the sequence 5′-AGGAGGU-3′. Experiments have shown that this sequence can base-pair with the 3′-end of the 16S rRNA in the small subunit of the ribosome (5′-ACCUCCU-3′). It is called the **ribosome binding site,** or **Shine–Dalgarno sequence.** It is thought to position the ribosome correctly with respect to the initiation codon.

Polysomes

When a ribosome has begun translating an mRNA molecule (see Topic Q2), and has moved about 70–80 nt from the initiation codon, a second ribosome can assemble at the ribosome-binding site and start to translate the mRNA. When this second ribosome has moved along, a third can begin and so on. Multiple ribosomes on a single mRNA are called **polysomes** (short for polyribosomes) and there can be as many as 50 on some mRNAs, although they cannot be positioned closer than about 80 nt.

Initiator tRNA

It has been shown that the first amino acid incorporated into a protein chain is methionine in both prokaryotes and eukaryotes, though in the former the Met has been modified to **N-formylmethionine**. In both types of organisms, the AUG initiation codon is recognized by a special **initiator tRNA**. The initiator tRNA differs from the one that pairs with AUG codons in the rest of the coding

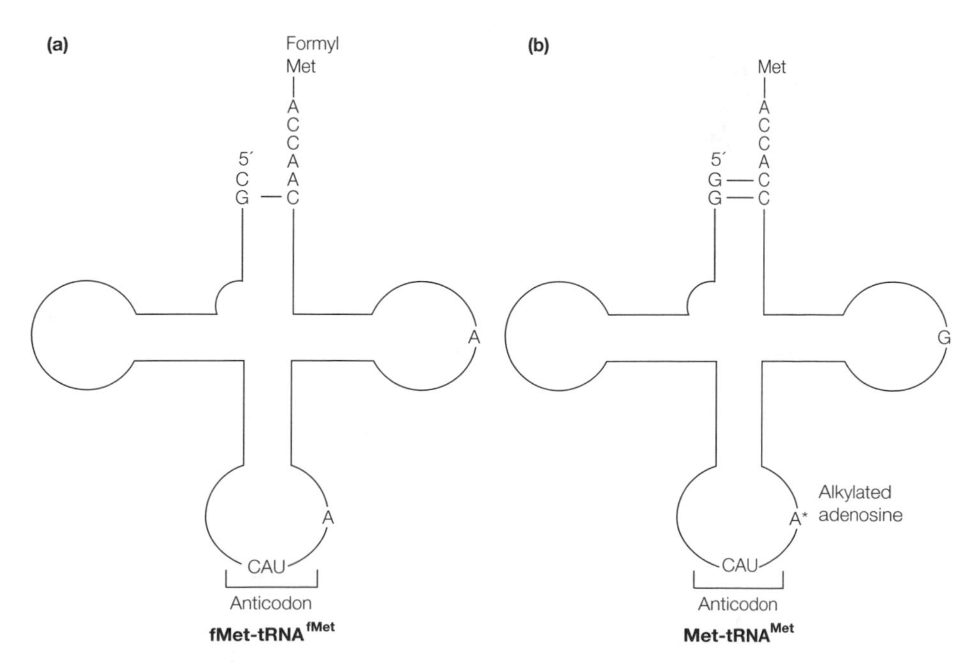

Fig. 2. The E. coli *methionine-tRNAs. (a) The initiator tRNA fMet-tRNA^{fMet}; (b) the methionyl-tRNA Met-tRNA^{Met}.*

region. In the prokaryote *E. coli*, there are subtle differences between these two tRNAs (*Fig. 2*). The initiator tRNA allows more flexibility in base pairing (wobble) because it lacks the alkylated A in the anticodon loop and hence it can recognize both AUG and GUG as initiation codons, the latter occurring occasionally in prokaryotic mRNAs. The noninitiator tRNA is less flexible and can only pair with AUG codons. Both tRNAs are charged with Met by the same **methionyl-tRNA synthetase** (see Topic P2) to give the methionyl-tRNA, but only the initiator methionyl-tRNA is modified by the enzyme **transformylase** to give *N*-formylmethionyl-tRNA^{fMet}. The *N*-formyl group resembles a peptide bond and may help this initiator tRNA to enter the P-site of the ribosome whereas all other tRNAs enter the A-site (see Topic Q2).

Q2 MECHANISM OF PROTEIN SYNTHESIS

Key Notes

Overview

There are three stages of protein synthesis:

- initiation – the assembly of a ribosome on an mRNA;
- elongation – repeated cycles of amino acid delivery, peptide bond formation and movement along the mRNA (translocation);
- termination – the release of the polypeptide chain.

Initiation

In prokaryotes, initiation requires the large and small ribosome subunits, the mRNA, the initiator tRNA, three initiation factors (IFs) and GTP. IF_1 and IF_3 bind to the 30S subunit and prevent the large subunit binding. IF_2 + GTP can then bind and will help the initiator tRNA to bind later. This small subunit complex can now attach to an mRNA via its ribosome-binding site. The initiator tRNA can then base-pair with the AUG initiation codon which releases IF_3, thus creating the 30S initiation complex. The large subunit then binds, displacing IF_1 and IF_2 + GDP, giving the 70S initiation complex which is the fully assembled ribosome at the correct position on the mRNA.

Elongation

Elongation involves the three factors (EFs), EF-Tu, EF-Ts and EF-G, GTP, charged tRNAs and the 70S initiation complex (or its equivalent). It takes place in three steps.

- A charged tRNA is delivered as a complex with EF-Tu and GTP. The GTP is hydrolyzed and EF-Tu·GDP is released which can be re-used with the help of EF-Ts and GTP (via the EF-Tu–EF-Ts exchange cycle).
- Peptidyl transferase makes a peptide bond by joining the two adjacent amino acids without the input of more energy.
- Translocase (EF-G), with energy from GTP, moves the ribosome one codon along the mRNA, ejecting the uncharged tRNA and transferring the growing peptide chain to the P-site.

Termination

Release factors (RF1 or RF2) recognize the stop codons and, helped by RF3, make peptidyl transferase join the polypeptide chain to a water molecule, thus releasing it. Ribosome release factor helps to dissociate the ribosome subunits from the mRNA.

Related topics

tRNA processing, RNase P and ribozymes (O2)
tRNA structure and function (P2)
Aspect of protein synthesis (Q1)
Initiation in eukaryotes (Q3)

Overview

The actual mechanism of protein synthesis can be divided into three stages:

- **initiation** – the assembly of a ribosome on an mRNA molecule;
- **elongation** – repeated cycles of amino acid addition;
- **termination** – the release of the new protein chain.

These are illustrated in *Figs 1–3* and involve the activities of a number of factors. In prokaryotes, the factors are abbreviated as IF or EF for **initiation** and **elongation factors** respectively, whereas in eukaryotes they are called eIF and eEF. There are distinct differences of detail between the mechanism in prokaryotes and eukaryotes, and most of these occur in the initiation stage. For this reason, this topic will describe the mechanism in prokaryotes and the following topic (Q3) will describe the differences in detail that occur in eukaryotes.

Initiation

The purpose of the initiation step is to assemble a complete ribosome on to an mRNA molecule at the correct start point, the **initiation codon**. The components involved are the large and small **ribosome subunits**, the **mRNA**, the **initiator tRNA** in its charged form, **three initiation factors** and **GTP.** The initiation factors **IF_1**, **IF_2** and **IF_3** are all just over one-tenth as abundant as ribosomes, and have masses of 9, 120 and 22 kDa respectively. Only IF_2 binds GTP. Although the finer details have yet to be worked out, the overall sequence of events (*Fig. 1*) is as follows:

- IF_1 and IF_3 bind to a free 30S subunit. This helps to prevent a large subunit binding to it without an mRNA molecule and forming an inactive ribosome.
- IF_2 complexed with GTP then binds to the small subunit. It will assist the charged initiator tRNA to bind.
- The 30S subunit attaches to an mRNA molecule making use of the ribosome-binding site (RBS) on the mRNA (see Topic Q1).
- The initiator tRNA can then bind to the complex by base pairing of its anticodon with the AUG codon on the mRNA. At this point, IF_3 can be released, as its roles in keeping the subunits apart and helping the mRNA to bind are complete. This complex is called the **30S initiation complex**.
- The 50S subunit can now bind, which displaces IF_1 and IF_2, and the GTP is hydrolyzed in this energy-consuming step. The complex formed at the end of the initiation phase is called the **70S initiation complex**.

As shown in *Figs 1–3*, the assembled ribosome has two tRNA-binding sites. These are called the **A- and P-sites**, for aminoacyl and peptidyl sites. The A-site is where incoming aminoacyl-tRNA molecules bind, and the P-site is where the growing polypeptide chain is usually found. These sites are in the cleft of the small subunit (see Topic O1) and contain adjacent codons that are being translated. One major outcome of initiation is the placement of the initiator tRNA in the P-site. It is the only tRNA that does this, as all others must enter the A-site.

Elongation

With the formation of the 70S initiation complex, the **elongation cycle** can begin. It can be subdivided into three steps as follows: (i) **aminoacyl-tRNA delivery**, (ii) **peptide bond formation** and (iii) **translocation** (movement). These are shown in *Fig. 2*, beginning where the P-site is occupied and the A-site is empty. It involves **three elongation factors EF-Tu, EF-Ts** and **EF-G** which all bind GTP or GDP and have masses of 45, 30 and 80 kDa respectively. EF-Ts and EF-G are about as abundant as ribosomes, but EF-Tu is nearly 10 times more abundant.

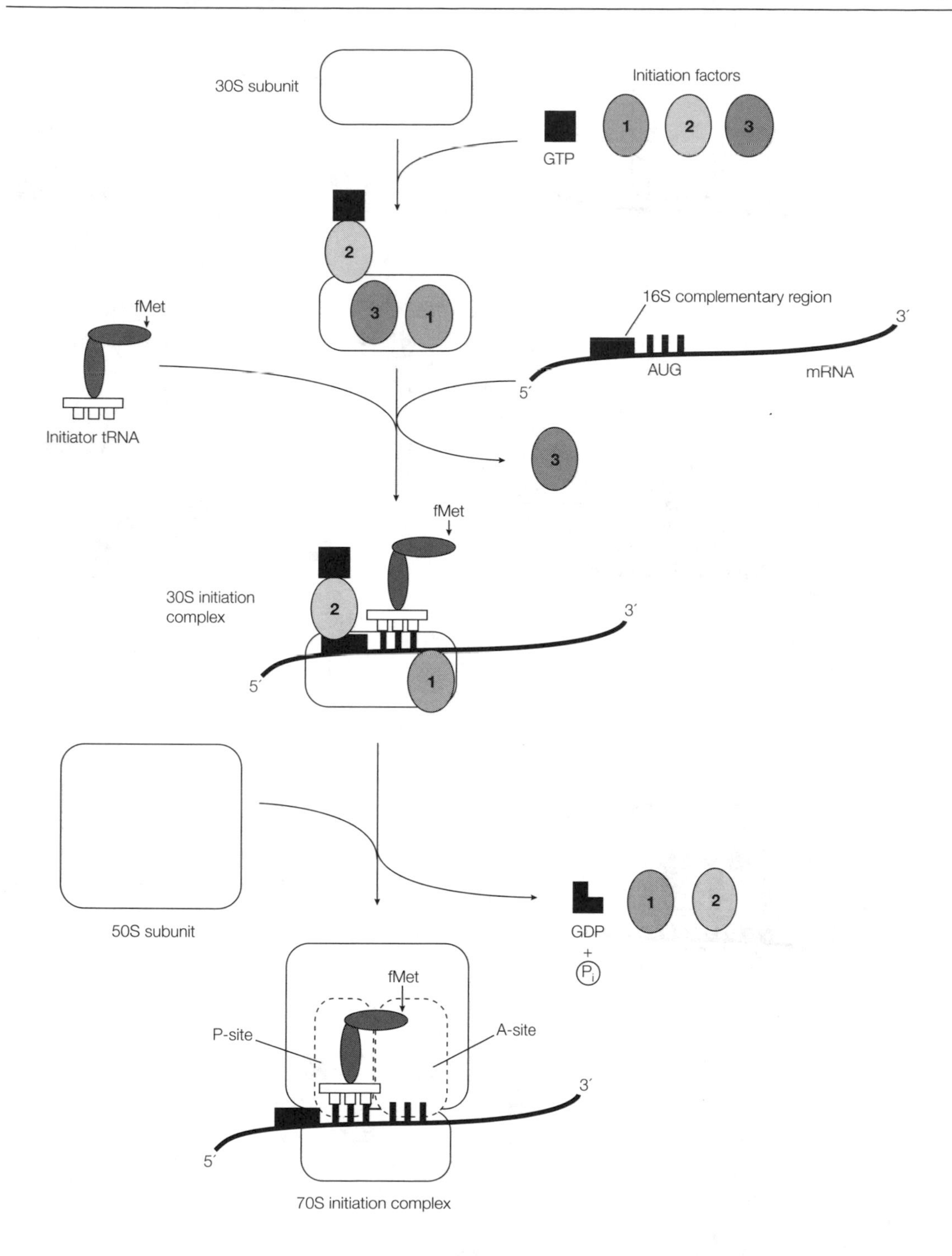

Fig. 1. Initiation of protein synthesis in the prokaryote E. coli.

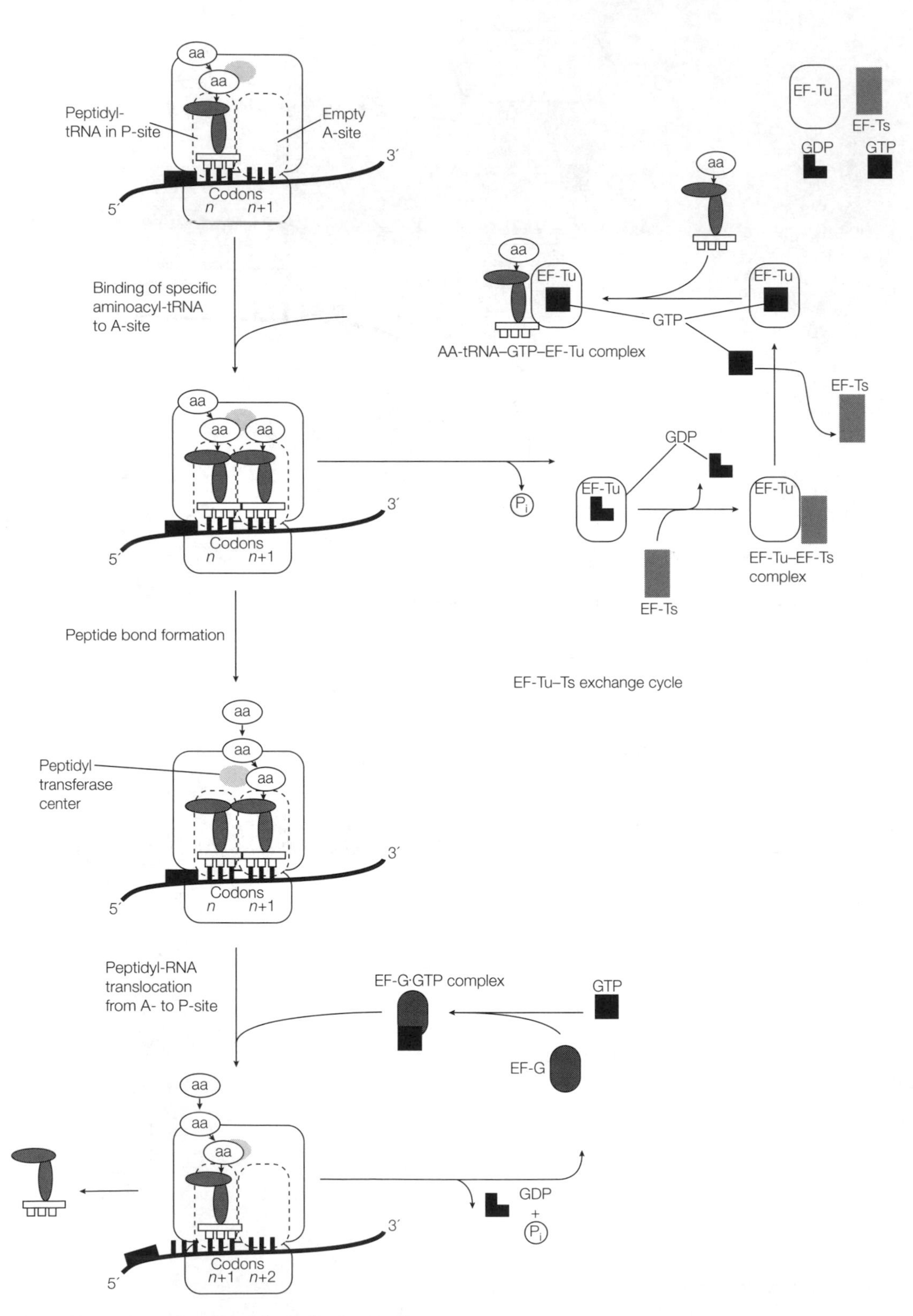

Fig. 2. Elongation stage of protein synthesis.

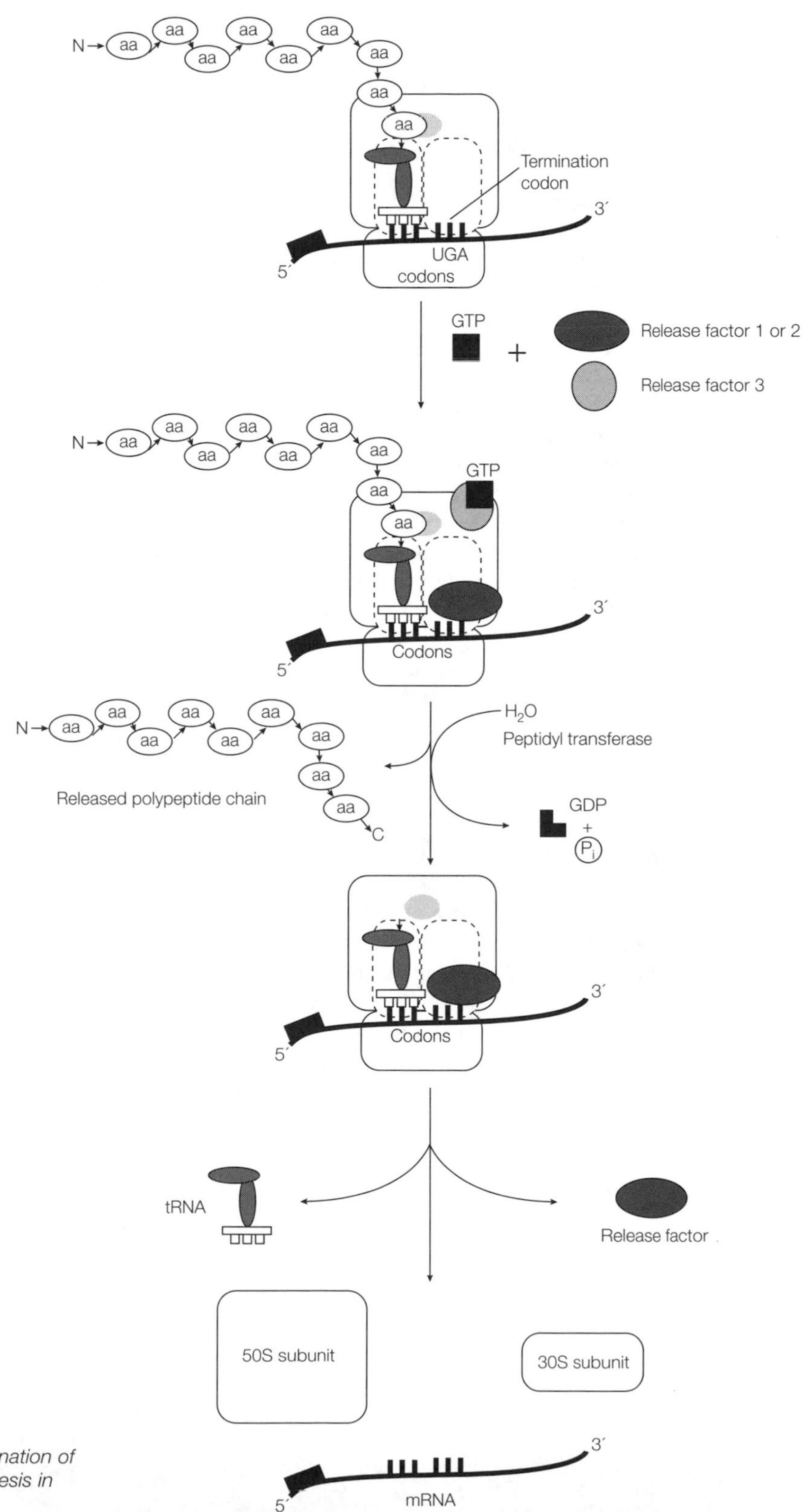

Fig. 3. Termination of protein synthesis in E. coli.

(i) Aminoacyl-tRNA delivery. EF-Tu is required to deliver the aminoacyl-tRNA to the A-site and energy is consumed in this step by the hydrolysis of GTP. The released **EF-Tu·GDP complex** is regenerated with the help of EF-Ts. In the **EF-Tu–EF-Ts exchange cycle**, EF-Ts displaces the GDP and subsequently is displaced itself by GTP. The resultant EF-Tu·GTP complex is now able to bind another aminoacyl-tRNA and deliver it to the ribosome. All aminoacyl-tRNAs can form this complex with EF-Tu, except the initiator tRNA.

(ii) Peptide bond formation. After aminoacyl-tRNA delivery, the A- and P-sites are both occupied and the two amino acids that are to be joined are in close proximity. The **peptidyl transferase** activity of the 50S subunit can now form a peptide bond between these two amino acids without the input of any more energy, since energy in the form of ATP was used to charge the tRNA (Topic P2).

(iii) Translocation. A complex of EF-G (**translocase**) and GTP binds to the ribosome and, in an energy-consuming step, the discharged tRNA is ejected from the P-site, the peptidyl-tRNA is moved from the A-site to the P-site and the mRNA moves by one codon relative to the ribosome. GDP and EF-G are released, the latter being re-usable. A new codon is now present in the vacant A-site. Recent evidence suggests that in prokaryotes the discharged tRNA is first moved to an **E-site** (**exit site**) and is ejected when the next aminoacyl-tRNA binds. In this way the ribosome maintains contact with the mRNA via 6 base pairs which may well reduce the chances of frameshifting (see Topic R4).

One cycle of the three-step elongation cycle has been completed, and the cycle is repeated until one of the three termination codons (stop codons) appears in the A-site.

Termination

There are no tRNA species that normally recognize stop codons. Instead, protein factors called release factors interact with these codons and cause release of the completed polypeptide chain. RF1 recognizes the codons UAA and UAG, and RF2 recognizes UAA and UGA. RF3 helps either RF1 or RF2 to carry out the reaction. The release factors make peptidyl transferase transfer the polypeptide to water rather than to the usual aminoacyl-tRNA, and thus the new protein is released. To remove the uncharged tRNA from the P-site and release the mRNA, EF-G together with ribosome release factor are needed for the complete dissociation of the subunits. IF_3 can now bind the small subunit to prevent inactive 70S ribosomes forming.

Q3 Initiation in eukaryotes

Key Notes

Overview

Most of the differences in the mechanism of protein synthesis between prokaryotes and eukaryotes occur in the initiation stage; however, eukaryotes have just one release factor (eRF). The eukaryotic initiator tRNA does not become *N*-formylated as in prokaryotes.

Scanning

The eukaryotic 40S ribosome subunit complex binds to the 5′-cap region of the mRNA complex and moves along it looking (scanning) for an AUG start codon. It is not always the first AUG, as it must have appropriate sequences around it.

Initiation

This is the major point of difference between prokaryotic and eukaryotic protein synthesis, there being at least nine eIFs involved. Functionally, these factors can be grouped. They either bind to the ribosome subunits or to the mRNA, deliver the initiator tRNA or displace other factors. In contrast to the events in prokaryotes, initiation involves the initiator tRNA binding to the 40S subunit before it can bind to the mRNA. Phosphorylation of eIF2, which delivers the initiator tRNA, is an important control point.

Elongation

This stage of protein synthesis is essentially identical to that described for prokaryotes (Topic Q2). The factors EF-Tu, EF-Ts and EF-G have direct eukaryotic equivalents called eEF1α, eEF1βγ and eEF2 respectively, which carry out the same roles.

Termination

Eukaryotes use only one release factor (eRF), which requires GTP, for termination of protein synthesis. It can recognize all three stop codons.

Related topics

tRNA processing, RNase P and ribozymes (O2)
tRNA structure and function (P2)
Mechanism of protein synthesis (Q2)

Overview

Apart from in the mitochondria and chloroplasts of eukaryotic cells (which are thought to originate from symbiotic prokaryotes; see Topic A2), details of the mechanism of protein synthesis differ from that described in Topic Q2. Most of these differences are in the initiation phase where a **greater number of eIFs** are involved. The method of finding the correct start codon involves a **scanning process** as there is no ribosome-binding sequence. Although there are two different tRNA species for methionine, one of which is the initiator tRNA, the attached methionine does **not** become converted to ***N*-formylmethionine**. A comparison of the factors involved in prokaryotes and eukaryotes is given in *Table 1*.

Scanning

Since there is no Shine–Dalgarno sequence in eukaryotic mRNA, the mechanism of selecting the start codon must be different. Kozak proposed a **scanning**

Table 1. Comparison of protein synthesis factors in prokaryotes and eukaryotes

Prokaryotic	Eukaryotic	Function
Initiation factors		
IF1, IF3	eIF3, eIF4C, eIF6	Binding to ribosome subunits
	eIF4B, eIF4F	Binding to mRNA
IF2	eIF2, eIF2B	Initiator tRNA delivery
	eIF5	Displacement of other factors
Elongation factors		
EF-Tu	eEF1α	Aminoacyl tRNA delivery to ribosome
EF-Ts	eEF1βγ	Recycling of EF-Tu or eEF1α
EF-G	eEF2	Translocation
Termination factors		
RF1		Polypeptide
RF2	eRF	chain
RF3		release

hypothesis in which the 40S subunit, already containing the initiator tRNA, attaches to the 5′-end of the mRNA and scans along the mRNA until it finds an appropriate AUG. This is not always the first one as it must be in the correct **sequence context** (5′-CCRCC**AUG**G-3′), where R = purine.

Initiation

Figure 1 shows the steps and factors involved in the initiation stage of protein synthesis in eukaryotes. Although there are at least nine reasonably well-defined initiation factors involved in eukaryotic protein synthesis, some have analogous functions to the three prokaryotic IFs. They can be grouped according to their functions as follows:

- those **binding to ribosomal subunits**, such as eIF6, eIF3 and eIF4C;
- those **binding to the mRNA** to recognize the 5′-cap and to melt secondary structure, such as eIF4B and eIF4F, which is a complex of a cap-binding protein, eIF4A and eIF4E;
- those **involved in initiator tRNA delivery**, such as eIF2 and eIF2B;
- those that **displace other factors**, such as eIF5 which releases two other factors so the 60S subunit can bind.

The following events take place, starting with a free **40S** subunit and a 5′-capped mRNA molecule. **eIF3** and **4C** bind to the 40S subunit which allows it to bind a complex of three components (**ternary complex**) – the **initiator tRNA, eIF2 and GTP**. Note this different order of assembly in eukaryotes where the initiator tRNA is bound to the small subunit before the mRNA binds (compare Topic Q2, *Fig. 1*). Before this large complex can bind to the mRNA, the latter must have interacted with **eIF4B** and **4F** (which recognizes the 5′-cap) and, using energy from ATP, have been unwound to remove secondary structure. When the 40S subunit complex has bound to the mRNA complex via the 5′-cap, ATP is used as the mRNA is scanned to find the AUG start codon. This is usually the first one. To allow the **60S** subunit to bind, **eIF5** must displace eIF2 and eIF3, and GTP is hydrolyzed. eIF4C is released when it has assisted 60S subunit binding to form the complete **80S initiation complex**.

The released eIF2·GDP complex is recycled by **eIF2B** and the rate of recycling (and hence the rate of initiation of protein synthesis) is regulated by

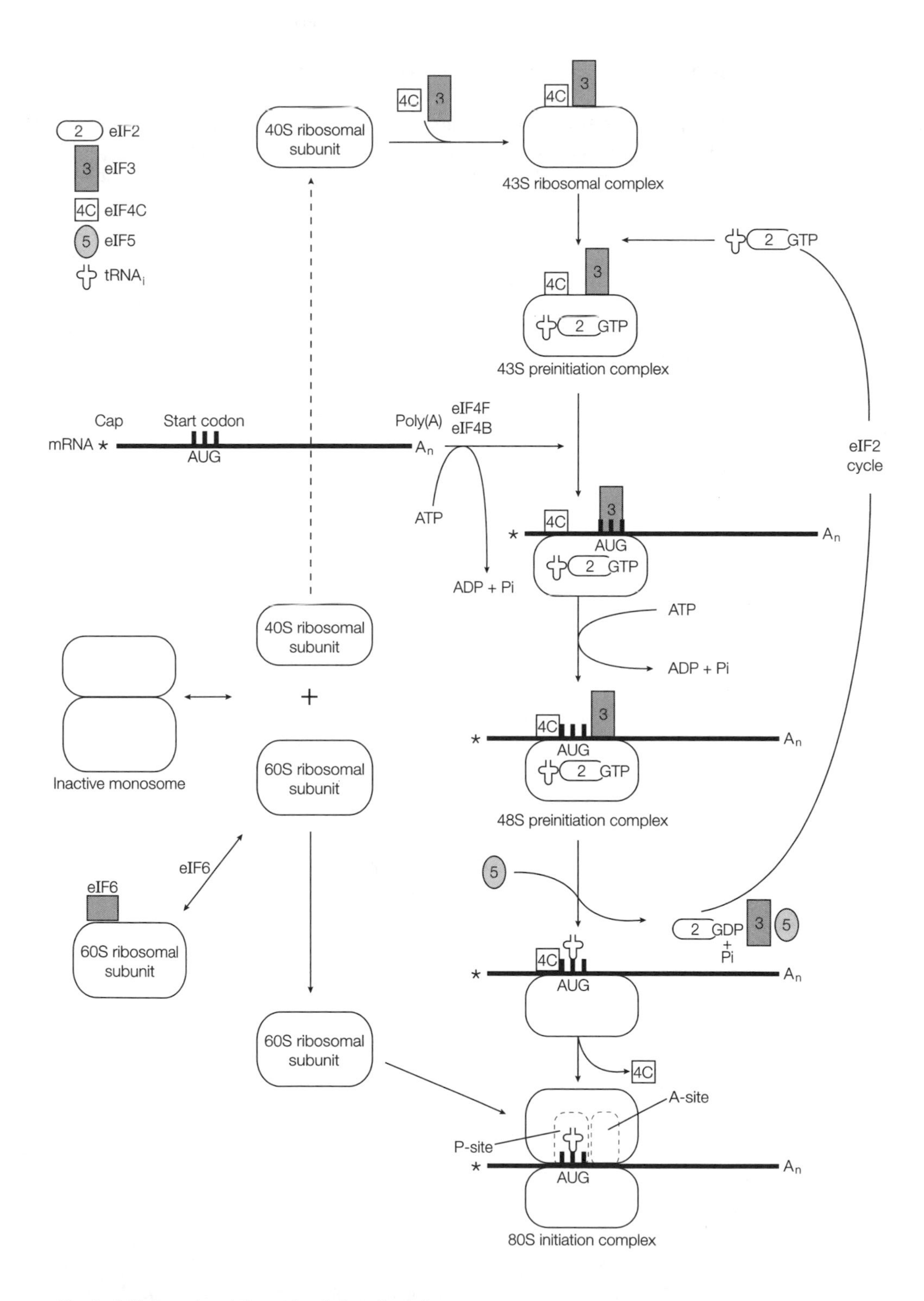

Fig. 1. Initiation of protein synthesis in eukaryotes.

phosphorylation of the α-subunit of eIF2. Certain events, such as viral infection and the resultant production of interferon, cause an inhibition of protein synthesis by promoting **phosphorylation of eIF2**.

Elongation

The protein synthesis elongation cycle in prokaryotes and eukaryotes is quite similar. Three factors are required with properties similar to their prokaryotic counterparts (*Table 1*). eEF1α, eEF1βγ and eEF2 have the roles described for EF-Tu, EF-Ts and EF-G respectively in Topic Q2.

Termination

In eukaryotes, a **single release factor, eRF**, recognizes all three stop codons and performs the roles carried out by RF1 (or RF2) plus RF3 in prokaryotes. eRF requires GTP for activity, but it is not yet clear whether there is a eukaryotic equivalent of ribosome release factor required for dissociation of the subunits from the mRNA.

Q4 TRANSLATIONAL CONTROL AND POST-TRANSLATIONAL EVENTS

Key Notes

Translational control

In prokaryotes, the level of translation of different cistrons can be affected by: (i) the binding of short antisense molecules, (ii) the relative stability to nucleases of parts of the polycistronic mRNA, and (iii) the binding of proteins that prevent ribosome access. In eukaryotes, protein binding can also mask the mRNA and prevent translation, and repeats of the sequence 5′-AUUUA-3′ can make the mRNA unstable and less frequently translated.

Polyproteins

A single translation product that is cleaved to generate two or more separate proteins is called a polyprotein. Many viruses produce polyproteins.

Protein targeting

Certain short peptide sequences in proteins determine the cellular location of the protein, such as nucleus, mitochondrion or chloroplast. The signal sequence of secreted proteins causes the translating ribosome to bind factors that make the ribosome dock with a membrane and transfer the protein through the membrane as it is synthesized. Usually the signal sequence is then cleaved off by signal peptidase.

Protein modification

The most common alterations to nascent polypeptides are those of cleavage and chemical modification. Cleavage occurs to remove signal peptides, to release mature fragments from polyproteins, to remove internal peptides as well as trimming both N-and C-termini. There are many chemical modifications that can take place on all but six of the amino acid side chains. Often phosphorylation controls the activity of the protein.

Protein degradation

Damaged, modified or inherently unstable proteins are marked for degradation by having multiple molecules of ubiquitin covalently attached. The ubiquitinylated protein is then degraded by a 26S protease complex.

Related topics

RNA Pol II genes: promoters and enhancers (M4)
mRNA processing, hnRNPs and snRNPs (O3)
Alternative mRNA processing (O4)
Mechanism of protein synthesis (Q2)
Initiation in eukaryotes (Q3)

Translational control

Because of the different natures of the mRNA in prokaryotes and eukaryotes (i.e. polycistronic vs. monocistronic; see Topic L1) and the absence of the nuclear membrane in the former, different possibilities exist for the **control of translation**. In prokaryotes, the structure formed by regions of the mRNA can obscure ribosome binding sites, thus reducing translation of some cistrons relative to others. The formation of stems and loops can **inhibit exonucleases** and give certain regions of the polycistronic mRNA a greater half-life (and hence a greater chance of translation) than others. Several operons encoding ribosomal proteins show an interesting form of translational control in *E. coli* where a region of the mRNA has a tertiary structure that resembles the binding site for a ribosomal protein encoded by the mRNA. If there is insufficient rRNA available for the translation product to bind to, it will bind to its own mRNA and prevent further translation. Prokaryotes sometimes make short **antisense RNA** molecules that form duplexes near the ribosome binding site of certain mRNAs, thus inhibiting translation.

Eukaryotes generally control the amount of specific proteins by varying the level of transcription of the gene (see Topic M4) and/or by RNA processing (see Topics O3 and O4), but some controls occur in the cytoplasm. The presence of multiple copies of **5′-AUUUA-3′**, usually in the 3′-noncoding region, marks the mRNA for rapid degradation and thus limited translation. Another form of translational control involves proteins binding directly to the mRNA and preventing translation. This RNA is called **'masked mRNA'**. In appropriate circumstances, the mRNA can be translated when the protein dissociates. Some noncoding sequences can cause mRNA to be located in a specific part of the cytoplasm and, when translated, can give rise to a gradient of protein concentration across the cell.

Polyproteins

Bacteriophage and viral transcripts (see Topic R2) and many mRNAs for hormones in eukaryotes (e.g. pro-opiomelanocortin) are translated to give a single polypeptide chain that is cleaved subsequently by specific proteases to produce multiple mature proteins from one translation product. The parent polypeptide is called a **polyprotein**.

Protein targeting

It has been discovered that the ultimate cellular location of proteins is often determined by specific, relatively short, amino acid sequences within the proteins themselves. These sequences can be responsible for proteins being secreted, imported into the nucleus or targeted to other organelles. The greater complexity of the eukaryotic cell (see Topic A1) means that there are more types of targeting in eukaryotes. **Protein secretion** in both prokaryotes and eukaryotes involves a **signal sequence** in the nascent protein and specific proteins or, in the latter, an RNP particle, **signal recognition particle (SRP)**, that recognizes it (*Fig. 1*).

If a cytosolic ribosome begins to translate an mRNA encoding a protein that is to be secreted, SRP binds to the ribosome and the emerging polypeptide and arrests translation. SRP is capable of recognizing ribosomes with a nascent chain containing a signal sequence (**signal peptide**) which is composed of about 13–36 amino acids having at least one positively charged residue followed by a hydrophobic core of 10–15 residues followed by a small, neutral residue, often Ala. SRP with the arrested ribosome binds to a receptor (**SRP receptor** or **docking protein**) on the cytosolic side of the **endoplasmic reticulum (ER)** (see Topic A2) and, when the ribosome becomes attached to **ribosome receptor proteins** on the ER, SRP is released and can be re-used. The ribosome is able to continue translation, and the nascent polypeptide chain is pushed through into the lumen of the ER. As it passes through, **signal peptidase** cleaves off the signal peptide. When the protein is released

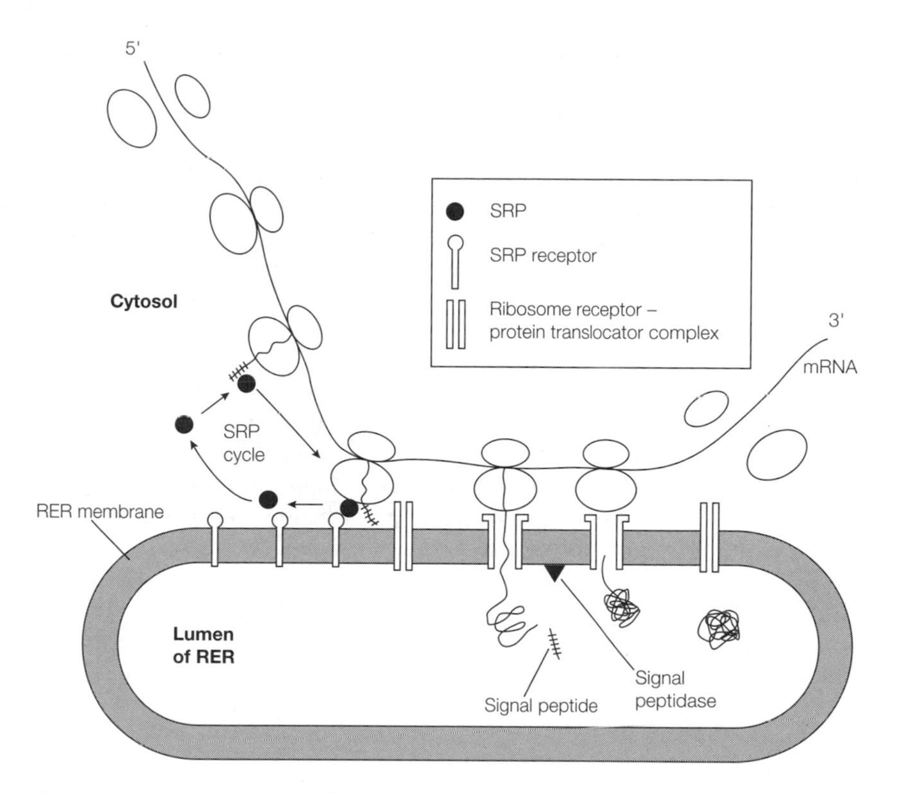

Fig. 1. Protein secretion in eukaryotes.

into the ER it is usually modified, often by **glycosylation**, and different patterns of glycosylation seem to control the final location of the protein.

Other peptide sequences in proteins are responsible for their cellular location. Different N-terminal sequences can cause proteins to be imported into mitochondria or chloroplasts, and the internal sequence -Lys-Lys-Lys-Arg-Lys, or any five consecutive positive amino acids, can be a **nuclear localization signal (NLS)** causing the protein containing it (e.g. histone) to be imported into the nucleus.

Protein modification

A newly translated polypeptide does not always immediately generate a functional protein (see Topic B2). Apart from correct folding and the possible formation of disulfide bonds, there are a number of other alterations that may be required for activity. These include **cleavage** and various covalent **modifications**. Cleavage is very common, especially trimming by **amino-** and **carboxypeptidases**, but the removal of internal peptides also occurs, as in the case of insulin. Signal sequences are also usually cleaved off secreted proteins and, where proteins are made as parts of polyproteins, they must be cleaved to release the component proteins. **Ubiquitin** is made as a polyprotein containing multiple copies linked end-to-end, and this must be cleaved to generate the individual ubiquitin molecules.

Chemical modifications are many and varied and have been shown to take place on the N and C termini, as well as on most of the 20 amino acid side chains, with the exception of Ala, Gly, Ile, Leu, Met and Val. The modifications include **acetylation, hydroxylation, phosphorylation, methylation, glycosylation** and even the **addition of nucleotides**. Hydroxylation of Pro is common in collagen, and some of the histone proteins are often acetylated (see Topic D3). The activity of many enzymes, such as glycogen phosphorylase and some transcription factors, is controlled by phosphorylation.

Protein degradation

Different proteins have very different half-lives. Regulatory proteins tend to turn over rapidly and cells must be able to dispose of faulty and damaged proteins. In eukaryotes, it has been discovered that the **N-terminal residue** plays a critical role in inherent stability. Eight N-terminal amino acids (Ala, Cys, Gly, Met, Pro, Ser, Thr, Val) correlate with stability ($t_{1/2} > 20$ hours), eight (Arg, His, Ile, Leu, Lys, Phe, Trp, Tyr) with short $t_{1/2}$ (2–30 min) and four (Asn, Asp, Gln, Glu) are destabilizing following chemical modification. A protein that is damaged, modified or has an inherently destabilizing N-terminal residue becomes **ubiquitinylated** by the covalent linkage of molecules of the small, highly conserved protein, ubiquitin, via its C-terminal Gly, to lysine residues in the protein. The ubiquitinylated protein is digested by a **26S protease complex (proteasome)** in a reaction that requires ATP and releases intact ubiquitin for re-use.

R1 Introduction to viruses

Key Notes

Viruses

Viruses are extremely small (20–300 nm) parasites, incapable of replication, transcription or translation outside of a host cell. Viruses of bacteria are called bacteriophages. Virus particles (virions) essentially comprise a nucleic acid genome and protein coat or capsid. Some viruses have a lipoprotein outer envelope, and some also contain nonstructural proteins essential for transcription or replication soon after infection.

Virus genomes

Viruses can have genomes consisting of either RNA or DNA, which may be double-stranded or single-stranded, and, for single-stranded genomes, positive, negative or ambi-sense (defined relative to the mRNA sequence). The genomes vary in size from around 1 kb to nearly 300 kb, and replicate using combinations of viral and cellular enzymes.

Replication strategies

Viral replication strategies depend largely on the type and size of genome. Small DNA viruses may make more use of cellular replication machinery than large DNA viruses, which often encode their own polymerases. RNA viruses, however, require virus-encoded RNA-dependent polymerases for their replication. Some RNA viruses use an RNA-dependent DNA polymerase (reverse transcriptase) to replicate via a DNA intermediate.

Virus virulence

Many viruses do not cause any disease, and often the mechanisms of viral virulence are accidental to the viral life cycle, although some may enhance transmission.

Related topics

Bacteriophages (R2)
DNA viruses (R3)
RNA viruses (R4)

Viruses

It is difficult to give a precise definition of a virus. The word originally simply meant a toxin, and was used by Jenner, in the 1790s, when describing the agents of cowpox and smallpox. Virus particles (**virions**) are sub-microscopic, and can replicate only inside a host cell. All viruses rely entirely on the host cell for translation, and some viruses rely on the host cell for various transcription and replication factors as well. Viruses of prokaryotes are called **bacteriophages** or **phages**.

Viruses essentially consist of a nucleic acid genome, of single- or double-stranded DNA or RNA surrounded by a virus-encoded protein coat, the **capsid**. It is the capsid and its interaction with the genome which largely determine the structure of the virus (see *Fig. 1* for examples of the structure of different viruses).

Viral capsids tend to be composed of protein subunits assembled into larger structures during the formation of mature particles, a process which may require interaction with the genome (the complex of genome and capsid is known as the **nucleocapsid**). The simplest models of this are some of the bacteriophages (e.g. the bacteriophage M13) and small mammalian viruses (e.g. poliovirus), but the same principle holds true even in larger and more complex viruses.

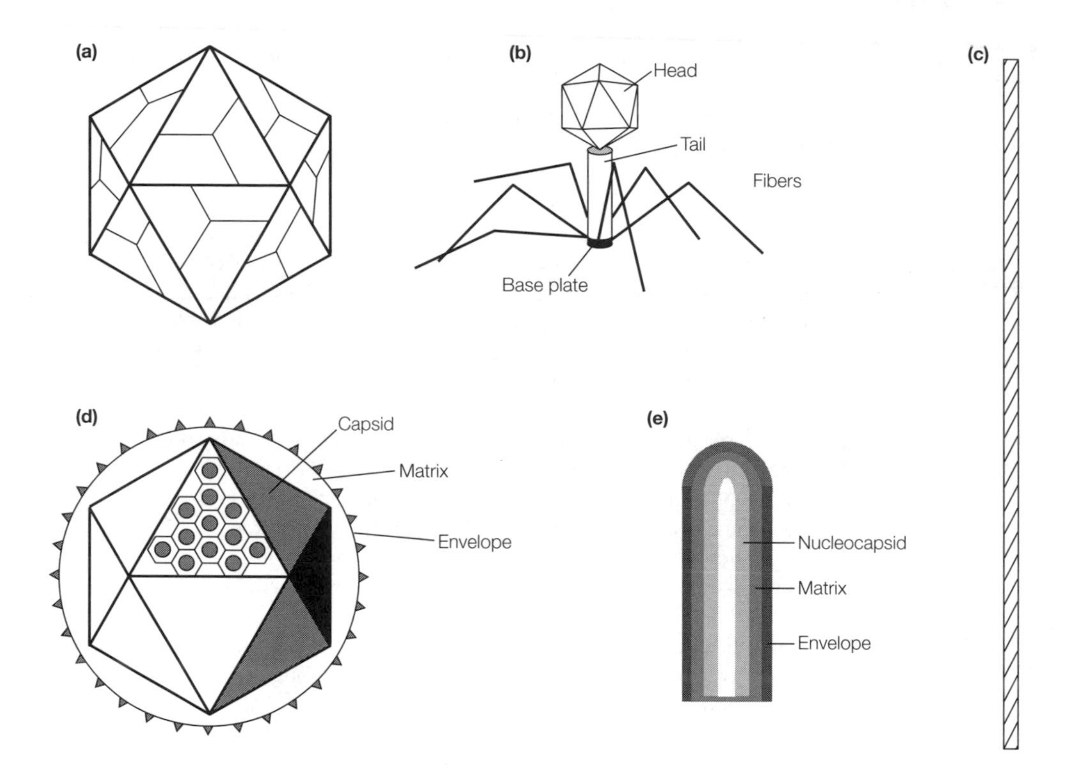

Fig. 1. Some examples of virus morphology (not drawn to scale). (a) Icosahedral virion (e.g. poliovirus); (b) complex bacteriophage with icosahedral head, and tail (e.g. bacteriophage T4); (c) helical virion (e.g. bacteriophage M13); (d) enveloped icosahedral virion (e.g. herpesviruses); (e) a rhabdovirus's typical bullet-shaped, helical, enveloped virion.

Many types of virus also have an outer bi-layer lipoprotein **envelope**. The envelope is derived from host cell membranes (by **budding**) and sometimes contains host cell proteins. Virus-encoded **envelope glycoproteins** are important for the assembly and structure of the virion. Envelope glycoproteins are often **receptor ligands** or **antireceptors** which bind to specific **receptors** on the appropriate host cell. **Matrix** proteins provide contact between the nucleocapsid and the envelope. Matrix and capsid proteins may have nonstructural roles in virus transcription and replication. The structural proteins are also often important antigens and, therefore, of great interest to those designing vaccines. Many RNA viruses, and some DNA viruses, also contain nonstructural proteins within the virion, necessary for immediate transcription or genome replication after infection of the cell.

Virus genomes

The genome of viruses is defined by its state in the mature virion, and varies between virus families. Unlike the genomes of true organisms, the virus genome can consist of DNA or RNA, which may be double- or single-stranded. In some viruses, the genome consists of a single molecule of nucleic acid, which may be **linear** or **circular**, but in others it is **segmented** or diploid. Single-stranded viral genomes are described as **positive sense** (i.e. the same nucleotide sequence as the mRNA), **negative sense** or **ambi-sense** (in which genes are encoded in both senses, often overlapping; see Topic P1).

Not all virions have a complete or functional genome, indeed the ratio of the number of virus particles in a virus preparation (as counted by electron microscopy) to the number of infectious particles (determined in cell culture) is usually greater than 100 and often many thousands. Genomes unable to replicate by themselves may be **rescued** during co-infection of a cell by the products of replication-competent **wild-type** genomes of **helper viruses**, or of genomes with different mutations or deletions in a process known as **complementation** (see Topic H2).

Replication strategies

The replication/transcription strategies of viruses vary enormously from group to group, and depend largely on the type of genome. DNA viruses may make more use of the host cell's nucleic acid polymerases than RNA viruses. DNA viruses with large genomes, such as **herpesvirus** (see Topic R3), are often more independent of host cell replication and transcription machinery than are viruses with small genomes, such as SV40, a **papovavirus** (see Topic R3). RNA viruses require RNA-dependent polymerases which are not present in the normal host cell and must, therefore, be encoded by the virus. Some RNA viruses such as the **retroviruses** encode a **reverse transcriptase** (an RNA-dependent DNA polymerase) to replicate their RNA genome via a DNA intermediate (see Topic R4).

The dependence of viruses on host cell functions for replication, and the requirements for specific cell-surface receptors determine host cell specificity. Cells capable of supplying the metabolic requirements of virus replication are said to be **permissive** to infection. Host cells which cannot provide the necessary requirements for virus replication are said to be **nonpermissive**. Under some circumstances, however, nonpermissive cells may be infected by viruses and the virus may have marked effects on the host cell such as cell **transformation** (see Topic R3 and Section S)

Virus virulence

Some viruses damage the cells in which they replicate, and if enough cells are damaged then the consequence is disease. It is important to realize that viruses do not exist in order to cause disease, but simply because they are able to replicate. In many circumstances, **virulence** (the capacity to cause disease) may be selectively disadvantageous (i.e. it may decrease the capacity for viral replication) but in others it may aid transmission. The evolution of virulence often results from a trade-off between damaging the host and maximizing transmission. The virulence mechanisms of viruses fall into six main categories:

(i) Accidental damage to cellular metabolism (e.g. competition for enzymes and nucleotides, or growth factors essential for virus replication).
(ii) Damage to the cell membrane during transmission between cells (e.g. cell lysis by many bacteriophages or cell fusion by herpesviruses).
(iii) Disease signs important for transmission between hosts (e.g. sneezing caused by common cold viruses, behavioral changes by rabies virus).

(iv) Evasion of the host's immune system, for example by rapid mutation.
(v) Accidental induction of deleterious immune responses directed at viral antigens (e.g. hepatitis B virus) or cross-reactive responses leading to autoimmune disease.
(vi) Transformation of cells (see Section S) and tumor formation (e.g. some papovaviruses such as SV40; see Topic R3).

R2 BACTERIOPHAGES

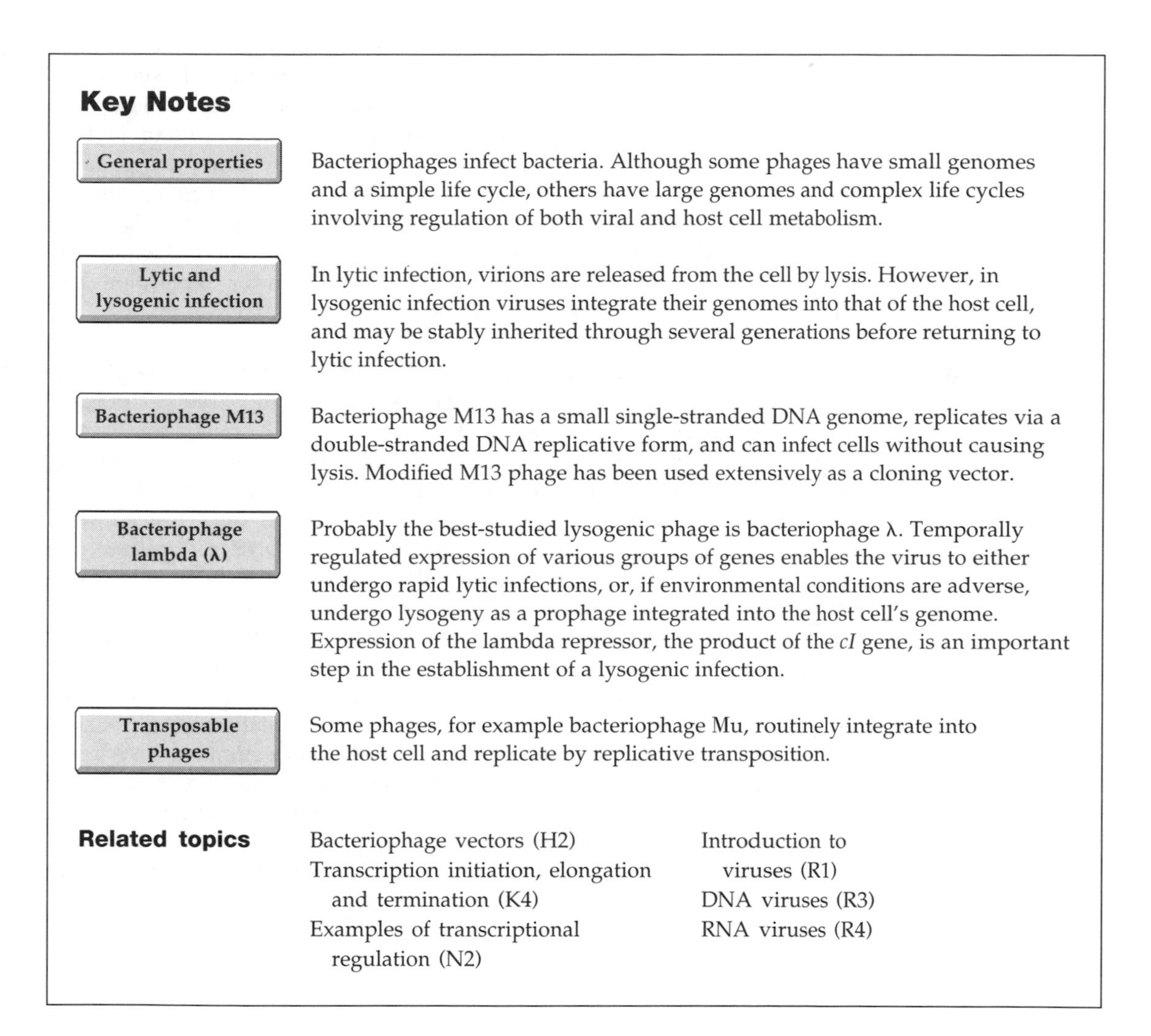

Key Notes

General properties

Bacteriophages infect bacteria. Although some phages have small genomes and a simple life cycle, others have large genomes and complex life cycles involving regulation of both viral and host cell metabolism.

Lytic and lysogenic infection

In lytic infection, virions are released from the cell by lysis. However, in lysogenic infection viruses integrate their genomes into that of the host cell, and may be stably inherited through several generations before returning to lytic infection.

Bacteriophage M13

Bacteriophage M13 has a small single-stranded DNA genome, replicates via a double-stranded DNA replicative form, and can infect cells without causing lysis. Modified M13 phage has been used extensively as a cloning vector.

Bacteriophage lambda (λ)

Probably the best-studied lysogenic phage is bacteriophage λ. Temporally regulated expression of various groups of genes enables the virus to either undergo rapid lytic infections, or, if environmental conditions are adverse, undergo lysogeny as a prophage integrated into the host cell's genome. Expression of the lambda repressor, the product of the *cI* gene, is an important step in the establishment of a lysogenic infection.

Transposable phages

Some phages, for example bacteriophage Mu, routinely integrate into the host cell and replicate by replicative transposition.

Related topics

Bacteriophage vectors (H2)
Transcription initiation, elongation and termination (K4)
Examples of transcriptional regulation (N2)
Introduction to viruses (R1)
DNA viruses (R3)
RNA viruses (R4)

General properties

Bacteriophages, or **phages**, are viruses which infect bacteria. Their genomes can be of RNA or DNA and range in size from around 2.5 to 150 kb. They can have simple lytic life cycles or more complex, tightly regulated life cycles involving **integration** in the host genome or even **transposition** (see Topic F4).

Bacteriophages have played an important role in the history of both virology and molecular biology; they were first discovered independently in 1915 and 1917 by Twort and d'Herelle. They have been studied intensively as model viruses and were important tools in the original identification of DNA as the genetic material, the determination of the genetic code, the existence of mRNA and many more fundamental concepts of molecular biology. Since phages parasitize prokaryotes, they often have significant sequence similarity to their hosts, and have, therefore, also been used extensively as simple models for various aspects of prokaryotic molecular biology. Some phages are also used as cloning vectors (see Topic H2). Bacteriologists also make use of strain-specific lytic phages to biologically type, and study the epidemiology of, various pathogenic bacteria.

Lytic and lysogenic infections

Some phages replicate extremely quickly: infection, replication, assembly and release by lysis of the host cell may all occur within 20 minutes. In such cases, replication of the phage genome occurs independently of the bacterial genome. Sometimes, however, replication and release of new virus can occur without lysis of the host cell (e.g. in bacteriophage M13 infection). Other phages alternate between a **lytic** phase of infection, with DNA replication in the cytosol, and a **lysogenic** phase in which the viral genome is integrated into that of its host (e.g. bacteriophage λ). Yet another group of phages replicate while integrated into the host cell genome via a combination of replication and **transposition** (e.g. bacteriophage Mu).

Bacteriophage M13

Bacteriophage M13 has a small (6.4 kb) single-stranded, positive-sense, circular DNA genome (*Fig. 1*). M13 particles attach specifically to *E. coli* sex pili (see Topic A1) (encoded by a plasmid called F factor), through a minor coat protein (g3p) located at one end of the particle. Binding of the minor coat protein induces a structural change in the major capsid protein. This causes the whole particle to shorten, injecting the viral DNA into the host cell. Host enzymes then convert the viral single-stranded genome into a dsDNA **replicative form (RF)**. The genome has 10 tightly packed genes and a small intergenic region which contains the origin of replication. Transcription occurs, again using host cell enzymes, from any of several promoters, and continues until it reaches one of two terminators (*Fig. 1*). This process leads to more transcripts being produced from genes closest to the terminators than from those further away, and provides the virus with its main method of regulation of expression. Multiple copies of the RF are produced by normal, double-stranded DNA replication, except that initiation of RF replication involves elongation of the 3′-OH group of a nick made in the (+) strand by a viral endonuclease (the product of gene 2), rather than RNA priming. Finally, multiple single-stranded (+) strands for packaging into new phage particles are made by continuous replication of

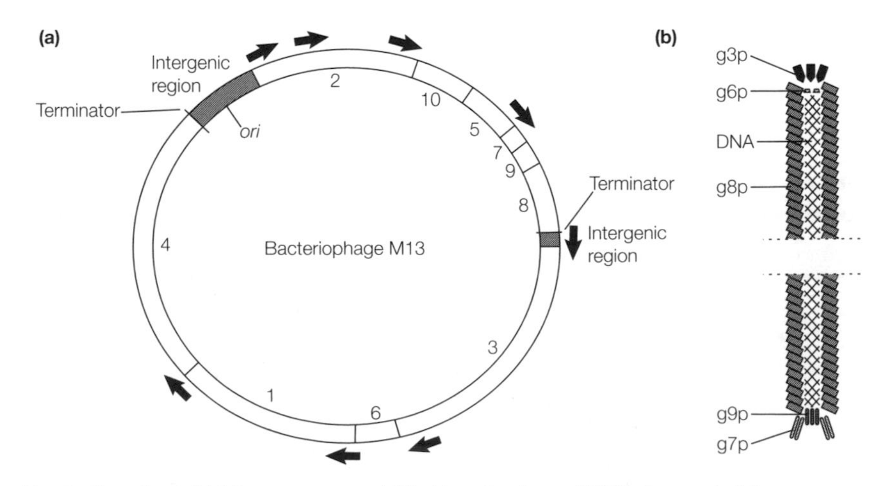

Fig. 1. Overview of (a) the genome and (b) virion structure of M13. Arrows in (a) are promoters. In (b), gene 3 protein = g3p, etc.

each RF, with the synthesis of the complementary (–) strand being blocked by coating the new (+) strands with the phage gene 5 protein. These packaging precursors are transported to the cell membrane and there, the DNA binds to the major capsid protein. At the same time, new virions are extruded from the cell's surface without lysis. M13-infected cells continue to grow and divide (albeit at a reduced rate), giving rise to generations of cells each of which is also infected and continually releasing M13 phage. What is more, the amount of DNA found in any particle is highly variable (giving rise to the variable length of the particles): virions containing multiple genomes and virions containing only partial genomes are found in any population.

Several peculiar properties of the M13 life cycle, described above, made it an ideal candidate for development as a cloning vector (see Topic H2). The double-stranded, circular RF can be handled in the laboratory just like a plasmid; furthermore, the lack of any strict limit on genome and particle size means that the genome will tolerate the insertion of relatively large fragments of foreign DNA. That the genome of the virion is single-stranded makes viral DNA an ideal template for sequencing reactions and, finally, the nonlytic nature of the life cycle makes it very easy to isolate large amounts of pure viral DNA.

Bacteriophage lambda (λ)

One of the best studied bacteriophages is bacteriophage λ, which has been much studied as a model for regulation of gene expression. Derivatives are commonly used as cloning vectors (see Topic H2). The λ phage virion consists of an icosahedral head containing the 48.5 kb linear dsDNA genome, and a long flexible tail. The phage binds to specific receptors on the outer membrane of *E. coli*, and the viral genome is injected through the phage's tail into the cell. Although the viral genome is linear within the virion, its termini are single-stranded and complementary. These are called **cos ends** (see Topic H2). The cohesive cos ends rapidly bind to each other once in the cell, producing a nicked circular genome which is repaired by cellular DNA ligase. Within the infected cell, the λ phage may either undergo **lytic** or **lysogenic** life cycles. In the lysogenic life cycle, the bacteriophage genome becomes integrated as a linear copy, or **prophage**, in the host cell's genome.

There are three classes of λ genes which are expressed at different times after infection. Firstly **immediate-early** and then **delayed-early** gene expression results in genome replication. Subsequently, **late** expression produces the structural proteins necessary for the assembly of new virus particles and lysis of the cell. The mechanisms by which the life cycle of phage λ is regulated are complex, and can only be described in outline here. A diagram of the phage λ genome is shown in *Fig. 2*.

Circularization of the genome is followed rapidly by the onset of immediate-early transcription. This is initiated at two promoters, *pL* and *pR*, and leads to the transcription of the immediate-early *N* and *Cro* genes. The two promoters are transcribed to the left (*pL*) and to the right (*pR*), using different strands of the DNA as template for RNA synthesis (*Fig. 2*). The terminators of both the *N* and *Cro* genes depend on transcription termination by rho (see Topic K4). The N protein acts as a transcription antiterminator, which enables the RNA polymerase to read through transcription termination signals of the *N* and *Cro* genes. As a result, mRNA transcripts are made from both *pL* and *pR*, which continue transcription, to the right through the replication genes *O* and *P* and into the *Q* gene and, to the left, through genes involved in recombination and

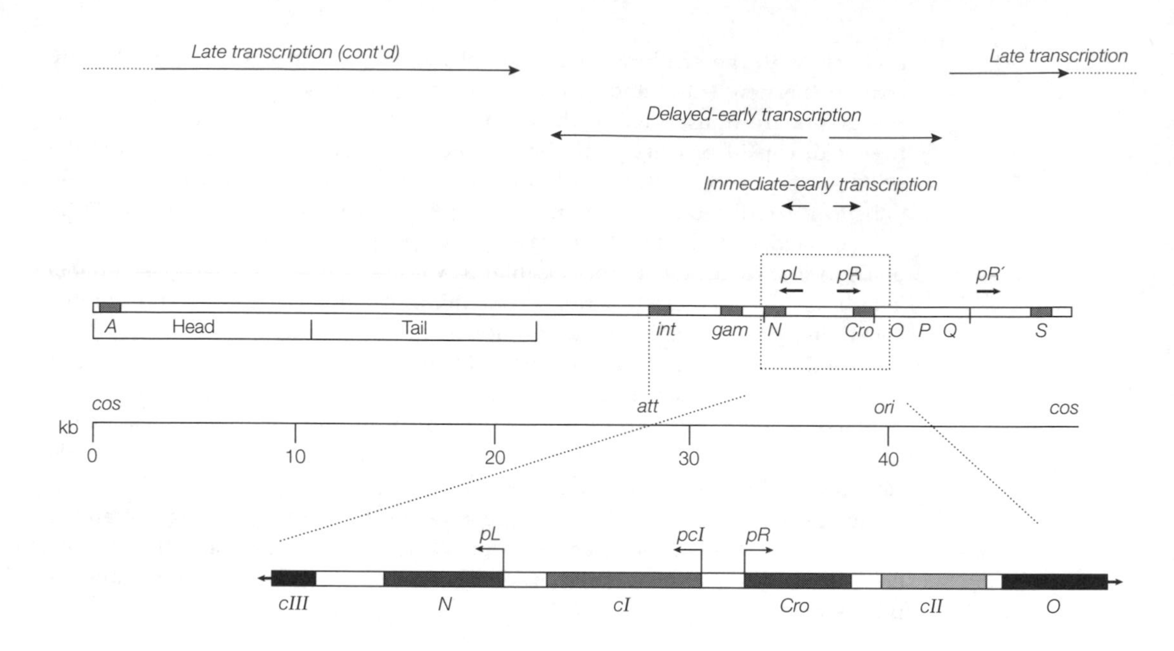

Fig. 2. Simplified map of the bacteriophage λ genome (linearized).

enhancement of replication. This leads to replication of the genome, which at this stage involves bi-directional DNA replication by host enzymes (see Topic E1), but initiated at the *ori* (origin of replication) site by a complex of the proteins pO and pP, and host cell helicase (DnaB). Later, build up of the *gam* gene product leads to conversion to **rolling circle replication** and the production of concatamers (mutiple length copies) of linear genomes (*Fig. 3*).

Transcription from *pR* results in the build-up of the Cro protein. Cro protein binds to sites overlapping *pL* and *pR*, and inhibits transcription from these promoters. As a result, early transcription is shut down. Q protein, like N protein, has an antitermination function (they can be compared with the HIV Tat protein, see Topic N2). The build-up of Q protein allows late transcription from *pR′* to occur (*Fig. 2*). The late genes encode the structural proteins of the virion head and tail, a protein to cleave the *cos* ends to produce a linear genome, and a protein which allows host cell lysis and viral release

The above lytic cycle can be completed in around 35 minutes, with the release of about 100 particles. For the lytic cycle to proceed, it requires the expression of the late genes. Lysogeny depends on the synthesis of a protein called the

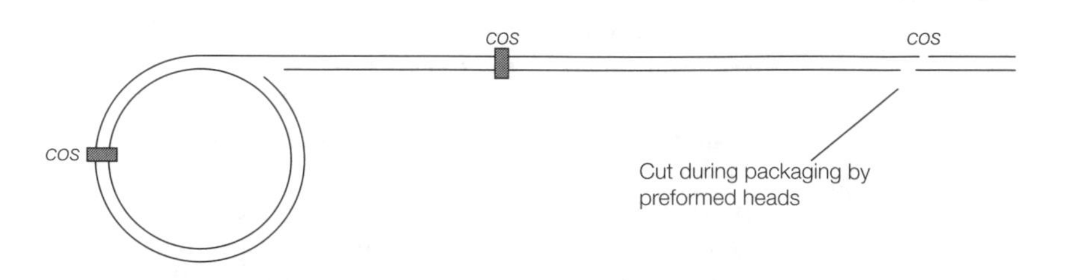

Fig. 3. Rolling circle replication of bacteriophage λ.

lambda repressor, which is the product of the *cI* gene. The delayed-early genes include replication and recombination genes, but also encode three regulators. One of these regulators, Q protein (described above), is responsible for the expression of the late genes. Two further regulator genes, *cII* and *cIII*, are expressed from *pR* and *pL* respectively. The *cII* gene product, which can be stabilized by the *cIII* gene product, binds to and activates the promoters of the *int* gene, which is responsible for λ integration into the host cell genome (required for lysogeny), and the *cI* gene, which encodes the λ repressor. The λ repressor represses both *pR* and *pL*, and thereby all early expression (including *Cro* and *Q* expression). Consequently, it represses both late gene expression and the lytic cycle, and this leads to lysogeny. The balance between the lytic and lysogenic pathways is determined by the concentration of the Cro protein (which inhibits early expression and *cI* expression) and Q protein (which activates the late genes) which favor lysis and, on the other hand, by the cII and cIII proteins and λ repressor protein which establish the lysogenic pathway.

Lysogenic infection can be maintained for many generations, during which the prophage is replicated like any other part of the bacterial genome. Transcription during lysogeny is largely limited to the *cI* gene: transcription of *cI* is from its own promoter which is enhanced by, but does not require, the cII protein, since the *cI* product can also regulate its own transcription. Escape from lysogeny occurs particularly in situations when the infected cell is itself under threat or if damage to DNA occurs (e.g. through ionizing radiation). Such situations induce the host cell to express *RecA* (see Topic F4) which cleaves the λ repressor, and enables progression into the lytic cycle.

Transposable phages

Transposable phages are found mainly in Gram-negative bacteria, particularly *Pseudomonas* species. One of the best-studied examples is bacteriophage mu (μ). Transposable phages have lytic and lysogenic phases of infection similar to those of λ phage, except that the method of genome replication is different. In the 'early' lytic phase, a complex process involving viral transposase, bacterial DNA polymerases and other viral and bacterial enzymes mediates both replication and transposition of the copy genome to elsewhere in the host cell's genome, without the original viral genome having to leave the host cell's genome (see Topic F4). Only after several rounds of **replicative transposition** are viral genomes, along with regions of adjacent cell DNA, excised from the cell's genome and encapsidated, causing degradation of the host cell's genome and lytic release of new phage particles.

R3 DNA VIRUSES

Key Notes

DNA genomes: replication and transcription

DNA virus genomes can be double-stranded or single-stranded. Almost all eukaryotic DNA viruses replicate in the host cell's nucleus and make use of host cellular replication and transcription as well as translation. Large dsDNA viruses often have more complex life cycles, including temporal control of transcription, translation and replication of both the virus and the cell. Viruses with small DNA genomes may be much more dependent on the host cell for replication.

Small DNA viruses

One example of a small DNA virus family is the Papovaviridae. Papovaviruses, such as SV40 and polyoma, rely on overlapping genes and splicing to encode six genes in a small, 5 kb double-stranded genome. These viruses can transactivate cellular replicative processes which mediate not only viral but cellular replication; hence they can cause tumors in their hosts.

Large DNA viruses

Examples of large DNA viruses include the family Herpesviridae. Herpesviruses infect a range of vertebrates, causing a variety of important diseases.

Herpes simplex virus-1

Herpes simplex virus-1 (HSV-1) has over 70 open reading frames (ORFs) and a genome of around 150 kb. After infection of a permissive cell, three classes of genes, the immediate-early (α), early (β) and late (γ) genes are expressed in a defined temporal sequence. These genes express a cascade of *trans*-activating factors which regulate viral transcription and activation. This virus has the ability to undergo latent infection.

Related topics

Eukaryotic vectors (H4)
Transcription in eukaryotes (Section M)
Introduction to viruses (R1)
Bacteriophages (R2)
RNA viruses (R4)
Tumor viruses and oncogenes (Section S)

DNA genomes: replication and transcription

Larger DNA viruses (e.g. herpesviruses or adenoviruses) have double-stranded genomes encoding up to 200 genes, and complex life cyles which can involve not only regulation of their own replication but sometimes that of the life cycle and functions of their host cells. At the other extreme, papovaviruses have an extremely small, double-stranded circular genome encoding only a few genes, and rely on their host for most replication functions.

Most eukaryotic DNA viruses replicate in the host cell's nucleus, where even the largest and most complex viruses can make use of cellular DNA metabolic pathways. This means that there is often considerable similarity between the sequences of viral promoters and those of their host cell. For this reason, the promoters of DNA viruses are often used in mammalian expression vectors (see Topic H4).

Small DNA viruses

The papovaviruses include **simian virus 40 (SV40)**, a simian (monkey) virus. This virus is well studied because it is a tumorigenic virus (see Section S). SV40 has a 5 kb, double-stranded circular genome, which is supercoiled and packaged with cell-derived histones within a 45 nm, icosahedral virus particle. In order to pack five genes into so small a genome, the genes are found on both strands and overlap each other. The genes are separated into two overlapping transcription units, the early genes and the late genes (*Fig. 1*). The different proteins are produced by a combination of the use of overlapping reading frames and differential splicing. SV40 depends on host cell enzymes for transcription and replication, but the early genes produce transcription activators (known as **large T-antigen** and **small t-antigen**) which stimulate both viral and host cell transcription and replication. These are responsible for the tumorigenic properties of this virus. The late genes produce three proteins, VP1, VP2 and VP3, which are required for virion production.

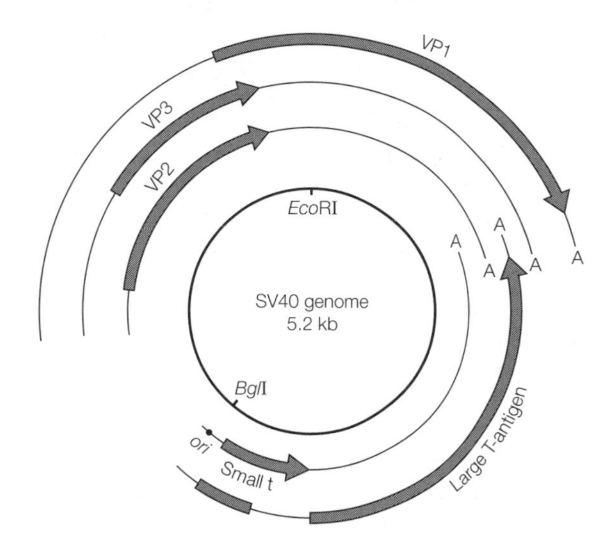

Fig. 1. Structure of the SV40 virus genome. Outer lines indicate transcripts and coding regions; A = poly (A) tail.

Large DNA viruses

Herpesviruses provide a good example of the complex ways in which a 'large' DNA virus and its host cell can interact. Herpesviruses infect vertebrates. The virions are large, icosahedral and enveloped (see Topic R1) and contain a double-stranded, linear DNA genome of up to 270 kb encoding around 100 open reading frames (ORFs). They are divided into three subfamilies, based on biological characteristics and genomic organization. There are well over 100 species, most of which are fairly host specific. Human examples include: herpes simplex virus-1 (HSV-1), the cause of cold sores; varicella zoster virus, the cause of chickenpox and shingles; and Epstein–Barr virus, a cause of infectious mononucleosis (glandular fever) and certain tumors.

Herpes simplex virus-1

HSV-1 is particularly well studied. Its genome is around 150 kb and contains over 70 ORFs. Genes can be found on both strands of DNA, sometimes overlapping each other. The genome (*Fig. 2*) can be divided into two parts ('short'

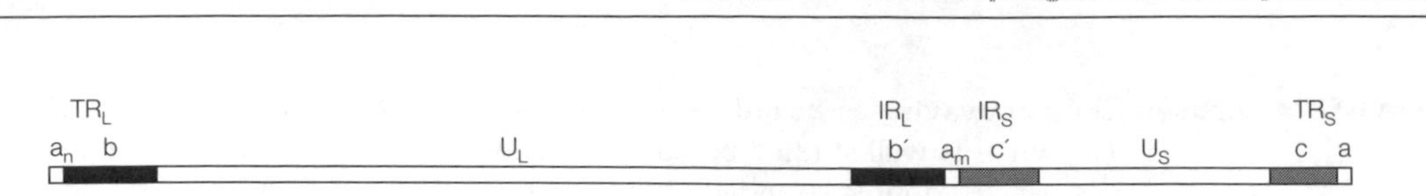

Fig. 2. Genome structure of herpes simplex virus-1.

and 'long'), each consisting of a unique section (U_L and U_S) with inverted repeats at the internal ends of the regions (IR_L and IR_S) and at the termini (TR_L and TR_S). These inverted repeats consist of sequences b and b′ and c and c′. In addition, a short sequence (a) is repeated a variable number of times (a_n and a_m).

The transcription and replication of the herpesvirus genome are tightly controlled temporally. After infection of a **permissive cell** (a cell which allows productive infection and virus replication, see Topic R1), the genome circularizes, and a group of genes located largely within the terminal repeat regions, the **immediate-early** or **α genes** are transcribed by cellular RNA polymerase II. Transcription of α genes is, however, *trans*-activated by a virus-encoded protein (**α-*trans*-inducing factor**, or **α-TIF**). In common with around one-third of the gene products of HSV-1, α-TIF is not essential for replication. Part of the mature virion's matrix, it interacts with cellular transcription factors after entry into the cell, binds to specific sequences upstream of α promoters and enhances their expression. The α mRNAs, some of which are spliced, encode *trans*-activators of the **early** or **β genes**. The β genes encode most of the nonstructural proteins used for further transcription and genome replication, and a few structural proteins. Early gene products include enzymes involved in nucleotide synthesis (e.g. thymidine kinase, ribonucleotide reductase), DNA polymerase, inhibitors of immediate-early gene expression and other products which can down-regulate various aspects of host cell metabolism. The promoters of β genes are similar to those of their hosts. They have an obvious TATA box 20–25 bases upstream of the mRNA transcription start site, and CCAAT box and transcription factor-binding sites – indeed herpesvirus β gene promoters function extremely well if incorporated into the host cell's genome, and have long been studied as model RNA polymerase II promoters.

Replication appears to be initiated at one of several possible origin (ORI) sites within the circular genome, and involves an ORI-binding protein, helicase–primase complexes and a polymerase–DNA-binding protein complex, which are all virus encoded. DNA synthesis is semi-discontinuous (see Section E) and results in concatamers (mutiple length copies) with multiple replication complexes and forks.

Late or **γ genes** (some of which can only be expressed after DNA replication) largely encode structural proteins, or factors which are included in the virion for use immediately after infection (such as α-TIF). Virus assembly takes place in the nucleus: empty capsids apparently associate with a free genomic terminal repeat 'a' sequence and one genome equivalent is packaged and cleaved, again at an 'a' sequence. The envelope is derived from modified nuclear membrane, and contains several viral glycoproteins important for attachment and entry.

In addition to this tightly regulated replication cycle, herpesviruses can undergo **latent infection**, that is they can down-regulate their own transcription to such an extent that the circular genome can persist extra-chromosomally in an infected cell's nucleus without replication. Latent infection can last the life of the cell, with only periodic reactivation (virus replication). HSV-1 undergoes latency mainly in neurons, but other herpesviruses can latently infect other tissues, including lymphoid cells. The precise mechanisms of herpesvirus latency and reactivation are still not fully understood, but may involve the transcription of specific RNAs (latency-associated transcripts; LATs) encoded within the terminal repeat regions and overlapping, but complementary to, some α genes.

Herpesviruses and other large DNA viruses (e.g. adenoviruses), by virtue of their large genomes and complex life cycles, contain many genes which, while important for optimal replication, are not essential, especially in cell culture (e.g. viral thymidine kinase genes). They are also able to withstand greater variation in genome size than smaller viruses without loss of stability. These features (and the relative ease of working with recombinant DNA rather than RNA) have made them prime candidates as vectors for foreign genes, both for use in the laboratory and as vaccines.

R4 RNA VIRUSES

Key Notes

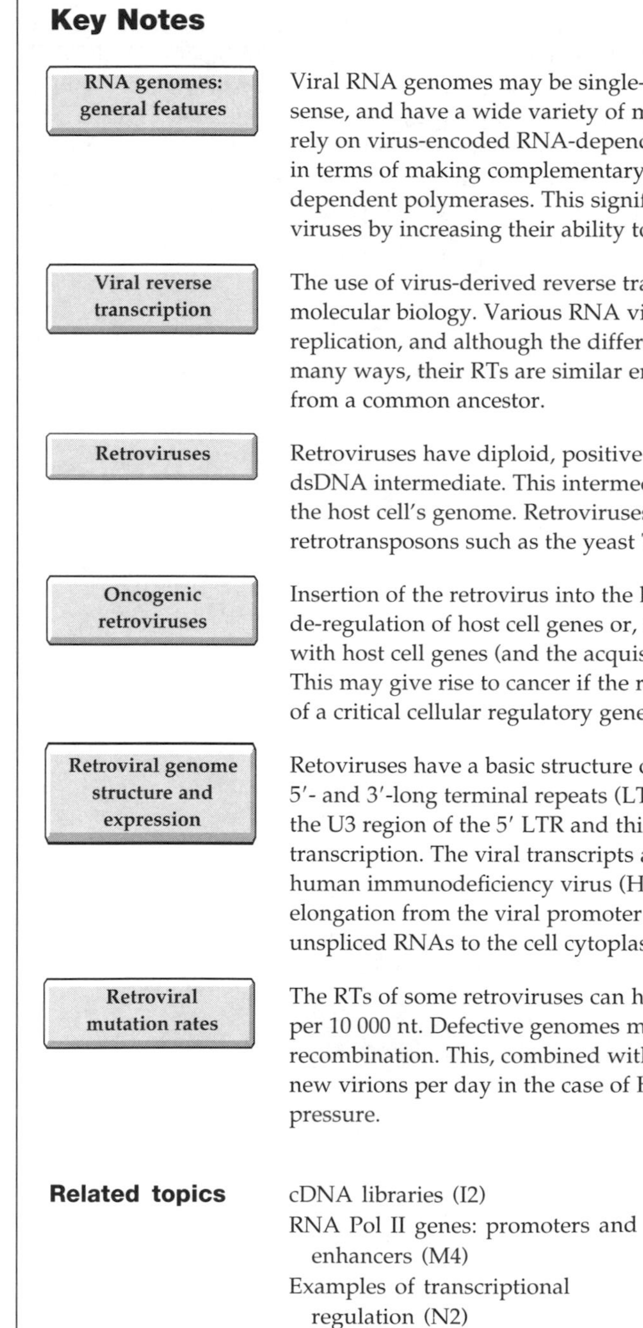

RNA genomes: general features

Viral RNA genomes may be single- or double-stranded, positive or negative sense, and have a wide variety of mechanisms of replication. All, however, rely on virus-encoded RNA-dependent polymerases, the inaccuracy of which in terms of making complementary RNA is much higher than that of DNA-dependent polymerases. This significantly affects the evolution of RNA viruses by increasing their ability to adapt, but limits their size.

Viral reverse transcription

The use of virus-derived reverse transcriptases (RTs) has revolutionized molecular biology. Various RNA viruses require reverse transcription for replication, and although the different virus families differ enormously in many ways, their RTs are similar enough to suggest that they have evolved from a common ancestor.

Retroviruses

Retroviruses have diploid, positive sense RNA genomes, and replicate via a dsDNA intermediate. This intermediate, called the provirus, is inserted into the host cell's genome. Retroviruses share many properties with eukaryotic retrotransposons such as the yeast Ty elements.

Oncogenic retroviruses

Insertion of the retrovirus into the host genome may cause either de-regulation of host cell genes or, occasionally, may cause recombination with host cell genes (and the acquisition of those genes into the viral genome). This may give rise to cancer if the retrovirus alters the expression or activity of a critical cellular regulatory gene called an oncogene.

Retroviral genome structure and expression

Retoviruses have a basic structure of *gag*, *pol* and *env* genes flanked by 5′- and 3′-long terminal repeats (LTRs). The retroviral promoter is found in the U3 region of the 5′ LTR and this promoter is responsible for all retroviral transcription. The viral transcripts are polyadenylated and may be spliced. In human immunodeficiency virus (HIV), Tat regulates transcriptional elongation from the viral promoter and Rev regulates the transport of unspliced RNAs to the cell cytoplasm.

Retroviral mutation rates

The RTs of some retroviruses can have a high error rate of up to one mutation per 10 000 nt. Defective genomes may be rescued by complementation and recombination. This, combined with the rapid turnover of virus (10^9–10^{10} new virions per day in the case of HIV), enables it to adapt rapidly to selective pressure.

Related topics

cDNA libraries (I2)
RNA Pol II genes: promoters and enhancers (M4)
Examples of transcriptional regulation (N2)
Introduction to viruses (R1)
DNA viruses (R3)
Tumor viruses and oncogenes (Section S)

RNA genomes: general features

Depending on the family, the RNA genome of a virus may be single- or double-stranded, and if single-stranded it may be positive or negative sense. As host cells do not contain RNA-dependent RNA polymerases, these must be encoded by the virus genome, and this 'polymerase' gene (*pol*) (which often also encodes other nonstructural functions) is often the largest gene found in the genome of an RNA virus. This makes RNA viruses true parasites of translation, and often totally independent of the host cell nucleus for replication and transcription. Thus, unlike eukaryote DNA viruses, many RNA viruses replicate in the cytoplasm.

RNA-dependent polymerases are not as accurate as DNA-dependent polymerases and, as a rule, RNA viruses are not capable of proofreading (see Topic F1); thus mutation rates of around 10^{-3}–10^{-4} (i.e. one mutation per 10^3–10^4 bases per replication cycle) are found in many RNA viruses compared with rates as low as 10^{-8}–10^{-11} in large DNA viruses (e.g. herpesviruses). This has three main consequences:

(i) mutation rates in RNA viruses are high and so, if the virus has a rapid replication cycle, significant changes in antigenicity and virulence can develop very rapidly (i.e. RNA viruses can evolve rapidly and quickly adapt to a changing environment, new hosts, etc.).

(ii) Some RNA viruses mutate so rapidly that they exist as **quasispecies**, that is to say as populations of different genomes (often replicating through complementation), within any individual host, and can only be molecularly defined in terms of a majority or average sequence.

(iii) As many of the mutations are deleterious to viral replication, the mutation rate puts an upper limit on the size of an RNA genome at around 10^4 nt, the inverse of the mutation rate (i.e the size which on average would give one mutation per genome).

Viral reverse transcription

The processes of **reverse transcription**, and the use of **reverse transcriptases (RTs)** in the laboratory, have been covered previously (see Topic I2). Although the different viruses which make use of reverse transcription can have very different morphologies, genome structures and life cycles, the amino acid sequences of their RTs and core proteins are sufficiently similar that it is thought they probably evolved from a common ancestor. Retroviral genomes have obvious structural similarities (and some sequence homology) to the retrotransposons found in many eukaryotes, such as the yeast Ty retrotransposons (see Topic F4).

Retroviruses

Retroviruses have a single-stranded RNA genome. Two copies of the sense strand of the genome are present within the viral particle. When they infect a cell, the single-stranded RNA is converted into a dsDNA copy by the RT. Replication and transcription occur from this dsDNA intermediate, the **provirus**, which is integrated into the host cell genome by a viral **integrase** enzyme. Retroviruses vary in complexity (*Fig. 1*). At one extreme there are those with relatively simple genomes which differ from retrotransposons essentially only by having an *env* gene, which encodes the envelope glycoproteins essential for infectivity. At the other extreme, lentiviruses, such as **the human immunodeficiency viruses (HIVs)**, have larger genomes which also encode various *trans*-acting factors and *cis*-acting sequences, which are active at different stages of the replication cycle, and involved in regulation of both viral and cellular functions.

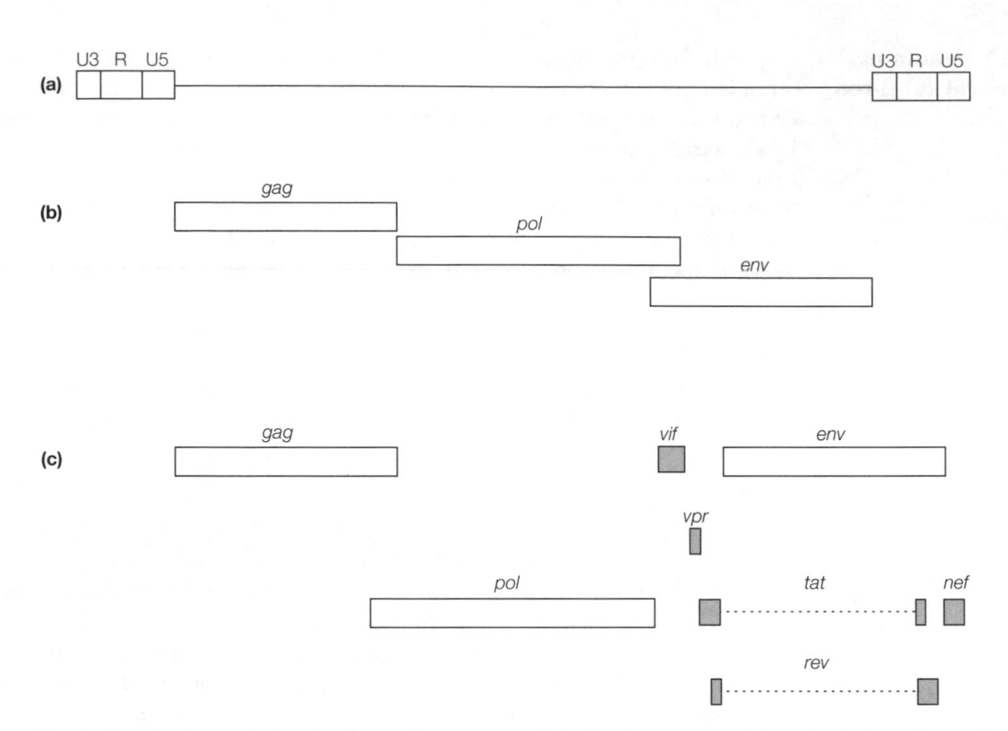

Fig. 1. The structure of retrovirus proviral genomes. (a) The basic structure of the retroviral provirus, with long terminal repeats (LTRs) consisting of U3, R and U5 elements; (b) avian leukosis virus (ALV), a simple retrovirus; (c) HIV-1, a complex retrovirus, with several extra regulatory factors.

Oncogenic retroviruses

Retroviruses sometimes recombine with host genomic material; such viruses are usually replication deficient. The retrovirus may either disrupt the regulated expression of a host gene, or it may recombine with the host DNA to insert the gene into its own genome. If the cellular gene encodes a protein involved in regulation of cell division, it may cause cancer and act as an **oncogene** (see Section S). Many oncogenic retroviruses expressing human oncogenes have been identified to date.

Retroviral genome structure and expression

Examples which represent the extreme ranges of retroviral genomes are shown in *Fig. 1*. All retroviral genomes have a similar basic structure. The *gag* gene encodes the core proteins of the icosahedral capsid, the *pol* gene encodes the enzymatic functions involved in viral replication (i.e. RT, RNase H, integrase and protease), and the *env* gene encodes the envelope proteins. At either end of the viral genome are unique elements (U5 and U3) and repeat elements which are involved in replication, host cell integration and viral gene expression.

Transcription of integrated provirus depends of the host's RNA polymerase II, directed by promoter sequences in the U3 region of the 5′ long terminal repeat (LTR). RNA transcripts are polyadenylated and may be spliced. Unspliced RNA is translated on cytoplasmic ribosomes to produce both a gag polyprotein and a gag–pol polyprotein (processed to pol proteins) through, for example, translational frameshifting (slippage of the ribosome on its template RNA by one reading frame to avoid a stop codon). The spliced mRNA is translated on membrane-bound ribosomes to produce the envelope glycoproteins. Some retroviruses, for example the lentiviruses, also produce other multiply or differently spliced RNAs which are translated to produce various factors such as Tat (see Topic N2) and Rev which regulate, respectively, transcription and mRNA processing. The Tat protein regulates transcription elongation in transcripts originating from the 5′ LTR of the HIV genome (see Topic N2). The Rev protein enhances nuclear to cytoplasmic export of unspliced viral mRNAs which can therefore synthesize a full range of structural viral proteins.

Retroviral mutation rates

The RTs of some retroviruses (e.g. HIV-1) have a high error rate of around 10^{-4}, that is an average of one per genome (as described above). Thus many genomes are replication defective, although this may be overcome in any virion through complementation by the second genome (two RNA genomes are packaged into each retroviral particle). Furthermore, recombination can occur between the two different genomes within any virion during reverse transcription. These two features, combined with the rapid turnover (10^9–10^{10} new virions per day) of HIV-1 enable it to adapt rapidly to new environments (e.g. under selective pressure from antibodies or drug treatment), a property which is increasingly recognized as central to our understanding of the pathogenesis of infections such as AIDS.

The ability of retroviruses to integrate into the host cell's genome, and the relative ease with which their ability to replicate can be genetically modified in the laboratory, has made them prime candidates as vectors both for the creation of novel cell lines and in whole organisms for gene therapy (see Topics H4 and J6).

S1 ONCOGENES FOUND IN TUMOR VIRUSES

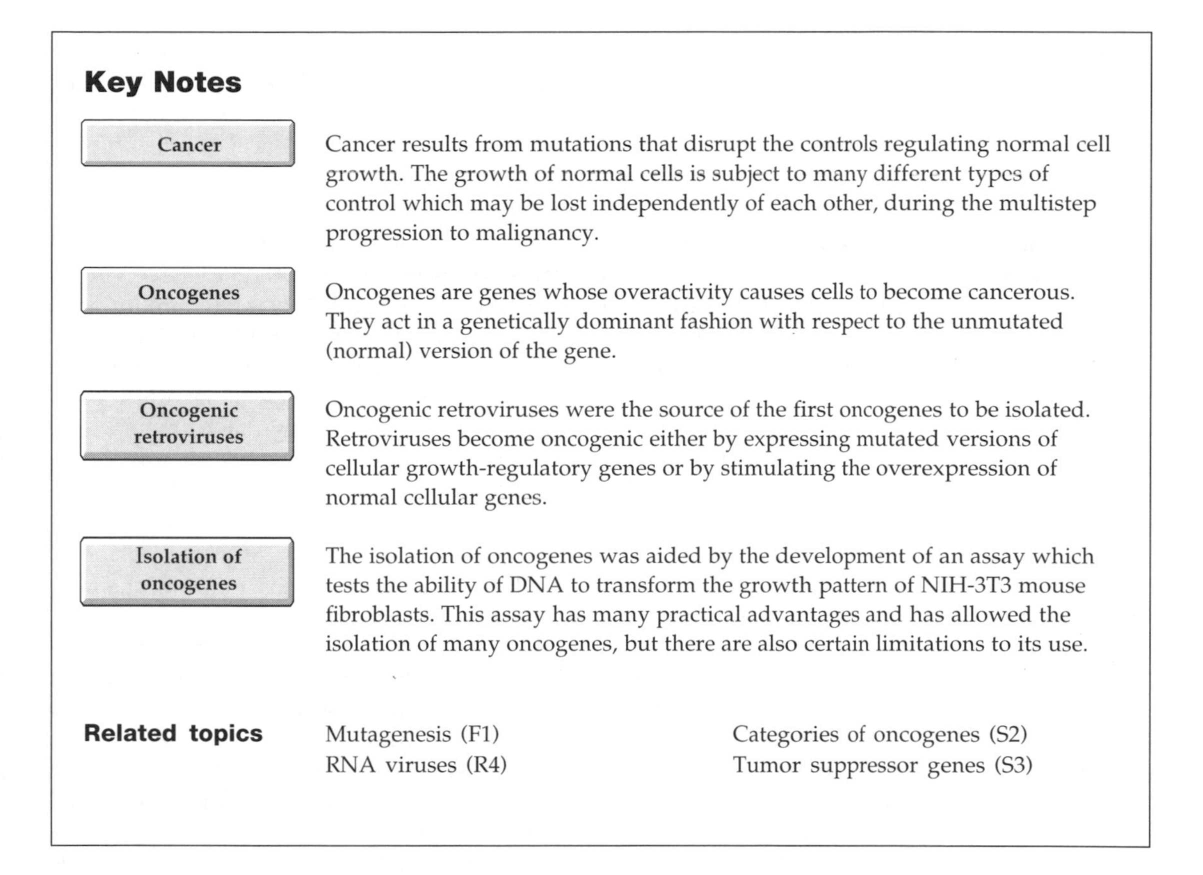

Key Notes

Cancer

Cancer results from mutations that disrupt the controls regulating normal cell growth. The growth of normal cells is subject to many different types of control which may be lost independently of each other, during the multistep progression to malignancy.

Oncogenes

Oncogenes are genes whose overactivity causes cells to become cancerous. They act in a genetically dominant fashion with respect to the unmutated (normal) version of the gene.

Oncogenic retroviruses

Oncogenic retroviruses were the source of the first oncogenes to be isolated. Retroviruses become oncogenic either by expressing mutated versions of cellular growth-regulatory genes or by stimulating the overexpression of normal cellular genes.

Isolation of oncogenes

The isolation of oncogenes was aided by the development of an assay which tests the ability of DNA to transform the growth pattern of NIH-3T3 mouse fibroblasts. This assay has many practical advantages and has allowed the isolation of many oncogenes, but there are also certain limitations to its use.

Related topics

Mutagenesis (F1)
RNA viruses (R4)
Categories of oncogenes (S2)
Tumor suppressor genes (S3)

Cancer

Cancer is a disease that results when the controls that regulate normal cell growth break down. The growth and development of normal cells are subject to a multitude of different types of control. A fully **malignant** cancer cell appears to have lost most, if not all, of these controls. However, conditions that seem to represent intermediate stages, when only some of the controls have been disrupted, can be detected. Thus the progression from a normal cell to a malignant cell is a **multistep process**, each step corresponding to the breakdown of a normal cellular control mechanism.

It has long been recognized that cancer is a disease with a genetic element. Evidence for this is of three types:

- the tendency to develop certain types of cancer may be inherited;
- in some types of cancer the tumor cells possess characteristically abnormal chromosomes;
- there is a close correlation between the ability of agents to cause cancer and their ability to cause mutations (see Topic F1).

It seems that normal growth controls become ineffective because of mutations in the cellular genes coding for components of the regulatory mechanism. Cancer can therefore be seen as resulting from the accumulation of a series of specific mutations in the malignant cell. In recent years, major progress has been made in identifying just which genes are mutated during carcinogenesis. The first genes to be identified as causing cancer were named **oncogenes**.

Oncogenes

Oncogenes are genes whose expression causes cells to become cancerous. The normal version of the gene (termed a **proto-oncogene**) becomes mutated so that it is overactive. Because of their overactivity, oncogenes are genetically dominant over proto-oncogenes, that is only one copy of an oncogene is sufficient to cause a change in the cell's behavior.

Oncogenic retroviruses

The basic concepts of oncogenes were given substance by studies on oncogenic viruses, particularly **oncogenic retroviruses** (see Topic R4). Oncogenic viruses are an important cause of cancer in animals, although only a few rare forms of human cancer have been linked to viruses. Retroviruses are RNA viruses that replicate via a DNA intermediate (the **provirus**) that inserts itself into the cellular DNA and is transcribed into new viral RNA by cellular RNA polymerase. Many oncogenic retroviruses were found to contain an extra gene, not present in closely related but nononcogenic viruses. This extra gene was shown to be an oncogene by transfecting it into noncancerous cells which then became tumorigenic. Different oncogenic viruses contain different oncogenes.

Isolation of oncogenes

The isolation of an oncogene, just like the isolation of any other biochemical molecule, depends upon the availability of a specific assay. The assay that has formed the backbone of oncogene isolation is DNA transfection of **NIH-3T3 cells**. These cells are a permanent cell line of mouse fibroblasts (connective tissue cells that grow particularly well *in vitro*) that do not give rise to tumors when injected into immune-deficient mice. The growth pattern of NIH-3T3 cells in culture is that of normal, noncancerous cells. NIH-3T3 cells are transfected with the DNA to be assayed (the DNA is poured on to the cells as a fine precipitate), so that a few cells will take up and express the foreign DNA. If this contains an oncogene, the growth pattern of the cells in culture will change to one that is characteristic of cancerous cells (*Fig. 1*). Advantages of this assay are:

- it is a cell culture rather than a whole animal test and so particularly suitable for screening large numbers of samples;
- results are obtained much more quickly than with *in vivo* tests;
- NIH-3T3 cells are good at taking up and expressing foreign DNA;
- it is a technically simple procedure compared with *in vivo* tests.

However, extensive use has revealed some drawbacks, both real and potential:

- some oncogenes may be specific for particular cell types and so may not be detected with mouse fibroblasts;
- large genes may be missed because they are less likely to be transfected intact;
- NIH-3T3 cells are not 'normal' cells since they are a permanent cell line and genes involved in early stages of carcinogenesis may therefore be missed;
- the assay depends upon the transfected gene acting in a genetically dominant manner and so will not detect tumor suppressor genes (see Topic S3).

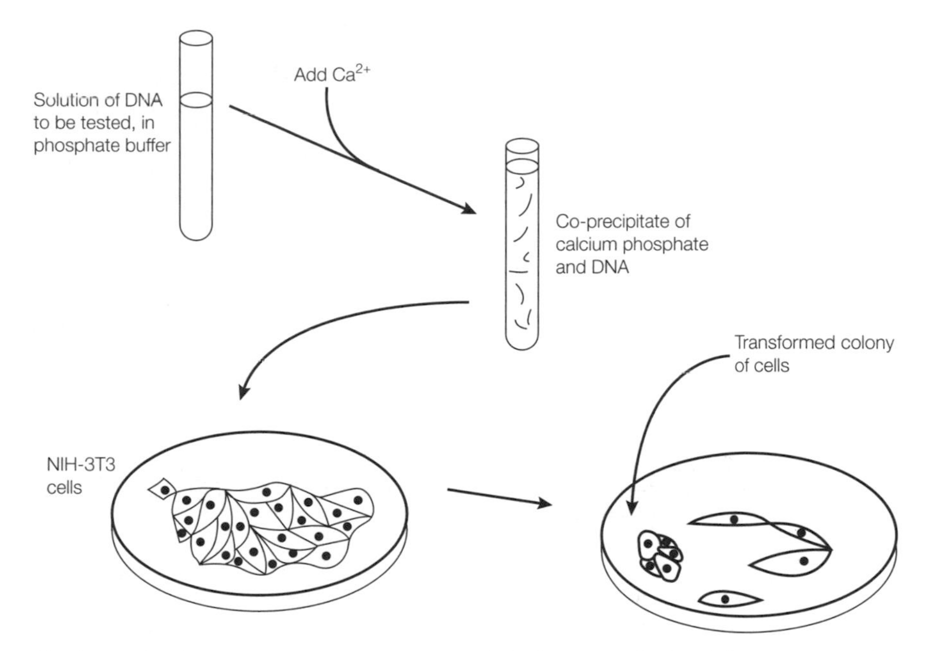

Fig. 1. Testing for the presence of an oncogene in DNA by revealing its ability to cause a change in the growth pattern of NIH-3T3 cells.

The first oncogenes to be isolated were those present in oncogenic retroviruses. When these had been cloned and were available for use as hybridization probes, an important discovery was made – genes with DNA sequences homologous to retroviral oncogenes were present in the DNA of normal cells. It was then realized that retroviral oncogenes must have originated as proto-oncogenes in normal cells and been incorporated into the viral genome when the provirus integrated itself nearby in the cellular genome. Subsequently, similar (sometimes the same) oncogenes were isolated from nonvirally caused cancers. In all cases, the oncogene differs from the normal proto-oncogene in important ways.

- A **quantitative difference**. The coding function of the gene may be unaltered but, because, for example, it is under the control of a viral promoter/enhancer or because it has been translocated to a new site in the genome, it is transcribed at a higher rate or under different circumstances from normal. This results in overproduction of a normal gene product, for example breast cancer in mice is caused by **mouse mammary tumor (retro) virus (MMTV)**. If the MMTV provirus inserts itself close to the mouse ***int*-2** gene (which codes for a growth factor), the viral enhancer sequences overstimulate the activity of *int*-2 (*Fig. 2*) and the excess of growth factor causes the cells to divide continuously.
- A **qualitative difference**. The coding sequence may be altered, for example by deletion or by point mutation, so that the protein product is functionally different, usually hyperactive. For example the ***erbB*** oncogene codes for a truncated version of a growth factor receptor. Because the missing region is responsible for binding the growth factor, the oncogene version is constitutively active, permanently sending signals to the nucleus (*Fig. 3*) instructing the cell to 'divide'.

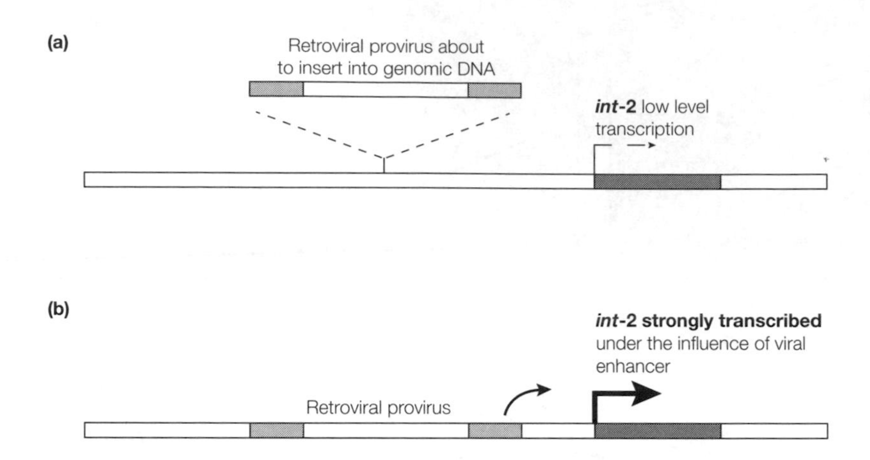

Fig. 2. The mechanism by which MMTV causes cancer in mouse mammary cells. (a) The mouse int-2 *gene before integration of the MMTV provirus; (b) integration of the provirus results in overexpression of* int-2.

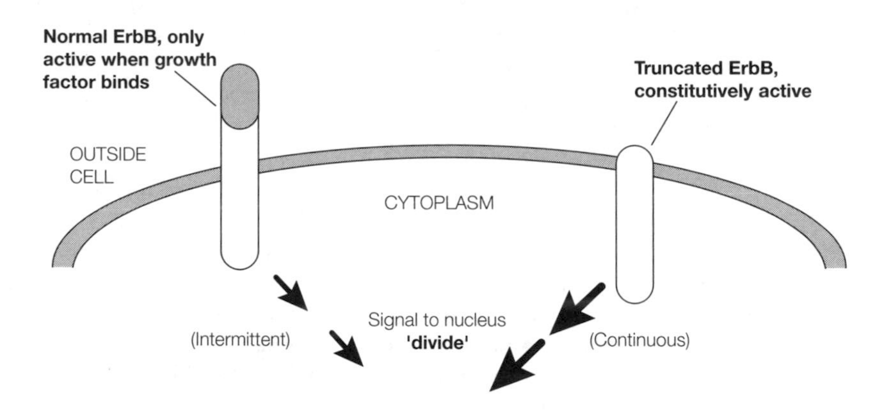

Fig. 3. The oncogene version of erbB *codes for a constitutively active growth factor receptor.*

S2 CATEGORIES OF ONCOGENES

Key Notes

Oncogenes and growth factors

Many oncogenes code for proteins that take part in various steps in the mechanism by which cells respond to growth factors. Such oncogenes cause the cancer cell to behave as though it were continuously being stimulated to divide by a growth factor.

Nuclear oncogenes

Other oncogenes code for nuclear DNA-binding proteins that act as transcription factors regulating the expression of other genes involved in cell division. As a result, the strict control that is exerted over the expression of genes required for cell division in normal cells no longer operates in the cancer cell.

Co-operation between oncogenes

When normal rat fibroblasts taken straight from the animal are transfected with oncogenes, it is found that both a growth factor-related and a nuclear oncogene are required to convert the cells to full malignancy. This reflects the fact that carcinogenesis *in vivo* requires more than one change to occur in a cell.

Related topics

Eukaryotic transcription factors (N1)

Oncogenes found in tumor viruses (S1)

Oncogenes and growth factors

The isolation of oncogenes has made it possible to investigate their individual roles in carcinogenesis. Many oncogenes seem to code for proteins related to various steps in the response mechanism for **growth factors**. Growth factors are peptides that are secreted into the extracellular fluid by a multitude of cell types. They bind to specific **cell-surface receptors** on nearby tissue cells (of the same or a different type from the secretory cell) and stimulate a response which frequently includes an increase in cell division rate. Oncogenes whose action appears to depend upon their growth factor-related activity include:

- the ***sis*** oncogene which codes for a subunit of **platelet-derived growth factor (PDGF)**. Overproduction of this growth factor autostimulates the growth of the cancer cell, if it possesses receptors for PDGF.
- The ***fms*** oncogene which codes for a mutated version of the receptor for **colony-stimulating factor-1 (CSF-1)**, a growth factor that stimulates bone marrow cells during blood cell formation. The 40 amino acids at the carboxyl terminus of the normal CSF-1 receptor are replaced by 11 unrelated amino acids in the Fms protein. As a result, the Fms protein is constitutively active regardless of the presence or absence of CSF-1 (*Fig. 1*).
- The various ***ras*** oncogenes which code for members of the **G-protein** family of plasma membrane proteins that transmit stimulation from many cell surface receptors to enzymes that produce **second messengers**. Normal

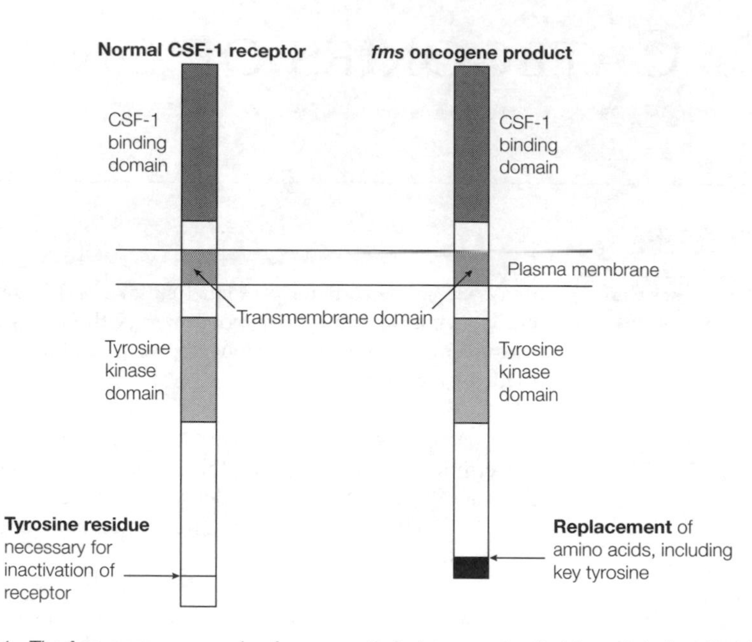

Fig. 1. The fms *oncogene codes for a growth factor receptor that is mutated so that it is constitutively active.*

G-proteins bind GTP when activated and are inactivated by their own GTPase activity. *ras* Oncogenes possess point mutations which inhibit their GTPase activity so that they remain activated for longer than normal (*Fig. 2*).

Nuclear oncogenes

Another group of oncogenes codes for nuclear DNA-binding proteins that act as transcription factors (see Topic N1) regulating the expression of other genes.

- The expression of the ***myc*** gene in normal cells is induced by a variety of mitogens (agents that stimulate cells to divide), including PDGF. The *myc*-encoded protein binds to specific DNA sequences and probably stimulates the transcription of genes required for cell division. Overexpression of *myc*

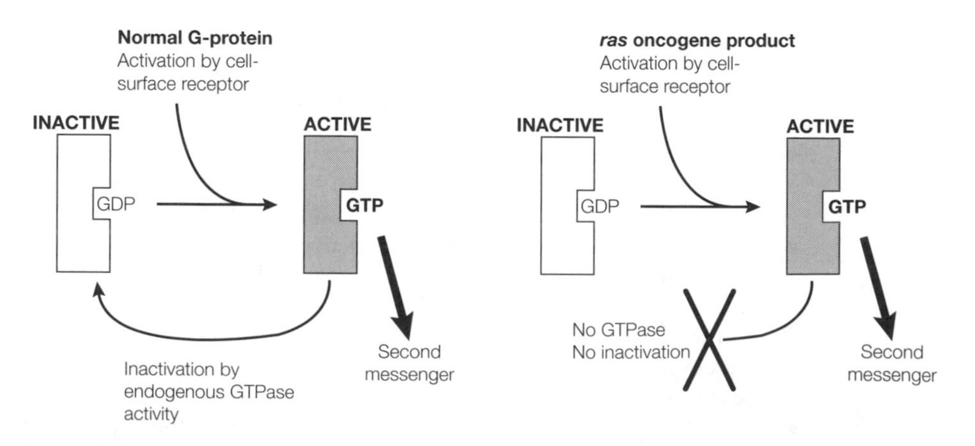

Fig. 2. The ras *oncogene codes for a signal transmission protein that is mutated so that it has lost the ability to inactivate itself.*

in cancer cells can occur by different mechanisms, either by increased transcription under the influence of a viral enhancer or by translocation of the coding sequence from its normal site on chromosome 8 to a site on chromosome 14, which places it under the control of the active promoter for the immunoglobulin heavy chain, or by deletion of the 5′-noncoding sequence of the mRNA, which increases the life time of the mRNA.

- The ***fos*** and ***jun*** oncogenes code for subunits of a normal transcription factor, **AP-1**. In normal cells, expression of *fos* and *jun* occurs only transiently, immediately after mitogenic stimulation. The normal cellular concentrations of the *fos* and *jun* gene products are regulated not only by the rate of gene transcription but also by the stability of their mRNA. In cancer cells, both processes may be increased.
- The ***erbA*** oncogene is a second oncogene (besides *erbB*) found in the avian erythroblastosis virus. It codes for a truncated version of the nuclear **receptor for thyroid hormone**. Thyroid hormone receptors act as transcription factors regulating the expression of specific genes, when they are activated by binding the hormone. The ErbA protein lacks the carboxyl-terminal region of the normal receptor so that it cannot bind the hormone and cannot stimulate gene transcription. However, it can still bind to the same sites on the DNA and appears to act as an antagonist of the normal thyroid hormone receptor.

Co-operation between oncogenes

The transformation of a normal cell into a fully malignant cancer cell is a multi step process involving alterations in the expression of several genes. Although transfection of a single one of the oncogenes described above will, in most cases, cause oncogenic transformation of cells of the NIH-3T3 cell-line, different results are seen when cultures of normal rat fibroblasts, taken straight from the animal, are substituted for the cell line. Neither the *ras* nor the *myc* oncogene on its own is able to induce full transformation in the normal cells, but simultaneous introduction of both oncogenes does achieve this. A variety of other pairs of oncogenes are able to achieve together what neither can achieve singly, in normal rat fibroblasts. Interestingly, to be effective, a pair must include one growth factor-related oncogene and one nuclear oncogene. It seems that any one activated oncogene is only capable of producing a subset of the total range of changes necessary to convert a completely normal cell into a fully malignant cancer cell.

S3 TUMOR SUPPRESSOR GENES

Key Notes

Overview

Tumor suppressor genes cause cells to become cancerous when they are mutated to become inactive. A tumor suppressor gene acts, in a normal cell, to restrain the rate of cell division.

Evidence for tumor suppressor genes

Evidence for tumor suppressor genes is varied and indirect. It includes the behavior of hybrid cells formed by fusing normal and cancerous cells, patterns of inheritance of certain familial cancers and '**loss of heterozygosity**' for chromosomal markers in tumor cells.

***RB1* gene**

The *RB1* gene was the first tumor suppressor gene to be isolated. It was shown to be the cause of the childhood tumor of the eye, retinoblastoma. Mutations in the *RB1* gene have also been detected in breast, colon and lung cancers.

***p53* gene**

p53 is the tumor suppressor gene that is mutated in the largest number of different types of tumor. When it was first identified, it appeared to have characteristics of both oncogenes and tumor suppressor genes. It is now known to be a tumor suppressor gene that may act in a **dominant-negative** manner to interfere with the function of a remaining, normal allele.

Related topics

Oncogenes found in tumor viruses (S1)

Categories of oncogenes (S2)

Overview

Tumor suppressor genes act in a fundamentally different way from oncogenes. Whereas proto-oncogenes are converted to oncogenes by mutations that increase the genes' activity, tumor suppressor genes become oncogenic as the result of mutations that eliminate their normal activity. The normal, unmutated version of a tumor suppressor gene acts to inhibit a normal cell from entering mitosis and cell division. Removal of this negative control allows a cell to divide. An important consequence of this mechanism of action is that both copies (alleles) of a tumor suppressor gene have to be inactivated to remove all restraint, that is tumor suppressor genes act in a genetically recessive fashion.

Evidence for tumor suppressor genes

Because there are no quick and easy assays for tumor suppressor genes, equivalent to the NIH-3T3 assay for oncogenes, fewer tumor suppressor genes have been isolated, and for a long time their basic importance for the development of cancer was not appreciated. Now that they are known to exist, more and more evidence for their involvement can be found.

- In the 1960s, it was shown that if a normal cell was fused with a cancerous cell (from a different species) the resulting **hybrid cell** was invariably noncancerous. Over subsequent cell generations, the hybrid cells lose chromosomes and may revert to a cancerous phenotype. Often, reversion can be correlated with the loss of a particular normal cell chromosome (carrying a tumor suppressor gene).

- Examination of the inheritance of certain **familial cancers** suggests that they result from recessive mutations.
- In many cancer cells, there has been a consistent loss of characteristic regions of certain chromosomes. This '**loss of heterozygosity**' is believed to indicate the loss of a tumor suppressor gene encoded on the missing chromosome segment (*Fig. 1*).

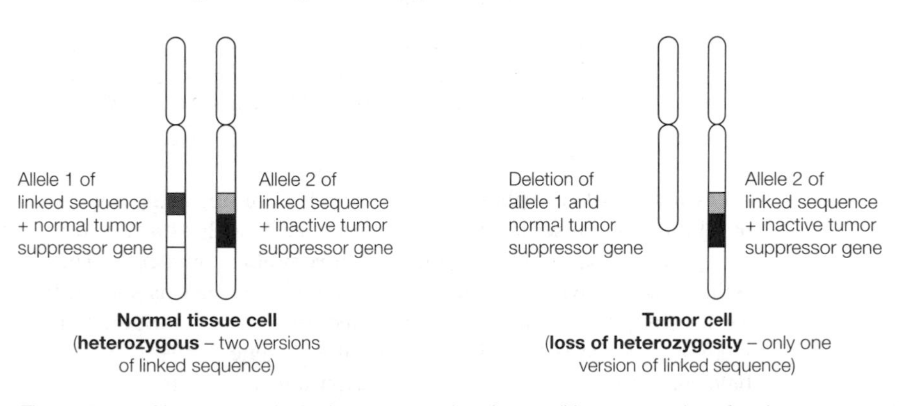

Fig. 1. Loss of heterozygosity is the process whereby a cell loses a portion of a chromosome that contains the only active allele of a tumor suppressor gene.

Retinoblastoma is a childhood tumor of the eye, and is the classic example of a cancer caused by loss of a tumor suppressor gene. Retinoblastoma takes two forms: familial (40% of cases), which exhibits the inheritance pattern for a recessive gene and which frequently involves both eyes; and sporadic, which does not run in families and usually only occurs in one eye. It was suggested that retinoblastoma results from two mutations which inactivate both alleles of a single gene. In the familial form of the disease, one mutated allele is inherited in the germ line. On its own this is harmless, but the occurrence of a mutation in the remaining normal allele, in a retinoblast cell, causes a tumor. Since there are 10^7 retinoblasts per eye, all at risk, the chances of a tumor must be relatively high. In the sporadic, noninherited form of the disease, both inactivating mutations have to occur in the same cell, so the likelihood is very much less and only one eye is usually affected (*Fig. 2*). It should be noted that whilst familial retinoblastoma constitutes the minority of cases, it is responsible for the majority of tumors. The 'two-hit' hypothesis for retinoblastoma was also supported by evidence for loss of heterozygosity. The retinoblastoma gene (***RB1***) was provisionally located on human chromosome 13, by analysis of the genetics of families with the familial disease. By using hybridization probes for sequences closely linked to *RB1*, it was possible to show that the retinoblastoma cells of patients who were heterozygous for the linked sequences had only a single copy of the sequence, that is there had been a deletion in the region of the supposed *RB1* gene in tumor cells, but not in nontumor cells.

RB1 gene

The *RB1* gene was then isolated by determining the DNA sequence of the region of chromosome 13 defined by the most tightly linked marker sequences (specific chromosomal DNA sequences that are most frequently inherited with *RB1*). *RB1*

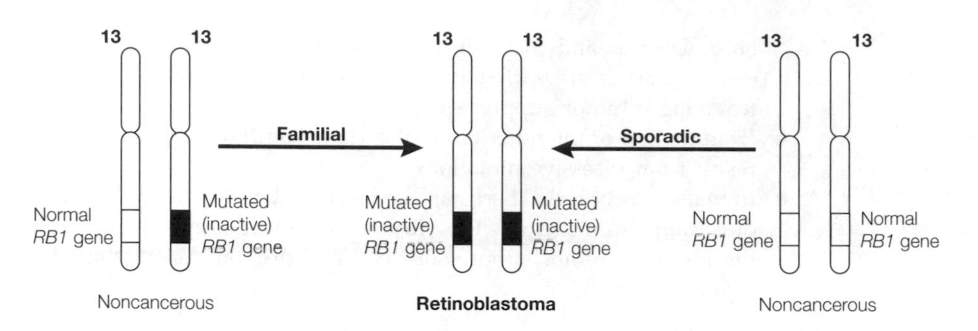

Fig. 2. Retinoblastoma results from the inactivation of both copies of the RB1 *gene on chromosome 13. This can occur by mutation of both normal copies of the gene (sporadic retinoblastoma) or by inheritance of one inactive copy followed by an acquired mutation in the remaining functional copy (familial retinoblastoma).*

codes for a 110 kDa phosphoprotein that binds to DNA, and has been shown to inhibit the transcription of proto-oncogenes such as *myc* and *fos*. *RB1* mRNA was found to be absent or abnormal in retinoblastoma cells. The role of *RB1* in retinoblastoma was established definitively when it was shown that retinoblastoma cells growing in culture reverted to a nontumorigenic state when they were transfected with a cloned, normal *RB1* gene. Unexpectedly, *RB1* mutations have also been detected in breast, colon and lung tumors.

p53 gene

Similar techniques have subsequently been used to identify/isolate tumor suppressor genes associated with other cancers, but the gene that really put tumor suppressors on the map is one called ***p53***. The gene for p53 is located on the short arm of chromosome 17, and deletions of this region have been associated with nearly 50% of human cancers. The mRNA for p53 is 2.2–2.5 kb and codes for a 52 kDa nuclear protein. The protein is found at a low level in most cell types and has a very short half-life (6–20 min). Confusingly, *p53* has some of the properties of both oncogenes and of tumor suppressor genes:

- many mutations (point mutations, deletions, insertions) have been shown to occur in the *p53* gene, and all cause it to become oncogenic. Mutant forms of *p53*, when co-transfected with the *ras* oncogene, will transform normal rat fibroblasts. In cancer cells, p53 has an extended half-life (4–8 h), resulting in elevated levels of the protein. All this seems to suggest that *p53* is an oncogene.
- A consistent deletion of the short arm of chromosome 17 has been seen in many tumors. In brain, breast, lung and colon tumors, where a *p53* gene was deleted, the remaining allele was mutated. This suggests that p53 is a tumor suppressor gene!

The explanation seems to be that p53 acts as a dimer. When a mutant (inactive) p53 protein is present it dimerizes with the wild-type protein to create an inactive complex (*Fig. 3*). This is known as a **dominant-negative** effect. However, inactivation of the normal *p53* gene by the mutant gene would not be expected to be 100%, since some normal–normal dimers would still form. Loss (by chromosomal deletion) of the remaining normal *p53* gene may, therefore, result in a more complete escape from the tumor suppressor effects of this gene.

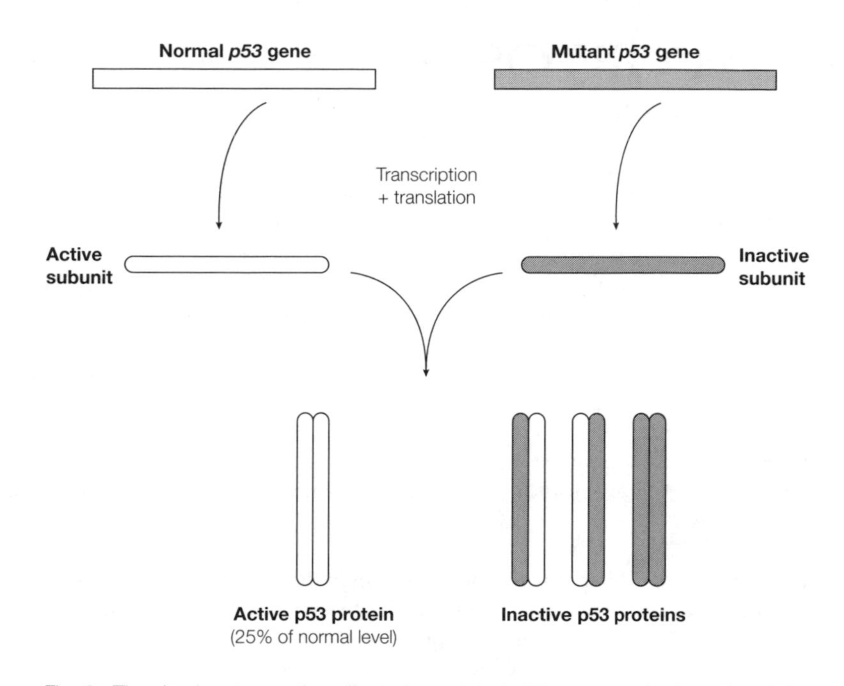

Fig. 3. The dominant–negative effect of a mutated p53 *gene results from the ability of the protein to dimerize with and inactivate the normal protein.*

S4 APOPTOSIS

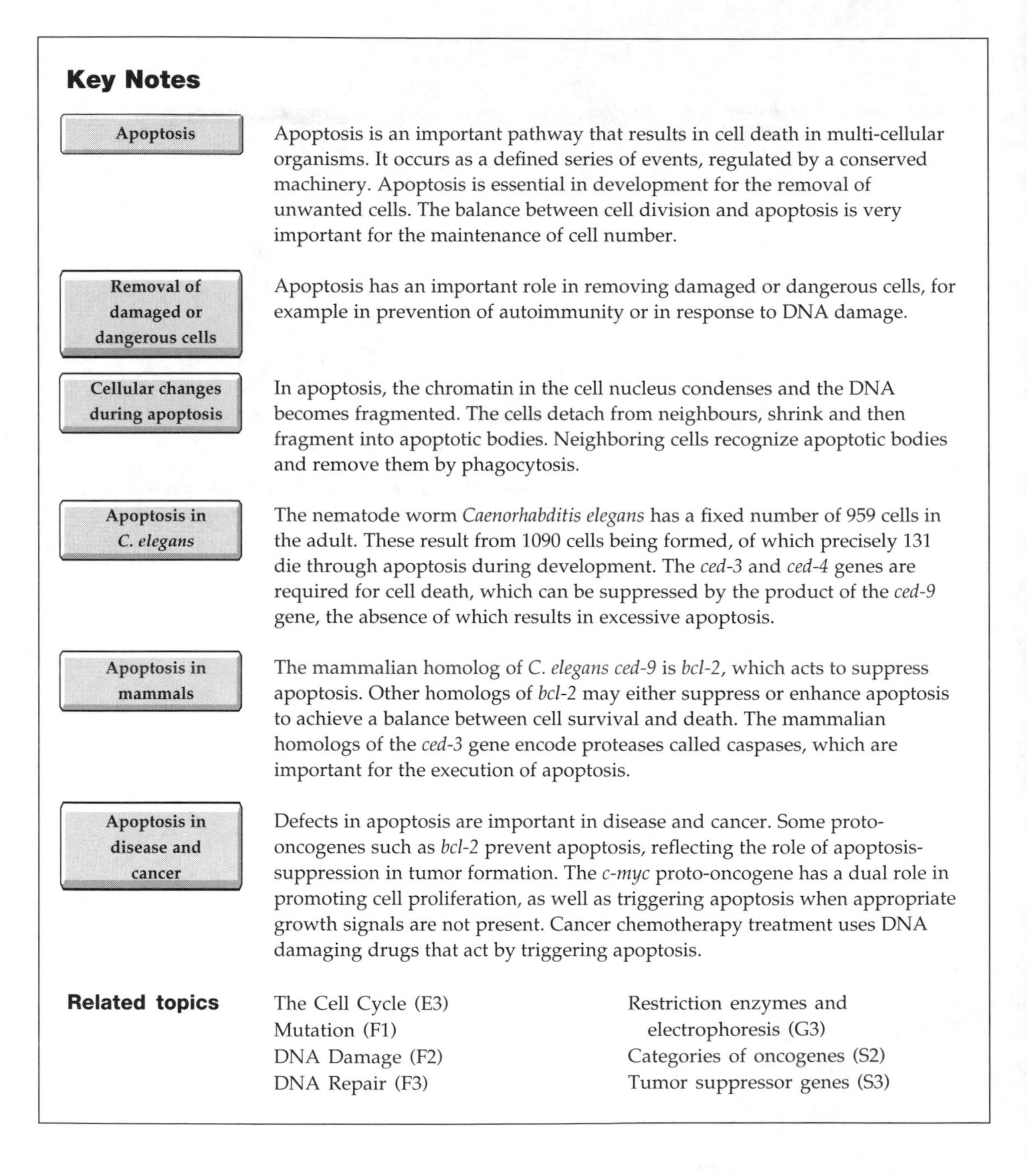

Key Notes

Apoptosis	Apoptosis is an important pathway that results in cell death in multi-cellular organisms. It occurs as a defined series of events, regulated by a conserved machinery. Apoptosis is essential in development for the removal of unwanted cells. The balance between cell division and apoptosis is very important for the maintenance of cell number.
Removal of damaged or dangerous cells	Apoptosis has an important role in removing damaged or dangerous cells, for example in prevention of autoimmunity or in response to DNA damage.
Cellular changes during apoptosis	In apoptosis, the chromatin in the cell nucleus condenses and the DNA becomes fragmented. The cells detach from neighbours, shrink and then fragment into apoptotic bodies. Neighboring cells recognize apoptotic bodies and remove them by phagocytosis.
Apoptosis in *C. elegans*	The nematode worm *Caenorhabditis elegans* has a fixed number of 959 cells in the adult. These result from 1090 cells being formed, of which precisely 131 die through apoptosis during development. The *ced-3* and *ced-4* genes are required for cell death, which can be suppressed by the product of the *ced-9* gene, the absence of which results in excessive apoptosis.
Apoptosis in mammals	The mammalian homolog of *C. elegans ced-9* is *bcl-2*, which acts to suppress apoptosis. Other homologs of *bcl-2* may either suppress or enhance apoptosis to achieve a balance between cell survival and death. The mammalian homologs of the *ced-3* gene encode proteases called caspases, which are important for the execution of apoptosis.
Apoptosis in disease and cancer	Defects in apoptosis are important in disease and cancer. Some proto-oncogenes such as *bcl-2* prevent apoptosis, reflecting the role of apoptosis-suppression in tumor formation. The *c-myc* proto-oncogene has a dual role in promoting cell proliferation, as well as triggering apoptosis when appropriate growth signals are not present. Cancer chemotherapy treatment uses DNA damaging drugs that act by triggering apoptosis.

Related topics

The Cell Cycle (E3)
Mutation (F1)
DNA Damage (F2)
DNA Repair (F3)
Restriction enzymes and electrophoresis (G3)
Categories of oncogenes (S2)
Tumor suppressor genes (S3)

Apoptosis

Apoptosis is the mechanism by which cells normally die. It involves a defined set of programmed biochemical and morphological changes. It is a frequent and widespread process in multi-cellular organisms, with an important role in the formation, maintenance and moulding of normal tissues. It occurs in almost all tissues during development and also in many adult tissues. For example, apoptosis is responsible for the gaps between the human fingers, which would otherwise be webbed. It is also responsible for the loss of the tadpole's tail during amphibian metamorphosis. It seems that in many cell-types, apoptosis is a built-in self-destruct pathway that is automatically triggered when growth signals are absent that would otherwise give the signal for the cell to survive. The balance between cell division and apoptosis is critical for maintaining homeostasis in cell number in the organism. Thus, tissue mass in an adult organism is maintained not only by the proliferation, differentiation and migration of cells, but also, in large part, by the controlled loss of cells through apoptosis.

Apoptosis is not the only pathway by which cells die. In **necrosis**, the cell membrane loses its integrity and the cell lyses releasing the cell contents. The release of the cell contents in an intact organism is generally undesirable. Apoptosis is therefore the major pathway of physiological cell death, and may also be called **programmed cell death** (and can be considered to be cell suicide).

Removal of damaged or dangerous cells

Apoptosis is responsible for the removal of damaged or dangerous cells. In the thymus, over 90% of cells of the immune system undergo apoptosis. This process is very important for the removal of self-reactive T lymphocytes, which would otherwise cause auto-immunity by turning the immune system against the organism's own cells. When T cells kill other cells, they do so by activating the apoptotic pathway and so induce the cells to commit suicide. Many virus-infected cells undergo apoptosis as a mechanism for limiting the spread of the virus within the organism. The tumor suppressor protein p53 (see Topic S3) can induce apoptosis in response to excessive DNA damage (see Topic F2), which the cell cannot repair (see Topic F3). The p53 protein therefore has a dual role in response to DNA damage, firstly in inhibiting the cell cycle (see Topic E3), and secondly in induction of apoptosis.

Cellular changes during apoptosis

During apoptosis, the nucleus shrinks and the chromatin condenses. When this happens, the DNA is often fragmented by nuclease-catalyzed cleavage between nucleosomes. This can be demonstrated by gel electrophoresis of the DNA (*Fig. 1a*; see Topic G3). The cell detaches from neighbors, rounds up, shrinks, and fragments into **apoptotic bodies** (*Fig. 1b*), which often contain intact organelles and an intact plasma membrane. Neighboring cells rapidly engulf and destroy the apoptotic bodies and it is thought that changes on the cell surface of the apoptotic bodies may have an important role in directing this phagocytosis. The morphological changes characteristic of apoptosis (*Fig. 1c*) may occur within half an hour of initiation of the process.

Apoptosis in *C. elegans*

In the nematode *Caenorhabditis elegans*, every adult hermaphrodite worm is identical and comprises exactly 959 cells. The cell lineage of *C. elegans* is closely regulated. During the worm's development, 1090 cells are formed of which 131 are removed by apoptosis. The products of two genes, *ced-3* and *ced-4*, are required for the death of these cells during development. If either gene is inactivated by mutation, none of the 131 cells die. The protein encoded by a further gene, *ced-9*, acts in an opposite way to suppress apoptosis. Inactivation of the *ced-9* by mutation results in excessive cell death, even in cells that do not normally die during development. Hence, *ced-9* is also required for the survival of cells that would not normally die and thus suppresses a general cellular program of cell death.

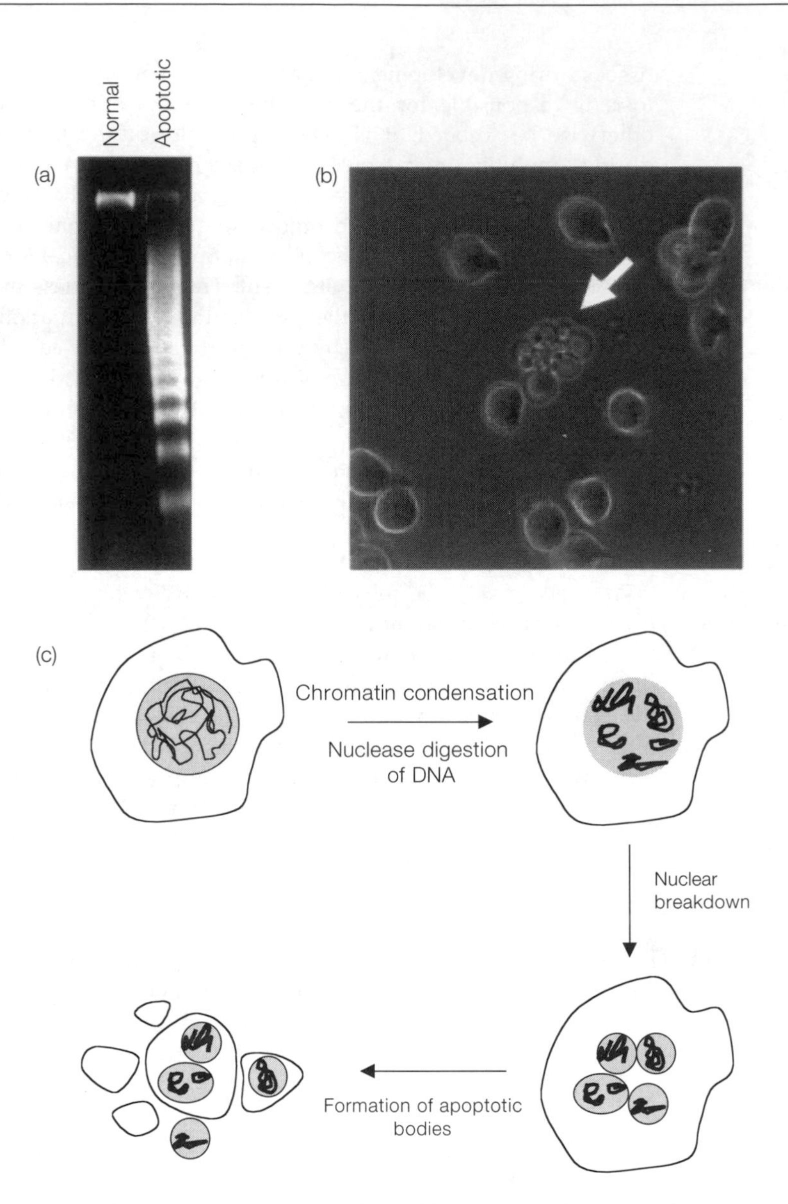

Fig. 1. Apoptosis. a) A characteristic DNA ladder from cells undergoing apoptosis. b) A microscopic image of T-cells showing apoptotic bodies (apoptotic cell marked with an arrow, original picture from Dr D. Spiller). c) A schematic diagram of the stages of apoptosis.

Apoptosis in mammals

The mammalian proto-oncogene ***bcl-2*** is homologous to the apoptosis-suppressing nematode *ced-9* gene. *bcl*-2 was the first of a novel functional class of proto-oncogene (see Topic S2) to be discovered whose members act to suppress apoptosis, rather than to promote cell proliferation. It is now known that a set of cell death-suppressing genes exist in mammals, several of which are homologous to *bcl-2*. Another group of proteins which include *bcl-2* homologs such as ***bax***, promote cell death. The bcl-2 and bax proteins bind to each other in the cell. It therefore seems that the regulation of apoptosis by cellular signaling pathways may occur in part through a change in the relative levels of cell death-suppressor and cell death-promoter proteins in the cell. Mammalian homologs of the nematode *ced-3* killer gene have also been identified. *ced-3* Encodes a polypeptide homologous to a family of cysteine-proteases of which the interleukin-1 β converting enzyme (ICE) is the archetype. These **ICE proteases** are also called **caspases** and they are responsible for the execution of apoptosis. Thus, it appears that apoptosis is a fundamentally important and evolutionarily conserved process.

Apoptosis in disease and cancer

Defects in the control of apoptosis appear to be involved in a wide range of diseases including neurodegeneration, immunodeficiency, cell death following a heart attack or stroke, and viral or bacterial infection. Most importantly, loss of apoptosis has a very important role in cancer. Cancer is a disease of multicellular organisms which is due to a loss of control of the balance between cell proliferation (see Topic E3) and cell death. Many proto-oncogenes (See Topic S2) regulate cell division, but others are known to regulate apoptosis (e.g. *bcl-2*), reflecting the importance of the balance between these processes. The proto-oncogene *c-myc* (see Topic S2) has a dual role since it stimulates cell division and can also act as a trigger of apoptosis. *c-myc* triggers apoptosis when growth factors are absent, or the cell has been subjected to DNA damage. Mutations in genes that result in the absence or relative down-regulation of the apoptosis pathway may therefore result in cancer. On the other hand, overexpression of genes, such as *bcl-2*, which normally inhibits the apoptotic pathway, may also result in cancer.

Most cancer treatment involves the use of DNA damaging drugs (for principles see Topic F2), that kill dividing cells. It has only recently been realized that rather than acting nonspecifically, these anti-cancer drugs act by triggering apoptosis in the cancer cells. One of the main mechanisms for the development of resistance to these drugs is suppression of the apoptotic pathway. This occurs in 50% of human cancers by mutation of the tumor suppressor *p53* (see Topic S3). *p53* has, as a result, been called 'the guardian of the genome'. In response to DNA damage, *p53* is central in triggering apoptosis, as well as switching on DNA repair (see Topic F3) and inhibiting progression through the cell cycle (see Topic E3). When the damage is too great for repair, this apoptosis-induction is important not only in removing cancer cells, but also in inhibiting proliferation of mutated cells in the cell population that might otherwise result in changes to the cancer cells which could give rise to a more dangerous tumor.

FURTHER READING

There are many comprehensive textbooks of molecular biology and biochemistry and no one book that can satisfy all needs. Different readers subjectively prefer different textbooks and hence we do not feel that it would be particularly helpful to recommend one book over another. Rather we have listed some of the leading books which we know from experience have served their student readers well.

General reading

Alberts, B., Bray, D., Lewis, J., Raff, M., Roberts, K. and Watson, J.D. (1994) *Molecular Biology of the Cell*, 3rd Edn. Garland Publishing, New York.

Brown, T.A. (1999) *Genomes.* BIOS Scientific Publishers, Oxford.

Freifelder, D.M. (1987) *Molecular Biology*, 2nd Edn. Jones and Bartlett, Boston, MA.

Lewin, B. (2000) Genes VII. Oxford University Press, Oxford.

Lodish, H., Baltimore, D., Berk, A., Zipursky, S.L., Matsudaira, P. and Darnell, J. (2000) *Molecular Cell Biology*, 4th Edn. Scientific American Books, W.H. Freeman, New York.

Twyman, R.M. (1998) *Advanced Molecular Biology*. BIOS Scientific Publishers, Oxford.

Voet, D. and Voet, J.G. (1995) *Biochemistry*, 2nd Edn. John Wiley & Sons, New York.

More advanced reading

The following selected articles are recommended to readers who wish to know more about specific subjects. In many cases they are too advanced for first year students but are very useful sources of information for subjects that may be studied in later years.

Section A

Bretscher, M.S. (1985) The molecules of the cell membrane. *Sci. Amer.* **253**, 86–90.

de Duve, C. (1996) The birth of complex cells. *Sci. Amer.* **274** (4), 38–45.

Gupta, R.S. and Golding, G.B. (1996) The origin of the eukaryotic cell. *Trends Biochem. Sci.* **21**, 166–171.

Pumplin, D.W. and Bloch, R.J. (1993) The membrane skeleton. *Trends Cell Biol.* **3**, 113-117.

Rothman, J.E. and Orci, L. (1996) Budding vesicles in living cells. *Sci. Amer.* **274** (3) 70–75.

Section B

Byard, E.H. and Lange, B.M.H. (1991) Tubulin and microtubules. *Essays Biochem.* **26**, 13–25.

Csermely, P. (1997) Proteins, RNAs and chaperones in enzyme evolution: a folding perspective. *Trends Biochem. Sci.* **22**, 147–149.

Darby, N.J. and Creighton, T.E. (1993) *Protein Structure: In Focus*. IRL Press, Oxford.

Doolittle, R.F. (1985) Proteins. *Sci. Amer.* **253** (4) 74–81.

Gahmberg, C.G. and Tolvanen, M. (1996) Why mammalian cell surface proteins are glycoproteins. *Trends Biochem. Sci.* **21**, 308-311.

Richards, F.M. (1991) The protein folding problem. *Sci. Amer.* **264** (1), 34–41.

Section C

Bates, A.D. and Maxwell, A. (1993) *DNA Topology: In Focus.* IRL Press, Oxford.

Drlica, K. (1990) Bacterial topoisomerases and the control of DNA supercoiling. *Trends Genet.* **6** (12), 433–437.

Maxwell, A. (1997) DNA gyrase as a drug target. *Trends Microbiol.* **5** (3), 102–109.

Neidle, S. (1994) *DNA Structure and Recognition: In Focus.* IRL Press, Oxford.

Roca, J. (1995) The mechanisms of DNA topoisomerases. *Trends Biochem. Sci.* **20** (4), 156–160.

Section D

Grosschedl, R., Giese, K. and Pagel, J. (1994) HMG domain proteins – architectural elements in the assembly of nucleoprotein structures. *Trends Genet.* **10** (3), 94–100.

Jones, P.A., Rideout, W.M., Shen, J.C., Spruck, C.H. and Tsai, Y.C. (1992) Methylation, mutation and cancer. *Bioessays* **14** (1), 33–36.

Roth, S.Y. and Allis, C.D. (1992) Chromatin condensation – does histone H1 dephosphorylation play a role? *Trends Biochem. Sci.* **17** (3), 93–98.

Travers, A.A. (1994) Chromatin structure and dynamics. *Bioessays* **16** (9), 657–662.

Wade, P.A., Pruss, D. and Wolffe, A.P. (1997) Histone acetylation: chromatin in action. *Trends Biochem. Sci.* **22** (4), 128–132.

Section E

Adams, R.L.P. (1991) *DNA Replication: In Focus.* IRL Press, Oxford.

Chong, J.P.J., Thommes, P. and Blow, J.J. (1996) The role of MCM/P1 proteins in the licensing of DNA replication. *Trends Biochem. Sci.* **21**, 102–106.

Diller, J.D. and Raghuraman, M.K. (1994) Eukaryotic replication origins – control in space and time. *Trends Biochem. Sci.* **19**, 320–325.

Greider, C.W. and Blackburn, E.H. (1996) Telomeres, telomerases and cancer. *Sci. Amer.* **274** (2) 80–85.

Hamlin, J.L. (1992) Mammalian origins of DNA replication. *Bioessays* **14**, 651–660.

Hozak, P. and Cook, P.R. (1994) Replication factories. *Trends Cell Biol.* **4**, 48–52.

Kornberg, A. and Baker, T. (1991) *DNA Replication,* 2nd Edn. W.H. Freeman, New York.

Murray, A.W. and Kirschner, M.W. (1991) What controls the cell cycle? *Sci. Amer.* **264** (3), 56–63.

Section F

Bridges, B.A. (1997) DNA turnover and mutation in resting cells. *Bioessays* **19**, 347–52.

Friedberg, E.C., Walker, G.C. and Siede, W. (1995) *DNA Repair and Mutagenesis* American Society for Microbiology, Washington, DC.

Shinagawa, H. and Iwasaki, H. (1996) Processing the Holliday junction in homologous recombination. *Trends Biochem. Sci.* **21**, 107–111.

Tanaka, K. and Wood, R.D. (1994) *Xeroderma pigmentosum* and nucleotide excision repair of DNA. *Trends Biochem. Sci.* **19**, 83–86.

Trends in Biochemical Sciences, Vol. 20, No. 10, 1995. Whole issue devoted to articles on DNA repair.

Section G

Brown T.A. (1995) *Gene Cloning: An Introduction*, 3rd Edn. Chapman and Hall, London.

Watson, J.D., Gilman, M., Witkowski, J. and Zoller, M. (1992) *Recombinant DNA*, 2nd Edn. W.H. Freeman, New York.

Section H

Anand, R. (1992) Yeast artificial chromosomes (YACs) and the analysis of complex genomes. *Trends Biotechnol.* **10** (1–2), 35–40.

Calos, M.P. (1996) The potential of extrachromosomal replicating vectors for gene-therapy. *Trends Genet.* **12** (11), 463–466.

Conzelmann, K.K. and Meyers, G. (1996) Genetic-engineering of animal RNA viruses. *Trends Microbiol.* **4** (10), 386–393.

Tinland, B. (1996) The integration of T-DNA into plant genomes. *Trends Plant Sci.* **1** (6), 178–184.

Walden, R. and Wingender, R. (1995) Gene-transfer and plant-regeneration techniques. *Trends Biotechnol.* **13** (9), 324–331.

Section I

Brown T.A. (1995) *Gene Cloning: An Introduction*, 3rd Edn. Chapman and Hall, London.

Watson, J.D., Gilman, M., Witkowski, J. and Zoller, M. (1992) *Recombinant DNA*, 2nd Edn. W.H. Freeman, New York.

Section J

Adams, R.L.P., Knowler, J.T. and Leader, D.P. (1992) *The Biochemistry of the Nucleic Acids*, 11th Edn. Chapman and Hall, London.

French Anderson, W. (1995) Gene therapy. *Sci. Amer.* **273**, 96–99.

Gurdon, J.B. and Colman, A. (1999) The future of cloning. *Nature* **402,** 743–746.

Heiter, P. (1999) What do yeast proteins do? *Nature* **402,** 362–363.

Mullis, K.B. (1990) The unusual origins of the polymerase chain reaction. *Sci. Amer.* **262**, 36–41.

Section K

Beebee, T.J.C. and Burke, J. (1992) *Gene Structure and Transcription: In Focus.* IRL Press, Oxford.

Ptashne, M. (1986) *A Genetic Switch.* Cell Press, Palo Alto, CA.

Section L

Beebee, T.J.C. and Burke, J. (1992) *Gene Structure and Transcription: In Focus.* IRL Press, Oxford.

Pugh, B.F. (1996) Mechanisms of transcription complex assembly. *Curr. Opin. Cell Biol.* **8**, 303–311.

Section M

Burley, S.K. (1996) The TATA box-binding protein. *Curr. Opin. Struct. Biol.* **6**, 69–75.

Chalut, C., Moncollin, V. and Egly, J.M. (1994) Transcription by RNA polymerase II. *Bioessays* **16**, 651–655.

Geiduschek, E.P. and Kassavetis, G.A. (1995) Comparing transcriptional initiation by RNA polymerase I and RNA polymerase III. *Curr. Opin. Cell Biol.* **7**, 344–351.

Roeder, R.G. (1996) The role of general initiation factors in transcription by RNA polymerase II. *Trends Biochem. Sci.* **21**, 327–334.

Verrijzer, C.P. and Tijan, R. (1996) TAFs mediate transcriptional activation and promoter selectivity. *Trends Biochem. Sci.* **21**, 338–342.

Section N

Laemmli, U.K. and Tijan, R. (1996) Nucleus and gene expression – a nuclear traffic jam – unraveling multicomponent machines and compartments. *Curr. Opin. Cell Biol.* 8, 299–302.

Maldonado, E. and Reinberg, D. (1995) News on initiation and elongation of transcription by RNA polymerase II. *Curr. Opin. Cell Biol.* **7**, 352–361.

Tijan, R. (1995) Molecular machines that control genes. *Sci. Amer.* **272**, 38–45.

Wolberger, C. (1996) Homeodomain interactions. *Curr. Opin. Struct. Biol.* **6**, 62–68.

Section O

Adams, M.D., Rudner, D.Z. and Rio, D.C. (1996) Biochemistry and regulation of pre-mRNA splicing. *Curr. Opin. Cell Biol.* **8**, 331–339.

Morrissey, J.P. and Tollervey, D. (1995) Birth of the snoRNPs – the evolution of RNase MRP and and the eukaryotic pre-rRNA-processing system. *Trends Biochem. Sci.* **20**, 78–82.

Nagai, K. (1996) RNA–protein complexes. *Curr. Opin. Struct. Biol.* **6**, 53–61.

Ross, J. (1996) Control of mRNA stability in higher eukaryotes. *Trends Genet.* **12**, 171–175.

Scott, W.G. and Klug, A. (1996) Ribozymes: structure and mechanism of RNA catalysis. *Trends Biochem. Sci.* **21**, 220–224.

Smith, H.C. and Snowden, M.P. (1996) Base modification mRNA editing through deamination – the good, the bad and the unregulated. *Trends Genet.* **12**, 418–424.

Tuschl, T., Thomson, J.B. and Eckstein, F. (1995) RNA cleavage by small catalytic RNAs. *Curr. Opin. Struct. Biol.* **5**, 296–302.

Wahle, E. and Keller, W. (1996) The biochemistry of polyadenylation. *Trends Biochem. Sci.* **21**, 247–250.

Section P

Arnstein, H.R.V. and Cox, R.A. (1992) *Protein Biosynthesis: In Focus.* IRL Press, Oxford.

Moras, D. (1992) Structural and functional relationships between aminoacyl-tRNA synthetases. *Trends Biochem. Sci.* **17**, 159–164.

Section Q

Garrett, R. (1999) Mechanism of the ribosome. *Nature* **400,** 811–812.

Liu, R. and Neupert, W. (1996) Mechanisms of protein import across the outer mitochondrial membrane. *Trends Cell Biol.* **6**, 56–61.

Martoglio, B. and Dobberstein, B. (1996) Snapshots of membrane-translocating proteins. *Trends Cell Biol.* **6**, 142–147.

Rassow, J. and Pfanner, N. (1996) Protein biogenesis – chaperones for nascent polypeptides, *Curr. Biol.* **6**, 115–118.

Riis, B., Rattan, S.I.S., Clark, B.F.C. and Merrick, W.C. (1990) Eukaryotic protein elongation factors. *Trends Biochem. Sci.* **15**, 420–424.

Section R

Cann, A.J. (1993) *Principles of Molecular Virology*, Academic Press, London.

Liljas, I. (1996) Viruses, *Curr. Opin. Struct. Biol.* **6**, 151–156.

Section S

Harris, C.C. (1993) p53: at the crossroads of molecular carcinogenesis and risk assessment. *Science* **26**, 1980–1981.

Macdonald, F. and Ford, C.H.J. (1991) *Oncogenes and Tumor Suppressor Genes.* BIOS Scientific Publishers, Oxford.

Weinberg, R.A. (1989) Oncogenes, antioncogenes and the molecular bases of multistep carcinogenesis. *Cancer Res.* **49**, 3713–3721.

Multiple choice questions

Section A – Cells and macromolecules

1. The glycosylation of secreted proteins takes place in the . . .

A mitochondria.
B peroxisomes.
C endoplasmic reticulum.
D nucleus.

2. Which of the following is an example of a nucleoprotein?

A keratin.
B chromatin.
C histone.
D proteoglycan.

3. Which of the following is not a polysaccharide?

A chitin.
B amylopectin.
C glycosaminoglycan.
D glycerol.

4. Transmembrane proteins . . .

A join two lipid bilayers together.
B have intra- and extracellular domains.
C are contained completely within the membrane.
D are easily removed from the membrane.

Section B – Protein structure

1. Which of the following is an imino acid?

A proline.
B hydroxylysine.
C tryptophan.
D histidine.

2. Protein family members in different species that carry out the same biochemical role are described as . . .

A paralogs.
B structural analogs.
C heterologs.
D orthologs.

3. **Which of the following is not a protein secondary structure?**

A α-helix.
B triple helix.
C double helix.
D β-pleated sheet.

4. **In isoelectric focusing, proteins are separated ...**

A in a pH gradient.
B in a salt gradient.
C in a density gradient.
D in a temperature gradient.

5. **Edman degradation sequences peptides ...**

A using a cDNA sequence.
B according to their masses.
C from the C-terminus to the N-terminus.
D from the N-terminus to the C-terminus.

Section C – Properties of nucleic acids

1. **The sequence 5′-AGTCTGACT-3′ in DNA is equivalent to which sequence in RNA?**

A 5′-AGUCUGACU-3′
B 5′-UGTCTGUTC-3′
C 5′-UCAGUCUGA-3′
D 5′-AGUCAGACU-3′

2. **Which of the following correctly describes A-DNA?**

A a right-handed antiparallel double helix with 10 bp/turn and bases lying perpendicular to the helix axis.
B a left-handed antiparallel double-helix with 12 bp/turn formed from alternating pyrimidine–purine sequences.
C a right-handed antiparallel double helix with 11 bp/turn and bases tilted with respect to the helix axis.
D a globular structure formed by short intramolecular helices formed in a single-stranded nucleic acid.

3. **Denaturation of double stranded DNA involves ...**

A breakage into short double-stranded fragments.
B separation into single strands.
C hydrolysis of the DNA backbone.
D cleavage of the bases from the sugar-phosphate backbone.

4. **Which has the highest absorption per unit mass at a wavelength of 260 nm?**

A double-stranded DNA.
B mononucleotides.
C RNA.
D protein.

5. **Type I DNA topoisomerases . . .**

A change linking number by ± 2.
B require ATP.
C break one strand of a DNA double helix.
D are the target of antibacterial drugs.

Section D – Prokaryotic and eukaryotic chromatin structure

1. **Which of the following is common to both *E. coli* and eukaryotic chromosomes?**

A the DNA is circular.
B the DNA is packaged into nucleosomes.
C the DNA is contained in the nucleus.
D the DNA is negatively supercoiled.

2. **A complex of 166 bp of DNA with the histone octamer plus histone H1 is known as a . . .**

A nucleosome core.
B solenoid.
C 30 nm fiber.
D chromatosome.

3. **In what region of the interphase chromosome does transcription take place?**

A the telomere.
B the centromere.
C euchromatin.
D heterochromatin.

4. **Which statement about CpG islands and methylation is not true?**

A CpG islands are particularly resistant to DNase I.
B CpG methylation is responsible for the mutation of CpG to TpG in eukaryotes.
C CpG islands occur around the promoters of active genes.
D CpG methylation is associated with inactive chromatin.

5. **Which of the following is an example of highly-repetitive DNA?**

A *Alu* element.
B histone gene cluster.
C DNA minisatellites.
D dispersed repetitive DNA.

Section E – DNA replication

1. **The number of replicons in a typical mammalian cell is . . .**

A 40–200.
B 400.
C 1000–2000.
D 50 000–100 000.

2. **In prokaryotes, the lagging strand primers are removed by . . .**

A 3′ to 5′ exonuclease.
B DNA ligase.
C DNA polymerase I.
D DNA polymerase III.

3. **The essential initiator protein at the *E. coli* origin of replication is . . .**

A DnaA.
B DnaB.
C DnaC.
D DnaE.

4. **Which phase would a cell enter if it was starved of mitogens before the R point?**

A G1.
B S.
C G2.
D G0.

5. **Which one of the following statements is true?**

A once the cell has passed the R point, cell division is inevitable.
B the phosphorylation of Rb by a G1 cyclin-CDK complex is a critical requirement for entry into S phase.
C phosphorylation of E2F by a G1 cyclin-CDK complex is a critical requirement for entry into S phase.
D cyclin D1 and INK4 p16 are tumor suppressor proteins.

6. **In eukaryotes, euchromatin replicates predominantly . . .**

A in early S-phase.
B in mid S-phase.
C in late S-phase.
D in G2-phase.

7. **Prokaryotic plasmids can replicate in yeast cells if they contain a cloned yeast . . .**

A ORC.
B CDK.
C ARS.
D RNA.

Section F – DNA damage, repair and recombination

1. **Per nucleotide incorporated, the spontaneous mutation frequency in *E. coli* is . . .**

A 1 in 10^6.
B 1 in 10^8.
C 1 in 10^9.
D 1 in 10^{10}.

2. **The action of hydroxyl radicals on DNA generates a significant amount of . . .**

A pyrimidine dimers.
B 8-oxoguanine.
C O^6-methylguanine.
D 7-hydroxymethylguanine.

3. **In methyl-directed mismatch repair in *E. coli*, the daughter strand containing the mismatched base is nicked by . . .**

A MutH endonuclease.
B UvrABC endonuclease.
C AP endonuclease.
D 3′ to 5′ exonuclease.

4. **Illegitimate recombination is another name for . . .**

A site-specific recombination.
B transposition.
C homologous recombination.
D translesion DNA synthesis.

5. **The excision repair of UV-induced DNA damage is defective in individuals suffering from . . .**

A hereditary nonpolyposis colon cancer.
B Crohn's disease.
C classical xeroderma pigmentosum.
D xeroderma pigmentosum variant.

Section G – Gene manipulation

1. **The presence of a plasmid in a bacterial culture is usually determined by . . .**

A blue–white screening.
B growth in the presence of an antibiotic.
C a restriction enzyme digest.
D agarose gel electrophoresis.

2. **The enzyme alkaline phosphatase . . .**

A joins double-stranded DNA fragments.
B adds a phosphate group to the 5′-end of DNA.
C cuts double-stranded DNA.
D removes a phosphate group from the 5′-end of DNA.

3. **Transformation is . . .**

A the take-up of a plasmid into a bacterium.
B the expression of a gene in a bacterium.
C the take-up of a bacteriophage into a bacterium.
D the isolation of a plasmid from a bacterium.

4. T4 DNA ligase . . .

A requires ATP.
B joins double-stranded DNA fragments with an adjacent 3′-phosphate and 5′-OH.
C requires NADH.
D joins single-stranded DNA.

5. In agarose gel electrophoresis . . .

A DNA migrates towards the negative electrode.
B supercoiled plamids migrate slower than their nicked counterparts.
C larger molecules migrate faster than smaller molecules.
D ethidium bromide can be used to visualize the DNA.

Section H – Cloning vectors

1. Blue–white selection is used . . .

A to test for the presence of a plasmid in bacteria.
B to reveal the identity of a cloned DNA fragment.
C to express the product of a cloned gene.
D to test for the presence of a cloned insert in a plasmid.

2. A multiple cloning site . . .

A contains many copies of a cloned gene.
B allows flexibility in the choice of restriction enzymes for cloning.
C allows flexibility in the choice of organism for cloning.
D contains many copies of the same restriction enzyme site.

3. Infection of *E. coli* by bacteriophage λ is normally detected by . . .

A resistance of the bacteria to an antibiotic.
B the growth of single bacterial colonies on an agar plate.
C the appearance of areas of lysed bacteria on an agar plate.
D restriction digest of the bacterial DNA.

4. Which vector would be most appropriate for cloning a 150 kb fragment of DNA?

A a plasmid.
B a λ vector.
C a BAC.
D a YAC.

5. Which vector would you choose to express a foreign gene in a plant?

A a baculovirus vector.
B a retroviral vector.
C a Yep vector.
D a T-DNA vector.

Section I – Gene libraries and screening

1. Which two of the following statements about genomic libraries are false?

A genomic libraries are made from cDNA.
B genomic libraries must be representative if they are to contain all the genes in an organism.
C genomic libraries must contain a minimum number of recombinants if they are to contain all the genes in an organism.
D the DNA must be fragmented to an appropriate size for the vector that is used.
E genomic libraries made from eukaryotic DNA usually use plasmid vectors.

2. Which statement correctly describes sequential steps in cDNA cloning?

A reverse transcription of mRNA, second strand synthesis, cDNA end modification, ligation to vector.
B mRNA preparation, cDNA synthesis using reverse transcriptase, second strand synthesis using terminal transferase, ligation to vector.
C mRNA synthesis using RNA polymerase, reverse transcription of mRNA, second strand synthesis, ligation to vector.
D double stranded cDNA synthesis, restriction enzyme digestion, addition of linkers, ligation to vector.

3. Which one of the following is not a valid method of screening a library?

A hybridization of colony/plaque-lifted DNA using a nucleic acid probe.
B using antibodies raised against the protein of interest to screen an expression library.
C screening pools of clones from an expression library for biological activity.
D hybridization of colony/plaque-lifted DNA using an antibody probe.

Section J – Analysis and uses of cloned DNA

1. A linear DNA fragment is (100%) labeled at one end and has 3 restriction sites for *Eco*RI. If it is partially digested by *Eco*RI so that all possible fragments are produced, how many of these fragments will be labeled and how many will not be labeled?

A 4 labeled; 6 unlabeled.
B 4 labeled; 4 unlabeled.
C 3 labeled; 5 unlabeled.
D 3 labeled; 3 unlabeled.

2. Which of the following are valid methods of labeling duplex DNA?

A 5′-end labeling with polynucleotide kinase.
B 3′-end labeling with polynucleotide kinase.
C 3′-end labeling with terminal transferase.
D 5′-end labeling with terminal transferase.
E nick translation.

3. Which one of the following statements about nucleic acid sequencing is correct?

A the Sanger method of DNA sequencing involves base specific cleavages using piperidine.
B the Maxam and Gilbert method of DNA sequencing uses a DNA polymerase and chain terminating dideoxynucleotides.
C enzymatic sequencing of RNA uses RNases A, T1, Phy M and *B. cereus* RNase.
D enzymatic sequencing of DNA uses a primer, which is extended by an RNA polymerase.
E enzymatic sequencing of RNA uses RNases T1, U2, Phy M and *B. cereus* RNase.

4. **Which one of the following statements about PCR is false?**

A the PCR cycle involves denaturation of the template, annealing of the primers and polymerization of nucleotides.
B PCR uses thermostable DNA polymerases.
C ideally, PCR primers should be of similar length and G+C content.
D PCR optimization usually includes varying the magnesium concentration and the polymerization temperature.
E. if PCR was 100% efficient, one target molecule would amplify to 2^n after n cycles.

5. **Which two of the following statements about gene mapping techniques are true?**

A S1 nuclease mapping determines the nontranscribed regions of a gene.
B primer extension determines the 3′-end of a transcript.
C gel retardation can show whether proteins can bind to and retard the migration of a DNA fragment through an agarose gel.
D DNase I footprinting determines where, on a DNA fragment, a protein binds.
E the function of DNA sequences in the promoter of a gene can be determined if they are ligated downstream of a reporter gene and then assayed for expression.

6. **Which one of these statements about mutagenesis techniques is false?**

A exonuclease III removes one strand of DNA in a 5′ to 3′ direction from a recessed 5′-end.
B exonuclease III removes one strand of DNA in a 3′ to 5′ direction from a recessed 3′-end.
C mutagenic primers can be used in PCR to introduce base changes.
D mutagenic primers can be used with a single stranded template and DNA polymerase to introduce base changes.
E deletion mutants can be created using restriction enzymes.

7. **Which one of these statements about the applications of gene cloning is false?**

A large amounts of recombinant protein can be produced by gene cloning.
B DNA fingerprinting is used to detect proteins bound to DNA.
C cloned genes can be used to detect carriers of disease-causing genes.
D gene therapy attempts to correct a disorder by delivering a good copy of a gene to a patient.
E genetically modified organisms have been used to produce clinically important proteins.

Section K – Transcription in prokaryotes

1. **Which two of the following statements about transcription are correct?**

A RNA synthesis occurs in the 3′ to 5′ direction.
B the RNA polymerase enzyme moves along the sense strand of the DNA in a 5′ to 3′ direction.
C the RNA polymerase enzyme moves along the template strand of the DNA in a 5′ to 3′ direction.
D the transcribed RNA is complementary to the template strand.
E the RNA polymerase adds ribonucleotides to the 5′ end of the growing RNA chain.
F the RNA polymerase adds deoxyribonucleotides to the 3′ end of the growing RNA chain.

2. **Which one of the following statements about *E. coli* RNA polymerase is false?**

A the holoenzyme includes the sigma factor.
B the core enzyme includes the sigma factor.
C it requires Mg^{2+} for its activity.
D it requires Zn^{2+} for its activity.

3. **Which one of the following statements is incorrect?**

A there are two α subunits in the *E. coli* RNA polymerase.
B there is one β subunit in the *E. coli* RNA polymerase.
C *E. coli* has one sigma factor.
D the β subunit of *E. coli* RNA polymerase is inhibited by rifampicin.
E the streptolydigins inhibit transcription elongation.
F heparin is a polyanion, which binds to the β' subunit.

4. **Which one of the following statements about transcription in *E. coli* is true?**

A the –10 sequence is always exactly 10 bp upstream from the transcription start site.
B the initiating nucleotide is always a G.
C the intervening sequence between the –35 and –10 sequences is conserved.
D the sequence of the DNA after the site of transcription initiation is not important for transcription efficiency.
E the distance between the –35 and –10 sequences is critical for transcription efficiency.

5. **Which one of the following statements about transcription in *E. coli* is true?**

A loose binding of the RNA polymerase core enzyme to DNA is non-specific and unstable.
B sigma factor dramatically increases the relative affinity of the enzyme for correct promoter sites.
C almost all RNA start sites consist of a purine residue, with A being more common than G.
D all promoters are inhibited by negative supercoiling.
E terminators are often A-U hairpin structures.

Section L – Regulation of transcription in prokaryotes

1. **Which two of the following statements are correct?**

A the double stranded DNA sequence that has the upper strand sequence 5′-GGATCGATCC-3′ is a palindrome.
B the double stranded DNA sequence that has the upper strand sequence 5′-GGATCCTAGG-3′ is a palindrome.
C the Lac repressor inhibits binding of the polymerase to the *lac* promoter.
D the *lac* operon is directly induced by lactose.
E binding of Lac repressor to allolactose reduces its affinity for the *lac* operator.
F IPTG is a natural inducer of the *lac* promoter.

2. **Which one of the following statements about catabolite-regulated operons is false?**

A cAMP receptor protein (CRP) and catabolite activator protein (CAP) are different names for the same protein.
B when glucose is present in the cell cAMP levels fall.
C CRP binds to cAMP and as a result activates transcription.
D CRP binds to DNA in the absence of cAMP.
E CRP can bend DNA, resulting in activation of transcription.

3. Which one of the following statements about the *trp* operon is true?

A the RNA product of the *trp* operon is very stable.
B the Trp repressor is a product of the *trp* operon.
C the Trp repressor, like the Lac repressor, is a tetramer of identical subunits.
D the Trp repressor binds to tryptophan.
E tryptophan activates expression from the *trp* operon.
F the *trp* operon is only regulated by the Trp repressor.

4. Which two of the following statements about attenuation at the *trp* operon are true?

A attenuation is rho-dependent.
B deletion of the attenuator sequence results in an increase in both basal and activated levels of transcription from the *trp* promoter.
C the attenuator lies upstream of the *trp* operator sequence.
D attenuation does not require tight coupling between transcription and translation.
E pausing of a ribosome at two tryptophan codons in the leader peptide when tryptophan is in short supply causes attenuation.
F a hairpin structure called the anti-terminator stops formation of the terminator hairpin, resulting in transcriptional read-through into the *trpE* gene, when tryptophan is scarce.

5. Which two of the following statements about sigma factors are false?

A the *E. coli* RNA polymerase core enzyme cannot start transcription from promoters in the absence of a sigma factor subunit.
B different sigma factors may recognize different sets of promoters.
C sigma factors recognize both the –10 and –35 promoter elements.
D heat shock promoters in *E. coli* have different –35 and –10 sequences and bind to a diverse set of 17 heat shock sigma factors.
E sporulation in *B. subtilis* is regulated by a diverse set of sigma factors.
F bacteriophage T7 expresses its own set of sigma factors as an alternative to encoding its own RNA polymerase.

Section M – Transcription in eukaryotes

1. Which one of the following statements about eukaryotic RNA polymerases I, II and III is false?

A RNA Pol II is very sensitive to α-amanitin.
B RNA Pol II is located in the nucleoplasm.
C RNA Pol III transcribes the genes for tRNA.
D eukaryotic cells contain other RNA polymerases in addition to RNA Pol I, RNA Pol II and RNA Pol III.
E each RNA polymerase contains subunits with homology to subunits of the *E. coli* RNA polymerase as well as additional subunits, which are unique to each polymerase.
F the carboxyl end of RNA Pol II contains a short sequence of only seven amino acids which is called the carboxyl-terminal domain (CTD) and which may be phosphorylated.

2. Which two of the following statements about RNA Pol I genes are true?

A RNA Pol I transcribes the genes for ribosomal RNAs.
B human cells contain 40 clusters of five copies of the rRNA gene.
C the 18S, 5.8S and 28S rRNAs are synthesized as separate transcripts.
D RNA Pol I transcription occurs in the nucleoplasm.
E RNA Pol I transcription occurs in the cytoplasm.
F rRNA gene clusters are known as nucleolar organizer regions.

3. **Which one of the following statements about RNA Pol I transcription is false?**

A in RNA Pol I promoters, the core element is 1000 bases downstream from the upstream control element (UCE).
B upstream binding factor (UBF) binds to both the UCE and the upstream part of the core element of the RNA Pol I promoter.
C selectivity factor SL1 stabilizes the UBF-DNA complex.
D SL1 contains several subunits including the TATA-binding protein TBP.
E in *Acanthamoeba* there is a single control element in rRNA gene promoters.

4. **Which two of the following statements about RNA Pol III genes are true?**

A the transcriptional control regions of tRNA genes lie upstream of the start of transcription.
B highly conserved sequences in tRNA gene coding regions are also promoter sequences.
C TFIIIC contains TBP as one of its subunits.
D TFIIIB is a sequence specific transcription factor on its own.
E in humans 5S rRNA genes are arranged in a single cluster of 2000 copies.

5. **Which one of the following statements is true?**

A RNA Pol II only transcribes protein-coding genes.
B the TATA box has a role in transcription efficiency but not in positioning the start of transcription.
C TBP binds to the TATA box.
D enhancers typically lie 100–200 bp upstream from the start of transcription.

6. **Which one of the following statements about general transcription factors is false?**

A TFIID binds to the TATA box.
B TFIID is a multiprotein complex consisting of TBP and TAF_{II}s.
C TBP is a common factor in transcription by RNA Pol I, RNA Pol II and RNA Pol III.
D TFIIB stabilizes the TFIID-DNA complex.
E TFIIE, TFIIH and TFIIJ associate with the transcription complex after RNA polymerase binding.
F TFIIH phosphorylates the CTD.

Section N – Regulation of transcription in eukaryotes

1. **Which two of the following statements about transcription factors are true?**

A the helix-turn-helix domain is a transcriptional activation domain.
B dimerization of transcription factors occurs through the basic domain.
C leucine zippers bind to DNA.
D it is often possible to get functional transcription factors when DNA binding domains and activation domains from separate transcription factors are fused together.
E the same domain of a transcription factor can act both as a repressor and as an activation domain.

2. **Which two of the following statements about transcriptional regulation are false?**

A SP1 contains two activation domains.
B steroid hormones regulate transcription through binding to cell surface receptors.
C phosphorylation of Stat1α leads to its migration from the cytoplasm to the nucleus.
D HIV Tat regulates RNA Pol II phosphorylation and processivity.
E the MyoD protein can form heterodimers with a set of other HLH transcription factors.
F the homeobox is a conserved DNA binding domain.

Section O – RNA processing and RNPs

1. Which of the following terms correctly describe parts of the *E. coli* large (50S) subunit?

A stalk, central protuberance, valley and cleft.
B upper third, lower third, valley and stalk.
C cleft, valley, stalk and small protuberance.
D stalk, polypeptide exit site, valley and central protuberance.

2. Which ribonucleases are involved in producing mature tRNA in *E. coli?*

A RNases A, D, E, and F.
B RNases D, E, F and H.
C RNases D, E, F and P.
D RNases A, D, H and P.

3. Most eukaryotic pre-mRNAs are matured by which of the following modifications to their ends?

A capping at the 3′-end, cleavage and polyadenylation at the 5′-end.
B addition of a GMP to the 5′-end, cleavage and polyadenylation to create the 3′-end.
C addition of a guanine residue to the 5′-end, cleavage and polyadenylation to create the 3′-end.
D addition of a GMP to the 5′-end, polyadenylation, then cleavage to create the 3′-end.

4. Which one of the following statements correctly describes the splicing process undergone by most eukaryotic pre-mRNAs?

A in a two-step reaction, the spliceosome removes the exon as a lariat and joins the two introns together.
B splicing requires conserved sequences, which are the 5′-splice site, the 3′-splice site, the branchpoint and the polypurine tract.
C the U1 snRNP initially binds to the 5′-splice site, U2 to the branchpoint sequence and then the tri-snRNP, U4, U5 and U6 can bind.
D in the first step of splicing the G at the 3′-end of the intron is joined to the 2′-hydroxyl group of the A residue of the branchpoint sequence to create a lariat.

Section P – The genetic code and tRNA

1. Which of the following list of features correctly apply to the genetic code?

A triplet, degenerate, nearly universal, comma-less, nonoverlapping.
B triplet, universal, comma-less, degenerate, nonoverlapping.
C overlapping, triplet, comma-less, degenerate, nearly universal.
D overlapping, comma-less, nondegenerate, nearly universal, triplet.

2. Which of the following statements about tRNAs is false?

A most tRNAs are about 76 residues long and have CCA as residues 74, 75 and 76.
B many tRNAs contain the modified nucleosides pseudouridine, dihydrouridine, ribothymidine and inosine.
C tRNAs have a common L-shaped tertiary structure with three nucleotides at one end able to base pair with an anticodon on a messenger RNA molecule.
D tRNAs have a common cloverleaf secondary structure containing three single stranded loops called the D-, T- and anticodon loops.

3. **Which three statements are true? The aminoacyl tRNA synthetase reaction . . .**

A joins AMP to the 3′-end of the tRNA.
B is a two step reaction.
C joins any amino acid to the 2′- or 3′-hydroxyl of the ribose of residue A76.
D is highly specific because the synthetases use identity elements in the tRNAs to distinguish between them.
E joins AMP to the amino acid to produce an intermediate.
F releases PPi in the second step.

Section Q – Protein synthesis

1. **Which statement about the codon-anticodon interaction is false?**

A it is antiparallel and can include nonstandard base pairs.
B inosine in the 5′-anticodon position can pair with A, C or U in the 3′-codon position.
C inosine in the 3′-anticodon position can pair with A, C or U in the 5′-codon position.
D A is never found in the 5′-anticodon position as it is modified by anticodon deaminase.

2. **Which one of the following statements correctly describes initiation of protein synthesis in *E. coli?***

A the initiator tRNA binds to the Shine-Dalgarno sequence.
B three initiation factors are involved and IF2 binds to GTP.
C the intermediate containing IF1, IF2, IF3, initiator tRNA and mRNA is called the 30S initiation complex.
D binding of the 50S subunit releases IF1, IF2, GMP and PPi.
E the initiation process is complete when the 70S initiation complex is formed which contains the initiator tRNA in the A site of the ribosome and an empty P site.

3. **Which statement about elongation of protein synthesis in prokaryotes is false?**

A elongation can be divided into three steps: peptidyl-tRNA delivery, peptide bond formation and translocation.
B the peptidyl transferase center of the large ribosomal subunit is responsible for peptide bond formation.
C in the EF-Tu-Ts exchange cycle, EF-Tu-GTP is regenerated by EF-Ts displacing GDP.
D EF-G is also known as translocase and uses GTP in its reaction.

4. ***E. coli* release factor 1 (RF1) recognizes which codons?**

A UAA only.
B UAG only.
C UGA only.
D UGA and UAA.
E UAG and UAA.
F UAG and UGA.

5. **Which two of the following statements about initiation of eukaryotic protein synthesis are true?**

A eukaryotes use a mRNA scanning method to locate the correct start codon.
B there are at least nine eukaryotic initiation factors (eIFs).
C eukaryotic initiation uses *N*-formylmethionine.
D the 80S initiation complex completes the initiation process and contains the initiator tRNA base-paired to the start codon in the A site.
E ATP is hydrolysed to AMP and PPi during the scanning process.
F the initiator tRNA binds after the mRNA has bound to the small subunit.

6. **Which of the following protein synthesis factors are not equivalent pairs in prokaryotes and eukaryotes?**

A EF-G; eEF2.
B EF-Tu; eEF1α.
C RF1 and RF3; eRF.
D EF-Ts; eEFαβ.

7. **Which statement about post-translational events is false?**

A some mRNAs encode polyproteins.
B protein targeting involves signal sequences in the nascent polypeptides.
C signal peptidase removes one or two amino acids from the amino terminus of some proteins.
D proteins can be modified by acetylation, phosphorylation and glycosylation.

Section R – Bacteriophages and eukaryotic viruses

1. **Which one of the following statements about viruses is false?**

A viruses can only replicate in a host cell.
B some viral envelopes contain host cell proteins.
C viral genomes may be double stranded or single stranded DNA or RNA.
D replication-defective viruses may be replicated through complementation.
E all viruses are dependent on the host cell replication and transcription machinery.
F some viruses use disease symptoms to aid their transmission between hosts.

2. **Which one of the following statements about M13 bacteriophage is true?**

A bacterophage M13 has a double stranded DNA genome.
B the M13 phage particle enters the *E. coli* host cell following binding to the sex pili.
C multiple copies of the M13 replicative form (RF) are produced by normal double stranded DNA replication using RNA priming.
D M13 phage particles are released by cell lysis.
E there is a highly variable amount of DNA in different M13 phage particles.

3. **Which three of the following statements about bacteriophage λ are true?**

A the bacteriophage λ has a double stranded DNA genome.
B λ phage particles bind to receptors on the *E. coli* outer membrane and inject the viral DNA into the cell.
C the lytic life cycle requires integration of the bacteriophage λ genome into the host cell genome.
D termination of the *N* and *Cro* genes is rho-independent.
E λ repressor acts to repress lysogeny.
F host cell stress tends to switch on the lytic cycle.

4. Which one of the following statements about DNA viruses is false?

A there is often sequence similarity between the promoters of DNA viruses and those of the host cell.
B the SV40 genome has overlapping reading frames.
C SV40 large T-antigen regulates both viral and host cell transcription and replication.
D the ability of DNA viruses to regulate host cell replication may cause tumors in host cells.
E SV40 proteins VP1, VP2 and VP3 are important for the tumorigenic properties of the virus.

5. Which one of the following statements about herpes viruses is true?

A herpes viruses are not associated with tumorigenesis.
B genes are only found on one strand of the herpes virus genome.
C HSV-1 encodes its own DNA polymerase.
D HSV-1 has a single origin of replication.
E latent infection by herpes viruses requires chromosomal integration of the viral genome.
F all of the herpes viral genes are required for viral replication.

6. Which one of the following statements about retroviruses is false?

A RNA-dependent polymerases are not as accurate as DNA-dependent polymerases.
B like retroviruses, yeast Ty transposons encode a reverse transcriptase enzyme.
C retroviruses have a single stranded RNA genome.
D the Rev protein of HIV regulates viral transcription.

Section S – Tumor viruses and oncogenes

1. Which one of the following statements does not support the view that cancer is a disease with a genetic element?

A some types of cancer are inherited.
B some cancer cells possess abnormal chromosomes.
C cancer is caused by carcinogens.
D many carcinogens cause mutations.

2. Which two of the following statements about the NIH-3T3 cell transfection assay for the isolation of oncogenes are false?

A it is technically simple compared to *in vivo* assays.
B it is quicker than *in vivo* assays.
C NIH-3T3 cells are normal, but can readily be transformed into tumor cells.
D NIH-3T3 cells are good at taking up and expressing foreign DNA.
E NIH-3T3 cells are stem cells that allow detection of cell-type specific oncogenes.

3. Which of the following protein groups contain no examples of oncogene products?

A transcription factors.
B cell surface receptors.
C protein kinases.
D lipases.
E peptide hormones.

4. **Which one of the following statements about tumor suppressor genes is false?**

A tumor suppressor genes act in a genetically dominant manner.
B there are fewer tumor suppressor genes known than oncogenes.
C the retinoblastoma gene is a tumor suppressor gene.
D tumor suppressor genes normally become oncogenic by mutations that eliminate their normal activity.
E tumor suppressor genes can be responsible for some familial cancers.

5. **Which three of the following could in theory enhance cancer cell formation or survival?**

A inactivation of one of the *bcl-2* gene family members.
B inactivation of one of the *bax* gene family members.
C over-expression of the *p53* gene.
D an increase in the cellular ratio of Bcl-2 protein over Bax protein.
E the removal of p53 protein.
F growth factor withdrawal.

6. **Which two of the following statements are true?**

A apoptosis is the only mechanism of cell death in multi-cellular organisms.
B the rate of apoptosis must equal the rate of cell division at all stages of development.
C apoptosis is thought to be the default pathway for many cells when they lack growth signals.
D apoptosis requires disruption of the integrity of the cell's plasma membrane.
E apoptosis involves breakdown of the nuclear DNA.

Answers

Section A
1C, 2B, 3D, 4B

Section B
1A, 2D, 3C, 4A, 5D

Section C
1A, 2C, 3B, 4B, 5C

Section D
1D, 2D, 3C, 4A, 5C

Section E
1D, 2C, 3A, 4D, 5B, 6A, 7C

Section F
1D, 2B, 3A, 4B, 5C

Section G
1B, 2D, 3A, 4A, 5D

Section H
1D, 2B, 3C, 4C, 5D

Section I
1A and E, 2A, 3D

Section J
1A, 2A, C and E, 3E, 4D, 5C and D, 6A, 7B

Section K
1B and D, 2B, 3C, 4E, 5B

Section L
1A and E, 2D, 3D, 4B and F, 5D and F

Section M
1F, 2A and F, 3A, 4B and E, 5C, 6D

Section N
1D and E, 2B and F

Section O
1D, 2C, 3B, 4C

Section P
1A, 2C, 3B, D and E

Section Q
1C, 2B, 3A, 4E, 5A and B, 6D, 7C

Section R
1E, 2E, 3A, B and F, 4E, 5C, 6D

Section S
1C, 2C and E, 3D and E, 4A, 5B, D and E, 6C and E

INDEX